ELSEVIER OCEANOGRAPHY SERIES, 22

Marine Gravity

FURTHER TITLES IN THIS SERIES

1 J.L. MERO
THE MINERAL RESOURCES OF THE SEA
2 L.M. FOMIN
THE DYNAMIC METHOD IN OCEANOGRAPHY
3 E.J.F. WOOD
MICROBIOLOGY OF OCEANS AND ESTUARIES
4 G. NEUMANN
OCEAN CURRENTS
5 N.G. JERLOV
OPTICAL OCEANOGRAPHY
6 V. VACQUIER
GEOMAGNETISM IN MARINE GEOLOGY
7 W.J. WALLACE
THE DEVELOPMENT OF THE CHLORINITY/SALINITY CONCEPT IN OCEANOGRAPHY
8 E. LISITZIN
SEA-LEVEL CHANGES
9 R.H. PARKER
THE STUDY OF BENTHIC COMMUNITIES
10 J.C.J. NIHOUL
MODELLING OF MARINE SYSTEMS
11 O.I. MAMAYEV
TEMPERATURE—SALINITY ANALYSIS OF WORLD OCEAN WATERS
12 E.J. FERGUSON WOOD and R.E. JOHANNES
TROPICAL MARINE POLLUTION
13 E. STEEMANN NIELSEN
MARINE PHOTOSYNTHESIS
14 N.G. JERLOV
MARINE OPTICS
15 G.P. GLASBY
MARINE MANGANESE DEPOSITS
16 V.M. KAMENKOVICH
FUNDAMENTALS OF OCEAN DYNAMICS
17 R.A. GEYER
SUBMERSIBLES AND THEIR USE IN OCEANOGRAPHY AND OCEAN ENGINEERING
18 J.W. CARUTHERS
FUNDAMENTALS OF MARINE ACOUSTICS
19 J.C.J. NIHOUL
BOTTOM TURBULENCE
20 P.H. LEBLOND and L.A. MYSAK
WAVES IN THE OCEAN
21 C.C. VON DER BORCH (Editor)
SYNTHESIS OF DEEP-SEA DRILLING RESULTS IN THE INDIAN OCEAN
22 P. DEHLINGER
MARINE GRAVITY

ELSEVIER OCEANOGRPAHY SERIES, 22

Marine Gravity

PETER DEHLINGER

Professor of Geophysics,
Department of Geology and Geophysics,
Marine Sciences Institute,
University of Connecticut,
Groton, Conn.,
U.S.A.

ELSEVIER SCIENTIFIC PUBLISHING COMPANY
Amsterdam – Oxford – New York 1978

ELSEVIER SCIENTIFIC PUBLISHING COMPANY
335 Jan van Galenstraat
P.O. Box 211, 1000 AE Amsterdam, The Netherlands

Distributors for the United States and Canada:

ELSEVIER NORTH-HOLLAND INC.
52, Vanderbilt Avenue
New York, N.Y. 10017

Library of Congress Cataloging in Publication Data

Dehlinger, Peter.
Marine gravity.

(Elsevier oceanography series ; 22)
Bibliography: p.
Includes index.
1. Gravity. 2. Marine geophysics. I. Title.
QB334.D43 526'.7'09162 78-3756
ISBN 0-444-41680-3

ISBN 0-444-41680-3 (Vol.22)
ISBN 0-444-41625-0 (Series)

Printed in The Netherlands

To Jean,
my center of gravity

PREFACE

Significant advances have been made during the past several decades in the knowledge of the earth's gravity field and in the determination of geologic structures which produce gravity anomalies of various wavelengths.

The advent of accurate sea gravimeters and accurate positioning at sea have made it possible to obtain reliable gravity measurements over a previously unsurveyed 70% of the earth's surface. The development of the theory of plate tectonics and determinations of the inhomogeneous nature of the crust and upper mantle (from deep seismic-refraction surveys) provide new approaches to interpretations of gravity anomalies. Digital computers provide means for rapidly analyzing anomalies which correspond to complex geologic models. New developments in inverse methods for analyzing sources of observed anomalies provide sophisticated approaches for rapid determinations of simplified structures. Falling-body experiments provide highly accurate determinations of absolute gravity. Trackings of orbital paths of artificial earth satellites have led to an improved reference earth spheroid, and measurements with highly accurate satellite altimeters have indicated local geoidal variations at sea. Finally, harmonic analysis of gravity anomalies obtained from perturbations in the paths of satellites, in addition to those obtained from surface measurements, have provided global free-air anomalies of a wide range of wavelengths which are associated with structures and tectonic processes at different depths.

The purpose of this book is to provide current information on the earth's gravity field, describe methods of interpreting gravity anomalies, and present free-air gravity anomalies over selected types of structures in marine areas, as may be useful to persons concerned with marine gravity: those designing field investigations, obtaining shipborne measurements, or making interpretations in the light of the new global tectonics. The developments of principles make the book a useful text for advanced undergraduate or first-year graduate students.

The book assumes knowledge of basic physics and the calculus. In several sections, where necessary, somewhat more advanced mathematics is used (including elements of spherical harmonics, differential and integral equations, Fourier transforms, and linear algebra). Most of the mathematical developments are presented in sufficient detail to be readily followed. Where the subject matter appears not to warrant development from basics, initial equations are presented without derivation, but developed such that the materials can be followed without extensive outside reading.

The cgs rather than the SI (rationalized mks) system of units is used since nearly all of the present literature in geophysics has used the cgs system: gravity is conventionally given in milligals (10^{-3} cm sec^{-2}), density in grams cm^{-3}, viscosity in poise. Appendix A provides conversions of units.

The plan of the book is to present an overview and a historical development of gravimetry (Chapter 1), develop principles of the earth's secular and time-varying gravity fields (Chapters 2 and 3), provide background information about the earth's interior, plate tectonics, and lithospheric flexuring and isostasy (Chapter 4), describe principles of gravity measuring instruments (Chapter 5), describe methods for reducing and correcting gravity measurements and computing various types of gravity anomalies (Chapter 6), outline methods for quantitative interpretations of gravity anomalies (Chapter 7), and present free-air anomalies of various types of oceanic and continental-margin areas (Chapter 8). Pertinent literature is referenced.

I thank the geophysicists who contributed anomaly maps, profiles, and associated crustal sections (especially Drs. R.W. Couch, A.B. Watts, P.D. Rabinowitz, and C.O. Bowin), and the graduate students who assisted in computing crustal sections. Thanks is also given to the ship crews who made gathering of the gravity measurements possible, to the government agencies which provided partial support for the field studies, and to The University of Connecticut for providing computer time.

During the first half of 1976 I was on sabbatic leave at the Lamont-Doherty Geological Observatory (L-DGO), Columbia University, and the Department of Geodesy and Geophysics of the University of Cambridge, writing part of the manuscript. I am particularly indebted to Drs. A.B. Watts, J.R. Cochran, and P.D. Rabinowitz at L-DGO, Sir Edward Bullard, Professor A.H. Cook and Dr. J.M. Woodside (then a visiting scientist) at Cambridge, Professor M.H.P. Bott of the University of Durham, and Dr. J.R. Gartner of the University of Connecticut for valuable discussions. Drs. A.B. Watts, R.W. Cochran, E.F. Chiburis, D.C. Skeels, and Messrs. J.H. Bodine and M. König reviewed selected parts of the manuscript.

P.D.
July 7, 1977

CONTENTS

Chapter 1

INTRODUCTION

A far-reaching revolution occurred in the earth sciences during the 1960's: the development and essential confirmation of the theory of plate tectonics. This theory is now generally accepted because many field tests based on widely different types of evidence have validated it well beyond the stage of hypothesis. The theory is revolutionary because it produced the first unified concept of tectonic processes occurring in the earth.

These are exciting times for earth scientists. Implications resulting from the new theory appear to be as revolutionary as those resulting from the breakthroughs in physics in the 1920's and 1930's, as those of genetic coding in the early 1960's, and possibly even as those of Darwinian evolution. Gravity has played a role, although not a dominant one, in the development of the theory.

According to the theory of plate tectonics, high-strength plates comprise the earth's lithosphere; these plates are in relative motion with respect to one another and to the underlying lower-strength asthenosphere. New lithosphere is generated at mid-ocean ridges, forming the trailing edges of plates; the lithosphere is consumed along subduction zones, near ocean trenches, where plates descend often to considerable depths into the upper mantle; plates are in relative motion along transform faults. Earthquakes characterize each of these three types of plate boundaries.

The theory of plate tectonics evolved from the Hess (1962) hypothesis of sea-floor spreading, which states that ocean floors spread laterally away from active mid-ocean ridges and ocean floors are therefore young relative to the continents. Extensive gatherings of marine geophysical and geological data interpreted in the light of this hypothesis, combined with locations and mechanisms of earthquake foci, led to the theory of plate tectonics. The proposed tectonic processes have been occurring at least for the past 200 million years, and very likely ever since the formation of a lithosphere, several billion years ago.

The theory confirms, in general, the Wegener (1912) hypothesis of continental drift, stating that continents are moving apart and are therefore in relative motion. Despite many striking but usually unrelated pieces of evidence which support continental drift, the concept was generally not accepted before the mid-1950's, when paleomagnetic measurements began to indicate that continents were not always in the same relative positions. Objections to drift centered primarily on the fact that a driving mechanism was not identified. Ironically, "drift" in the form of plate motions is now accepted, even though (at the time of this writing) driving mechanisms which cause plate motions are still not well understood.

Gravity anomalies, particularly at sea, are now interpreted in light of the theory of plate tectonics. Moreover, the theory led to gravimetric studies of new areas: examples are continental shelves for petroleum exploration, passive continental margins to determine contacts between basement rocks of older continental and younger oceanic crusts, mid-ocean ridges and plate boundaries to determine subcrustal densities, oceanic trenches where great thicknesses of sediments are piled up, and island arcs where ore deposits may be formed as a consequence of plate subduction.

SOME SCIENTIFIC CONTRIBUTIONS FROM GRAVIMETRY

Early geodesists provided two distinct breakthroughs which contributed significantly to understanding structures and related tectonics in the crust and upper mantle (or lithosphere and asthenosphere). The first was the establishment of isostatic compensation, proposed by Pratt (1855) and Airy (1855) and confirmed by many later investigators. Compensation means that weights of topographic masses are counterbalanced by underlying light-weight materials. In conjunction with observed surface geologic variations, isostatic compensation indicates a laterally inhomogeneous crust.

The second breakthrough occurred in the 1920's when Vening Meinesz (1929) made the first pendulum measurements in submarines across the Sunda trench south of Java. Unexpectedly large-amplitude negative gravity anomalies were observed paralleling the trench axis. These anomalies were too large, it was reasoned, to be produced by only crustal structures; therefore, mantle structures were assumed to be the source. This deduction resulted in a revision of the then held view that the mantle (including the upper mantle) is laterally homogeneous and is overlain by a laterally inhomogeneous crust. Island arc—trench systems were viewed as regions where crustal material is forced down into the upper mantle. More recently, many other types of measurements, mostly seismic, have established that the upper mantle is indeed generally inhomogeneous, and not just in island arc—trench areas.

Investigations of the gravity field and gravity anomalies around the world have resulted in additional new concepts about the structures and tectonic processes in the crust and upper mantle. The more prominent of these concepts include the following.

(1) Isostatic compensation is world-wide; about 85% of the earth's surface is estimated to be in equilibrium. However, the compensation appears to be regional more than local, the lithosphere bending down like an elastic plate in its support of long-duration topographic loads. Regional down-bending is consistent with a high-strength lithosphere.

(2) Along active oceanic trenches characteristically large negative gravity anomalies occur, which parallel positive anomalies over adjacent island arcs. The negative anomalies persist so long as the leading edge of a lithospheric plate plunges down into the mantle, but the anomalies appear to die out after active subduction ceases.

(3) Along mid-ocean ridges free-air gravity-anomaly values are uniformly positive (+20 to +30 mgal), but small for the mountainous height of the ridges above adjacent abyssal plains. The small anomaly values indicate that the high ridges are essentially compensated at depth; this implies that ridges are formed in response to passive processes (isostatic readjustment) and are not likely to be produced by processes actively forcing plates apart.

(4) Different wavelength components of gravity anomalies have been correlated (with varying degrees of correlation) with: (a) topographic terrain; (b) crustal and subcrustal structures; (c) regional elevations; (d) possible asthenospheric configurations and associated flow patterns. Correlations between anomaly values and topographic terrain are fairly good in local, uncompensated areas, but appear to be generally poor to fair over the larger, compensated provinces. Anomalies of very long wavelength extend across continental—oceanic boundaries and, except island arc—trench systems, across many plate boundaries, indicating that sources may be several hundred kilometers deep (thus providing possible further means for investigating deep inhomogeneities in the upper mantle).

(5) The ellipticity of the earth spheroid was determined with fair accuracy from gravity measurements before the advent of artificial earth satellites (circa 1960). (Orbital paths of satellites have led to a more accurate value of ellipticity.)

(6) The geoidal shape of the earth (described in Chapter 2) has been determined from computed undulations of the geoid and deflections of the vertical in those areas of the world where gravity anomalies were observed, in effect providing data for a map of the geoid.

(7) A world-geodetic system has been established with the advent of gravity measurements on all continents and oceans, making it possible to tie together the various independent geodetic systems that had been established around the world.

(8) A significantly improved geoid has been obtained (as recently as 1976) from combined data on perturbations of satellite paths and observed gravity anomalies. A geoid described to spherical harmonics degree and order 18 includes wavelengths over 4,000 km derived primarily from satellite data, and under 4,000 km derived primarily from averaged gravity anomalies. Free-air anomaly maps similarly obtained from satellite and gravity measurements provide data for correlating anomalies with various types of regional features (e.g., mid-ocean ridges, oceanic trenches, archipelagoes, abyssal plains).

PURPOSE OF MAKING GRAVITY MEASUREMENTS

The objective of obtaining gravity measurements is to locate and determine anomalous masses in the outer earth. In geology, anomalous masses are translated into density contrasts and then into geologic structures (including topographic terrain and regional elevation), possibly with inferences on tectonic processes. In geodesy, anomalous masses are translated into geoidal undulations and deflections of the vertical, to describe the shape of the geoid better and thus improve geodetic surveys and positioning by navigational methods.

A few examples will illustrate the wide range in types of features which can be analyzed by their anomaly fields. Local and regional structures, as in prospecting for economic deposits, are evaluated for structural trends, to locate specific geologic features, and to determine dimensions of larger structures in the upper crust (e.g., the thickness of a sedimentary basin, size and configuration of a salt dome, average density of a seamount). Anomalies over larger and deeper structures are analyzed to determine crustal and subcrustal layering and associated densities, as across continental margins, mid-ocean ridges, oceanic trenches, mountains, or archipelagoes. The nature of regional isostatic compensation and lithospheric flexuring can be studied by long-wavelength anomalies, with inferences on lithospheric thickness, strength, and elastic-viscoelastic properties. The longest-wavelength anomalies can be analyzed to study deep processes, such as possible flow in the asthenosphere or mineral phase changes.

Free-motion apparatuses were developed within the past decade which make it possible to measure absolute gravity within a few parts in 10^{-9} g (a few microgals). These apparatuses, which are free of the instrument drift affecting all gravimeters, have sufficiently high precision that new types of geophysical studies can be undertaken. One example is measuring secular changes in gravity associated with long-duration tectonic adjustments, and even those associated with the 18.6-year nutation of the earth's axis. Another is investigating possible time variance of the gravitational constant, which, since the time of Newton, has been assumed to be invariant.

EVALUATION AND ANALYSIS OF GRAVITY MEASUREMENTS

Evaluation of anomalous masses requires determination of gravity-anomaly fields. Anomalies are the difference between gravity on the geoid and the earth spheroid (discussed in Chapter 2), that on the geoid representing observed gravity reduced to its sea-level equivalent, and that on the spheroid representing normal gravity assuming uniform layers of constant density (no mass anomalies). To compute an anomaly, observed gravity is therefore first reduced to its sea-level equivalent, for measurements made above sea level. Measurements made on a moving platform are corrected to an equivalent stationary value.

Gravity reductions follow standard procedures to obtain one of three specifically defined types of anomalies: the free-air, Bouguer, and isostatic (described in Chapter 6). Each anomaly type has particular uses, limitations, and complexity of computation. Hence, care should be taken in obtaining the most appropriate type of anomaly. Anomaly maps are usually constructed from the computed anomalies; they are inspected for anomaly trends, for highs and lows, and for changes in gradient. Where it is desirable to discard large fractions of anomalies as not being of interest, regional anomalies are determined, as described in Chapter 7, and resultant residuals are analyzed in terms of plausible structures. Where long-wavelength anomalies are of interest, averaged values are determined.

Gravity anomalies cannot provide unique solutions to mass distributions. Assumptions must therefore be made for quantitative interpretations. Usually a hypothetical model of a possible structure which fits the observed anomaly (or residual) is constructed, and the model adjusted until the computed anomaly satisfactorily matches the observed. Digital computers are commonly used to make these computations. To constrain the model as much as possible, it is made to conform with existing seismic, borehole, and other available data.

A more sophisticated and possibly more rapid analysis is based on direct methods of interpretation, particularly inverse methods described in Chapter 7. These analyses solve equations which describe the observed anomaly, assuming either the shape of the body (usually an equivalent layer) and solving for the density distribution, or assuming a density contrast and solving for the shape of the body.

COMPONENTS OF THE EARTH'S GRAVITY FIELD

The earth's field consists of two secular and one time-varying component. The secular components result from the earth's mass attraction, which produces more than 99.4% of the total field, and the earth's rotation. The time-varying component, produced by lunar and solar tides, is relatively very small.

The component of the gravity vector produced by the mass attraction acts vertically downward, and that produced by the centrifugal acceleration outward normal to the earth's rotation axis. The total vector thus acts downward (essentially vertical), and its vertical component is slightly smaller than the attraction produced by the mass alone. The centrifugal acceleration, which causes the equatorial bulge to be formed, is greatest at the equator, 3.6 gal (cm $\sec^{-2}$) or $3.6 \cdot 10^{-3}\ g$ (where g is gravity), and vanishes at the poles.

Gravity at the equator is 5.2 gal ($5.2 \cdot 10^{-3}$ g) less than at the poles as a result of the increased radius, the mass of the bulge, and the outward centrifugal acceleration. The increased radius, 21.3 km, itself causes gravity to decrease 6.3 gal. With the further decrease of 3.6 gal from the rotational effect, the resultant decrease is 9.9 gal, which is 4.7 gal larger than the observed 5.2 gal. The attraction produced by the mass of the bulge is thus about 4.7 gal at the equator.

The tides produce very small variations in the direction of the total gravity vector (described in Chapter 3). Magnitudes of the attractions produced by the tides range up to 0.3 mgal ($3 \cdot 10^{-7}$ g). Although small, the tidal attractions are measurable with gravimeters, especially the highly sensitive tidal gravimeters, and the attractions must be corrected for (eqs. 3.30a,b) if anomalies are to be accurate within a tenth of a milligal.

Anomalous mass distributions in the outer earth produce secular gravity anomalies which are essentially constant, except for mass redistributions (produced by earthquake displacements, creep, subsidence, lithospheric plate movements, and isostatic readjustment), which are usually difficult to measure gravimetrically. Values of secular gravity anomalies range up to about 400 mgal ($4 \cdot 10^{-4}$ g) These anomalies are thus orders of magnitude smaller than the change in gravity from pole to equator, and orders of magnitude larger than tidal attractions.

HISTORICAL DEVELOPMENT OF GRAVIMETRY

Galileo Galilei, in about 1590, established that the force of gravity is the same in all bodies. He showed that materials of different densities fall equal distances during equal times. It was nearly a century earlier (circa 1500), however, that Leonardo da Vinci had already deduced that topographic masses must be at least partly compensated. One of his notebooks records that densities of mountains must be lower than those of level lands (Delaney, 1940).

In the latter half of the 17th century it was observed that pendulum clocks require adjustment of their pendulum lengths when the clocks are moved from one part of the earth to another. In 1672 the French astronomer Richer found that the length of a second-period pendulum in a clock moved from Paris to Cayenne (West Indies) had to be shortened about 0.1 inch (Todhunter, 1962, I, pp. 30, 39), and in 1677 the British astronomer Halley (Todhunter, I, p. 41) found that the length of a pendulum clock had to be shortened when it was moved from England to St. Helena (Richer's earlier observations were apparently not then known to Halley). Since the period of oscillation varies inversely as the square root of gravity, these (and later) pendulum observations demonstrated that gravity varies throughout the earth's surface.

Sir Isaac Newton is the founder of the science of attracting masses. He established that at a given site gravity is constant to about 1 part in 10^3 (using gold, silver, lead, glass, sand, salt, wood, water and wheat in his experiments). Later experiments by Bessel, Eötvös and others demonstrated that gravity is constant to more than 1 part in 10^8, and recently it was found to be constant to better than 1 part in 10^{10} (see, e.g., Cook, 1969, p. 89).

On the basis of extensive researches on attractions and by applying Kepler's laws of planetary motions, Newton derived the celebrated law of universal gravitation (i.e., the inverse-square law between two spherical homogeneous masses). These experiments and analyses are described in the *Philosophiae Naturalis Principia Mathematica* (1687). Newton stated that gravity is produced by the earth's mass attraction and its centrifugal acceleration (due to rotation), and weight is the effect produced by gravity.

Newton also showed that variations in gravity indicated by the pendulum observations made at different latitudes can be explained by allowing for the equatorial bulge, produced by the earth's rotation. He deduced that the earth has the shape of an oblate spheroid, with a radius about 17 miles greater at the equator than at the poles (the actual increase is 13 miles); the spheroid thus has an ellipticity of 1/230 (Table 1.I). He also concluded that gravity increases as the square of the sine of latitude between the earth's equator and pole.

Huygens (1693) developed the principle that the sea surface everywhere is perpendicular to the plumb line. Using accounts of pendulum experiments at different locations, he deduced a ratio of polar to equatorial axes of 577/578 (Todhunter, 1962, I, p. 30) and obtained an equation approximating the

TABLE 1.I

Historical estimates of the earth's ellipticity e

Source	Date	e
Newton	1687	$(230)^{-1}$
Legendre	1789	$(318)^{-1}$
Bessel	1841	$(299)^{-1}$
Clarke	1866	$(295)^{-1}$
Helmert	1901	$(298.2)^{-1}$
Hayford	1909	$(297.0)^{-1}$
Bowie	1928	$(297.5)^{-1}$
Heiskanen	1928	$(297.1)^{-1}$
International ellipsoid	1930	$(297.0)^{-1}$
Krassovsky	1938	$(298.3)^{-1}$
Jeffreys	1948	$(297.1)^{-1}$
Hydrostatic earth	1965	$(299.8)^{-1}$
Satellite determined	1967	$(298.247)^{-1}$
Satellite determined	1974	$(298.256)^{-1}$

earth's surface. He calculated that the centrifugal force at the equator is 1/289 that of the mass attraction, and therefore that the plumb line at Paris deviates about 6 minutes of arc from the direction in a nonrotating earth.

An arc of meridian was first measured by Picard in France in 1669 (Picard, 1675; also Todhunter, 1962, I, p. 69). Cassini and his son (see Todhunter, I, p. 51), on the basis of triangulation arcs measured in France in the latter 17th century, concluded that the shape of the earth is a prolate spheroid, not the oblate spheroid previously deduced by Newton and by Huygens. De Maupertius of the Paris Academy of Sciences (1735; also Todhunter, 1962, I, p. 69) showed that the axes of an earth ellipsoid of revolution can be calculated from lengths of a degree of latitude obtained by meridian-arc measurements.

The Paris Academy undertook to answer the question whether the earth is an oblate or prolate ellipsoid. It sent an expedition to Peru in 1735 to measure a length of arc at a low latitude, and shorly thereafter (1736) another expedition to Lapland to make similar measurements at a high latitude. Bouguer was a principal participant in the Peru expedition, who also made numerous contributions published in the Paris *Memoirs* and in his book *La Figure de la Terre* (1749). The Bouguer correction (eq. 6.10), which is applied to gravity measurements to compute the Bouguer anomaly, is widely used today. The Lapland expedition, under the direction of De Maupertius and with Clairaut as participant, was particularly successful. It demonstrated that the earth's shape is indeed an oblate spheroid. Clairaut had been conducting studies on the theory of fluid equilibrium and the effects of variations in the lengths of meridian arcs at different latitudes, which were published in the Paris *Memoirs* between about 1733 and 1741 (Todhunter, 1962, I, p. 83) and in his *Théorie de la Figure de la Terre* (1743). He derived the famous Clairaut's theorem (eq. 2.58) relating the earth's geometric ellipticity (flattening) to its centrifugal acceleration and gravitational flattening.

Many renowned mathematicians made theoretical contributions to the knowledge of the shape of the earth. MacLaurin (1742; also Todhunter, 1962, I, pp. 133—175) investigated attractions on ellipsoids of revolutions, giving exact values of the attractions at the poles and equator, and also studied tidal attractions. D'Alembert (Todhunter, 1962, I, pp. 249—304) investigated the problem of equilibrium of a fluid on a nearly spherical ellipsoid, obtaining formulas for the attraction at any point on the ellipsoid. Jacobi (Todhunter, I, p. 261) developed a theorem which showed that a rotating fluid ellipsoid, even with three different axes, can be in equilibrium.

Laplace published extensively in the Paris *Memoirs* beginning in 1776. Particularly relevant is his introduction of potentials and analyses of spherical harmonics and their coefficients, which are fundamental to the analysis of the earth's shape, and were used, for example, in the development of Laplace's tidal equation (3.26). His book *Théorie du Mouvement et de la*

Figure Elliptique des Planets (1784) is a major contribution to the theory of the attraction of ellipsoids at both internal and external points. Legendre (1785) developed the potential as the sum of elements of a body divided by their distances from a fixed point (e.g., eq. 2.11), which is basic in analyses of spherical harmonics, and he showed that attractions can be obtained as gradients of this potential. He also calculated the ellipticity of the earth, obtaining a value of 1/318.

Euler (1736), the most prolific writer in mathematics (Bell, 1937, p. 163), developed the theory of rotation of solid bodies, including ellipsoids of revolution and triaxial ellipsoids. He showed that even a slight deviation in the axes of rotation from the direction of a principal moment of inertia in an ellipsoid produces a gyroscopic wobble or free nutation, which is governed by the difference in the moments of inertia about the principal (polar and equatorial) axes. For a rigid earth, this Eulerian period is 305 days. Chandler (1891) discovered that the earth has a wobble period of 430 days, and Newcomb (1892) showed that the Chandler period or wobble is indeed a free nutation, but that the period is greater than for a rigid earth because the earth is actually elastic. The excitation of the wobble has remained a long-standing problem; recent calculations indicate (O'Connell and Dziewonski, 1976) that atmospheric motions and large earthquakes can account for the observed wobble amplitudes.

The first laboratory measurements of the universal gravitational constant (eq. 2.2) were made by Cavendish (1798) by measuring the attraction which large lead spheres exert on smaller masses. Cavendish measured the deflection of a torsion balance devised by the Rev. John Mitchell (Todhunter, 1962, II, p. 164), who died before he could conduct his own experiment with it. Cavendish obtained a value of $6.674 \cdot 10^{-8}$ cm^3 g^{-1} sec^{-2}, compared with $6.6720 \cdot 10^{-8}$, a value recommended by the International Council of Scientific Unions in 1973. Many determinations of the constant were made between 1798 and 1973 (values are listed in Nettleton, 1976, p. 13). Using his value of the constant, Cavendish computed the average density of the earth to be 5.48 g cm^{-3} (compared with the current value of 5.515).

Various investigators have calculated the ellipticity or flattening of the earth (Table 1.I) since Newton's first approximation. Good values were estimated in the 19th century, particularly by Bessel in 1841 and by Clarke in 1866, and more accurate values were obtained in the present century. Helmert (1901) determined a value which is almost identical to that now used, although astro-geodetic data available for this determination were limited. Hayford (1909) obtained an ellipticity of 1/297, based on available astro-geodetic data, particularly in North America. The Hayford value was officially adopted by the International Geodetic Association in 1924 and was used until improved ellipticities were determined from the regression of the nodes and perturbations of orbital paths of artificial earth satellites in the 1960's. Values nearly equal to Hayford's were obtained later by Bowie (1924), Heiskanen (1928), and Jeffreys (1948), using European and African as well as North American data. Measurements along a different astro-geodetic arc (Europe and Asia) led Krassovsky (1938) to obtain a value for ellipticity that is very near Helmert's earlier figure and the later satellite-determined values. Ellipticities determined from satellite orbital paths have accuracies of about two orders of magnitude greater than those from astro-geodetic data.

Stokes (1849) initiated studies on the form of the geoid using spherical harmonics to calculate undulations of the geoid from a reference ellipsoid using observed gravity anomalies throughout the earth. These investigations led to the famous Stokes' theorem (eq. 2.78), which makes it possible to map the geoid relative to a reference ellipsoid when sufficient numbers of gravity anomalies are known. Deflections of the vertical (eqs. 2.83a, b) can be readily computed from the undulations. Numerous investigators have used Stokes' theorem to determine the shape of the geoid, where the principal limiting factor has been gravity anomalies in unsurveyed areas. Hayford and Bowie (1912), and later Heiskanen (1924, 1938a), Vening Meinesz (1929, 1941a), Uotila (1960), and others have applied Stokes' theorem using available gravity data around the world to compute geoidal undulations and deflections of the vertical, particularly in areas of large gravity anomalies. Insufficient numbers of gravity-anomaly values were known over the world's oceans before about 1970, however, to compute undulations accurately. Since the early 1970's many new gravity values have been measured at sea, as a result of increasing numbers of reliable sea gravimeters and satellite-navigation receivers operating on research vessels. Geoidal shapes have also been determined from orbital paths of artificial earth satellites (e.g., King-Hele, 1958; Kaula, 1966; Gaposchkin, 1974; Wagoner, et al., 1976) Currently described geoids are based on values of spherical harmonic coefficients in which the longer-wavelength components are based on perturbations of satellite orbital paths and the shorter wavelengths on averages of observed gravity anomalies.

The first geodesist to write on isostasy appears to have been Bouguer (1749), who commented that the gravitational attraction produced by the Andes Mountains is considerably smaller than expected for the amount of mass above sea level.

The principal founders of isostasy are Pratt and Airy, who, working independently of one another, published their theories in the same year. Pratt (1855) in India had analyzed then recent triangulation measurements obtained by George Everest in northern India in which there appeared to be a 5″.24 relative deflection of the vertical between two stations, one about 700 km further from the Himalayan Mountains than the other. He concluded that the difference in deflections is real and results from the attraction of the adjacent Himalayas. Upon computing deflections which the topographic effects of the mountains should produce at the two stations, however, Pratt obtained a difference about three times greater than that observed. Thus, two-thirds of the horizontal attraction of the Himalayas at the two stations must be compensated by mass deficiencies, presumably located beneath the mountains. These analyses led Pratt to his concept of isostatic compensation. Airy (1855) in England reasoned that mountains, such as the Himalayas, should exhibit at most a small attraction at relatively large horizontal distances because the weight of a mountain mass could not be maintained by the crust. He calculated the weight to be so large that all or part of the mountain would break through the crust and sink into the underlying dense "lava," deducing therefore that the mountain mass must be supported from below. Airy proposed his concept of roots of mountains, which he considered analogous to a raft of timber floating on water, a mechanism which he discussed in detail.

Both the Pratt and Airy isostatic concepts explain compensations of mountain masses, although in radically different ways. In the Pratt view, the higher the uplift of the mountain, the smaller its mean density. In the Airy view, a constant-density mountain sinks into the denser underlying layer.

Other geodesists have later confirmed the compensation established from deflections of the vertical and from gravity values in India. Helmert (1884) in Germany maintained that continents are compensated by underlying, lower-density material in the crust. Hayford (1909, 1910) in the United States made extensive studies of the effects of large topographic masses on deflections of the vertical, from which he concluded that isostatic compensation is widespread. Heiskanen concluded from his isostatic anomaly determinations that about 85% of the earth's surface is essentially in isostatic equilibrium.

The first geologist to discuss isostatic compensation in detail was Dutton (1889) in the United States. He proposed the word "isostasy" (from the Greek "isostasis"); he would have preferred "isobary," as more descriptive of equal pressure, but this word already was an accepted meteorological term.

Methods for computing isostatic compensation, as indicated by gravity anomalies, were developed principally by Hayford (1909, 1910) based on the Pratt concept of compensation, and by Heiskanen (1924, 1938b) based on the Airy concept. Both methods assume that compensation occurs directly beneath the topographic masses. Vening Meinesz developed a regional method of compensation, with downbending of a relatively rigid lithosphere. A two-dimensional, pseudo-isostatic computational procedure has recently been used by Rabinowitz (1974) and co-workers at the Lamont-Doherty Geological Observatory of Columbia University (New York) to evaluate gravity profiles across continental margins.

The principal components of the earth tides produced by the attraction of the sun and moon can be obtained from Laplace's tidal equation (eq. 3.26), which applies to a rigid earth. In 1876 Kelvin indicated that tides deform the earth and that the earth cannot, therefore, be treated as a completely rigid body. Shortly thereafter Darwin (1882) showed that amplitudes of solid-earth tides are about one-third as large as those of ocean tides, from which it was deduced that the rigidity of a homogeneous earth is about that of steel. Love (1909) introduced the Love numbers h and k, which characterize the over-all elastic response of the earth, as can be determined from measurements with tidal gravimeters, coast tide gages, tilt meters, and strain meters. The observations can provide average values of Love numbers for the earth, and also geographic variations in their values, which can indicate lateral and vertical variations in the earth's elastic properties (provided the measurements have been corrected for effects of ocean loading and of fractures of all sizes in the vicinity of the station).

Deviations in the vertical due to tidal forces were measured with horizontal pendulums by Paschwitz in 1891 and later by Schweydar (1905), and corresponding earth deformations, also measured with horizontal pendulums, by Hecker (1907) and Orlov (1915). Tilting motions of the earth's crust caused by tidal loading were studied by Takahasi (1929). Doodson (1921) introduced a tidal constant, the Doodson constant, which is used in the analyses of tide-generating potentials and in predicting tidal forces. Tomaschek and Schaffernicht (1932) initiated gravimetric measurements of earth tides, studies which were also conducted by Wykoff (1936), Melchior (1959), and many others (Melchior, 1966). Tidal gravimeters with sensitivities of 1 μgal were constructed by the 1950's, which made it possible to study amplitudes of the separate tidal components observed at a land station.

Pendulums are the first instruments used to measure gravity accurately. In 1583 Galileo Galilei (Fleet, 1961, p. 43) observed that the to-and-fro swing of a lamp on a long chain in the Pisa Cathedral might be a way of measuring time, but it was not until 1657 that Christian Huygens (see Fleet, same page) successfully constructed a pendulum clock. Huygens (1693) and Jacob and Johannes Bernoulli were the first to make theoretical investigation on the operation of the pendulum, and Euler (1736) developed equations which related pendulum swing periods with amplitudes. Kater (1818; see also Cook, 1965) constructed a reversible pendulum with which he obtained the first accurate measurement of absolute gravity. In 1826 Bessel analyzed effects of knife edges and the lengths of the simple second-period pendulum in the determination of gravity. In 1898 Helmert described in detail the theory of the pendulum in a memoir which formed the basis of the Kühnen and Furtwängler (1906) construction of a reversible pendulum that was used to measure absolute gravity in Potsdam, Germany. This absolute value became the world gravity reference and Potsdam the base station until the 1960's, when improved values of gravity were obtained with falling-body experiments, showing the Potsdam value to be 14 mgal too high.

An excellent discussion of methods used for determining absolute gravity has been given by Cook (1965). After the 1906 gravity determination, reversible pendulums were used to provide accurate measurements of gravity in Washington, D.C. (Heyl and Cook, 1936), in Teddington, England (Clark, 1939), and in Leningrad (Yanovsky, 1958; see, e.g., Cook, 1965). These measurements already indicated that the earlier Potsdam value is too high. Later, free-fall apparatuses were built which provided more accurate measurements of absolute gravity. Preston-Thomas et al. (1960) in Ottawa, Canada, Thulin (1961) in Sèvres (Paris), France, and Faller (1965) in Princeton, New Jersey, conducted free-fall measurements. Cook (1967) constructed the first symmetric free-motion apparatus, in which a glass ball is projected upward and allowed to fall back under the force of gravity. The symmetric motion method is not subject to the systematic errors inherent in the free-fall measurements, and Cook obtained a value of gravity with an uncertainty not exceeding 0.5 mgal. Sakuma (1971) in Sères (Paris) perfected a symmetric free-motion apparatus that by 1969 measured absolute gravity with an uncertainty of a few microgals, a two order-of-magnitude increase in accuracy. There appears to be little need for improving upon the present accuracy.

Gravity differences, the difference in the value of gravity at two points, were first measured with a four-pendulum, half-second period apparatus constructed by Sterneck in 1887, who used the instrument to make numerous field measurements in Europe. Modifications of this pendulum were used extensively to establish geodetic base stations. Relative-gravity measuring pendulums were also one of the first types of geophysical instruments used in petroleum exploration. These land pendulums were modified for operation at sea (in submerged submarines) by Vening Meinesz (1929), who constructed a three-pendulum-type apparatus in which two pendulums swing in opposition to eliminate first-order effects of horizontal acceleration. Browne (1937) introduced corrections for second-order horizontal and vertical accelerations, thereby improving the accuracy of sea-pendulum measurements. Vening Meinesz made approximately a thousand measurements of gravity in the Atlantic, Pacific, and Indian oceans. The advent of the accurate, rapid-reading land gravimeters in the mid-1930's made land pendulums obsolete for relative-gravity determinations; similarly, the development of sea gravimeters by 1960 for measurements aboard surface ships made the sea-pendulum apparatus obsolete.

Eötvös (1896) constructed a torsion balance in the 1880's to make field measurements of curvatures in the earth's gravity field. Considerable time was required to obtain a gradient measurement, so this instrument became obsolete by the mid-1930's when the more useful gravimeters became available.

Land gravimeters, with the short times required to obtain highly sensitive field measurements, were developed in various forms in the 1930's by different persons and organizations in the United States and Europe. These instruments are similar to long-period seismographs; both consist essentially of a mass on a spring acted upon by gravity, where the spring provides the restoring force. (Chapter 5 shows operating principles of several types of gravimeters.) Early gravimeters exhibited relatively large amounts of drift and had limited ranges over which measurements could be made without resetting the reading dial. LaCoste and Romberg introduced geodetic gravimeters in the 1950's, which have low drift rates, so that these meters could be used anywhere on the earth's surface without resetting the reading dial, thereby greatly facilitating gravity measurements between distant stations.

Relative gravity has been measured at sea by instruments based on different principles: barometric pressure, springs, and vibrating strings, as well as pendulums. Hecker (1903) made the earliest measurement of gravity at sea, between Rio de Janeiro and Lisbon, using a gas-pressure device; the measurements were not satisfactory, however. Haalck (1931) used an improved meter of this type aboard ship between Hamburg and Bremen. Although he obtained measurements accurate to about 5 mgal in calm seas, work on this meter was not continued.

The earliest method used to measure gravity in shallow waters was a land gravimeter mounted on a tripod whose legs rested on the bottom, with the meter above the water surface. Pepper (1941) developed an underwater housing in which a gravimeter is lowered to the bottom in coastal regions, where leveling of the sensing unit is obtained with a remote-control system. LaCoste and Romberg (LaCoste, 1952a, b) constructed a stabilized, servo-controlled elevator system for use with an underwater meter to average out disturbing microseismic accelerations on the shelf bottom. Diving bells were also designed for submergence to shallow bottoms, housing both the operator and meter (Frowe, 1947).

A vibrating-string-type gravimeter was constructed by Gilbert (1949) and tested in a submarine. Other investigators subsequently modified this Gilbert meter for use aboard surface ships, mounting it on a stabilized platform (e.g., Lozinskaya, 1959; Tsuboi et al., 1961; Tomoda and Kanomori, 1962). Nonlinearities inherent in the system were essentially eliminated by mounting two oppositely directed vibrating accelerometers (Wing, 1969).

Askania Werke in Germany and LaCoste and Romberg in the United States modified their beam-type land gravimeters by 1957 for operation on surface ships by increasing the meter damping. The Askania meter operated on a stabilized platform and the LaCoste and Romberg meter in a gimbal suspension. Initial meter systems were only partly successful. Damping in the Askania meters was greatly increased in the early 1960's, and by 1965 the LaCoste and Romberg meters were modified to operate on a stabilized platform, so as to measure gravity more satisfactorily at larger ship accelerations. Both meters are subject to a cross-coupling of vertical and horizontal components of ship accelerations when operated on a stabilized platform. To correct for such coupling, horizontal accelerations acting on the meter are measured, the cross-coupling correction computed, and then subtracted from the measured gravity. The Bell-Aerospace Company constructed a force-balanced accelerometer in which a mass is constrained in a null position by means of a servo loop. This system is not subject to appreciable cross-coupling; however, the system has not been widely used.

In the early 1970's Askania developed an axially symmetric, spring-type gravimeter for use at sea which is not affected by cross-coupling accelerations. LaCoste and Romberg is also developing an axially symmetric unit. Axially symmetric meters should provide more reliable measurements at sea than the beam-type meters; cross-coupling corrections are difficult to determine accurately at large ship accelerations encountered during high-sea states.

With the use of satellite receivers, developed in the 1960's to provide quite accurate navigation positioning, many gravity measurements obtained in the world oceans by the late 1960's and the 1970's have uncertainties of only a few milligals. At the time of this writing, regional gravity has been measured on much of the earth's land and ocean surfaces. To obtain complete world coverage, additional surveys are needed, particularly in relatively inaccessible areas (including the Arctic, Antarctic, and Central Asian highlands).

Airborne measurements (both in airplanes and in helicopters) have been attempted with sea gravimeters since the 1960's. Results have not been encouraging, however, primarily because long-period vertical craft acceleration could not be accurately compensated by the meter systems. Additional problems encountered in airplanes include large Eötvös corrections (described in Chapter 6) for even small errors in flight course or velocity, and appreciable amplitude diminutions as well as time shifts in anomaly profiles, produced in rapid transit across an anomaly by a meter which itself averages values over periods of minutes (a consequence of heavy meter damping).

Prospects for airborne measurements nevertheless appear to be improving, particularly with slow-flying planes or helicopters (capable of covering substantial distances in otherwise inaccessible areas) which have auto-pilots and automatic height stabilizers, or in which vertical velocities can be maintained uniform for a duration several times that of the meter-averaging period. Satellite data now make it possible to determine ground elevations accurately (to within a fraction of a meter) and also changes in sea-level heights (as across a continental margin). Height above ground can be measured accurately with modern radar. Such data should make it possible to measure accurately changes in plane altitude. If these changes can be kept linear for times long compared with meter averaging time, it should be possible to correct for long-period altitude changes.

Airborne measurements would be enhanced by decreasing meter averaging times from those in the present sea gravimeters (designed for heavy sea acceleration); then altitude changes during shorter time periods could be applied to measured gravity. Measurements would also be improved if made only when a craft can be maintained at a uniform altitude or when altitude changes are linear during required periods.

Efforts have been underway to develop an airborne gradiometer, which measures gravity gradients. Gravity gradiometers are essentially unaffected by vertical craft accelerations; they therefore have appeal for airborne measurements. Gradiometers do not yet have the sensitivity to provide gravity determinations accurate to about 1 mgal. Gradiometers do have application in satellite measurements.

Chapter 2

GRAVITATIONAL ACCELERATION AND POTENTIAL—GEOID AND REFERENCE SPHEROID

The earth's gravity field is usually described by its gravitational potential; components of gravity are spacial derivatives of the potential. Earth potentials, geopotentials, are commonly described as constant or equipotential surfaces produced by the earth's mass (including irregular mass distributions) and its rotation. The constant geopotential corresponding to mean sea level over the oceans is the geoid. The geoid is thus an equipotential surface which, to a first approximation, is represented by mean sea level.

Irregular mass distributions in the earth, particularly in the outer layers, affect the shape of the geoid and, in fact, cause the geoid to be sufficiently irregular that it cannot readily be described mathematically. The general shape of the geoid (Fig. 2.1) can be approximated by an oblate spheroid, for which formulas are developed.

To sufficient accuracy, the spheroid is approximated by an ellipsoid of revolution with a specific geometric ellipticity or flattening $e = (a - c)/a$, where a and c are the equatorial and polar radii. The best value of e is 1/298.256, which has been determined from the regression of the nodes and perturbations of orbital paths of artificial earth satellites. Before the satellite era, the earth's ellipticity was obtained first theoretically, and later from

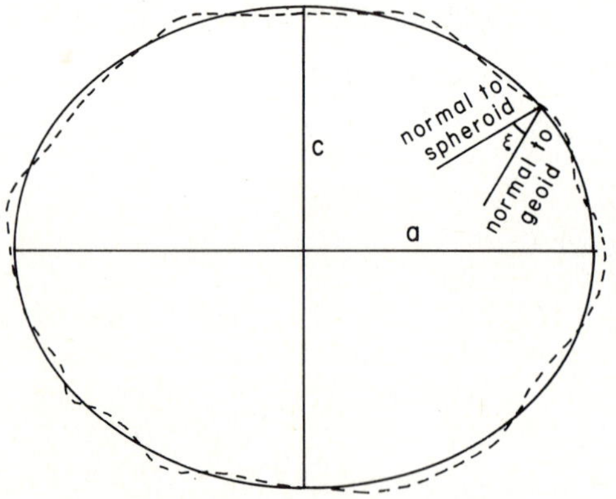

Fig. 2.1. Earth spheroid (solid curve), with equatorial and polar axes a and c, and hypothetical geoid (dashed curve).

gravity measurements made on long astro-geodetic arcs on different continents. Ellipticities determined from regressions of the nodes of satellites are considerably more accurate than those obtained from arcs. Table 1.I lists various estimated values of ellipticity.

GRAVITATIONAL ACCELERATION

The force of attraction between two point masses or two homogeneous spheres is described precisely by Newton's law of gravitation:

$$F = G\frac{mm'}{q^2} \tag{2.1}$$

in which m and m' are the point or spherical masses separated by a distance q. G is the gravitational constant of proportionality, which is:

$$G = 6.6720 \cdot 10^{-8}\ \mathrm{cm}^3\ \mathrm{g}^{-1}\ \mathrm{sec}^{-2} \tag{2.2}$$

The force at m is identical to that at m', but has the opposite sense. Acceleration is the force per unit mass. Hence, the acceleration of the mass m' (point P, Fig. 2.2) towards m is:

$$g = \frac{F}{m'} = G\frac{m}{q^2} \tag{2.3}$$

The acceleration of m' toward a generalized mass (Fig. 2.3) is the acceleration produced by all point masses. Accelerations due to the individual masses are summed geometrically, because they do not generally act in the same direction; components in the same direction are added algebraically. In calculating gravity at a point P, the vertical component is

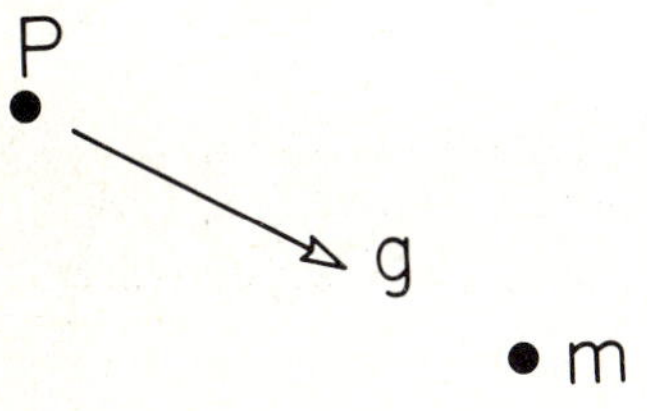

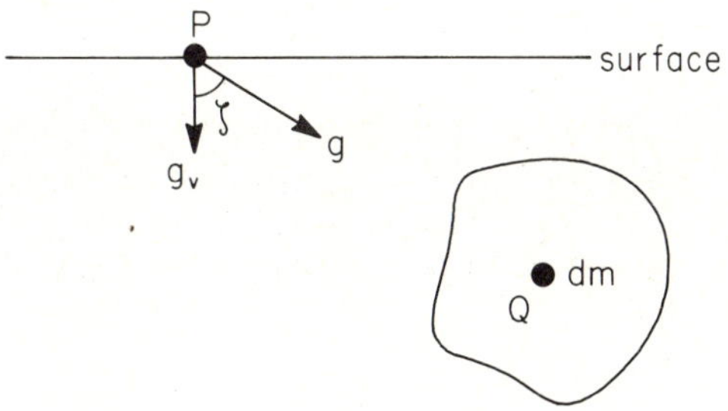

Fig. 2.2. Acceleration g at point P produced by a point mass m.

Fig. 2.3. Acceleration g and its vertical component g_v at a point P produced by a generalized mass m; element dm at point Q.

usually of interest since it acts in the same general direction as the earth's field. The vertical component g_v at P (Fig. 2.3) is given by:

$$g_v = G \int_\tau \frac{\rho \cos \zeta \, d\tau}{q^2} \tag{2.4}$$

in which ρ is the density in the body with volume τ, q is the vector from P (where gravity is measured) to Q, $d\tau$ an element of volume, and ζ the angle q makes with the vertical.

GRAVITATIONAL POTENTIAL

A particle of unit mass at a very great distance from a mass m, when released, will move freely toward m because of the work done by the gravity field produced by m. The work done in moving the unit mass is the acceleration over the distance moved, that is, from infinity to a point at distance q from the center of m. Thus:

$$V = G \int_\infty^q \frac{m \, dq}{q^2} = -G \frac{m}{q} \tag{2.5}$$

in which V is the potential of mass m at distance q. The sign of V is negative, a convention used in physics. In geodesy and astronomy (see, e.g., Kaula, 1963, p. 508), and also in geophysics, the sign convention is to take the gravitational potential positive. We will write V positive throughout the book; thus:

$$V = G \frac{m}{q} \tag{2.6}$$

Potential energy is the capacity to do work, whereas gravitational potential is the work done by gravity. Gravitational potential is the negative of potential energy per unit mass.

The work gravity does on the unit mass as it travels a distance dq is $-g dq$ (where g is taken positive in the direction toward the mass m and dq is positive in the opposite direction). This work is the change from potential V to potential $V + dV$ (Fig. 2.4). Thus:

$$dV = -g \, dq \tag{2.7}$$

or

$$g = -\frac{\partial V}{\partial q} \tag{2.8}$$

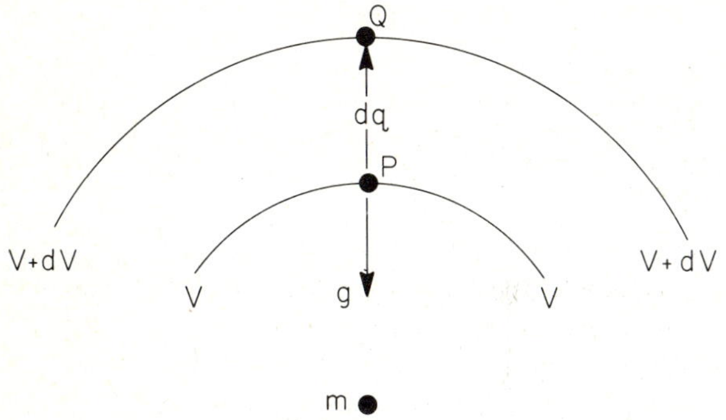

Fig. 2.4. Potential surfaces V at point P and $V + V$ at Q produced by point mass m; g acts toward m; q is distance outward from m.

The negative sign indicates that gravity decreases with increasing distance from the attracting mass.

The potential produced by a distributed mass is the sum of the individual potentials at point P, that is:

$$V = \sum_i G \frac{m_i}{q_i} = G \int_m \frac{dm}{q} = G \int_\tau \frac{\rho \, d\tau}{q} \tag{2.9}$$

The component of gravity at P in any direction u is:

$$g_u = -\frac{\partial V}{\partial u} = -\frac{\partial V}{\partial q}\frac{\partial q}{\partial u} \tag{2.10}$$

GEOPOTENTIALS AND THE GEOID

The geopotential at a point P is the sum of the potentials resulting from the earth's mass attraction and its rotation. The part produced by the attraction of the earth's mass M is, similar to eq. 2.9:

$$V = G \int_M \frac{dM}{q} \tag{2.11}$$

in which the origin of coordinates is at the earth's center of gravity (Fig. 2.5), and:

$$q = (r_0^2 + r^2 - 2r_0 r \cos \psi)^{1/2} = [(x_0 - x)^2 + (y_0 - y)^2 + (z_0 - z)^2]^{1/2} \tag{2.12}$$

is the distance from a point $P(x_0, y_0, z_0)$ to an element of mass $dM(x, y, z)$.

The part of the geopotential produced by rotation is obtained from the centrifugal acceleration acting at P (Fig. 2.6). This is $-\omega^2 r_0 \cos \phi$, in which

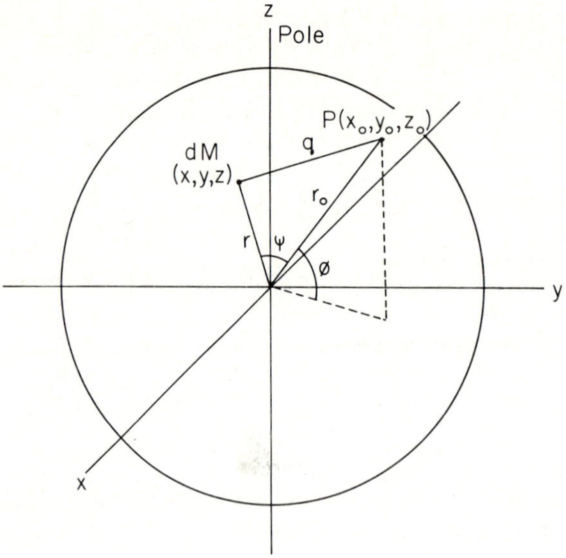

Fig. 2.5. Coordinates at point P on or outside the earth's surface and an element of mass dM; q is distance from P to dM.

r_0 is the radius to P, $\omega(7.29211 \cdot 10^{-5}\,\text{rad sec}^{-1})$ the earth's angular velocity, and ϕ the geocentric latitude at P. The negative sign indicates that the acceleration acts outward from the earth; the radial component is cos ϕ of the centrifugal acceleration. The rotational potential is related to the radial component of centrifugal acceleration by:

$$-\frac{\partial}{\partial r}\left(\tfrac{1}{2}\,\omega^2 r_0^2 \cos^2\phi\right) = -\,\omega^2 r_0 \cos^2\phi \qquad (2.13)$$

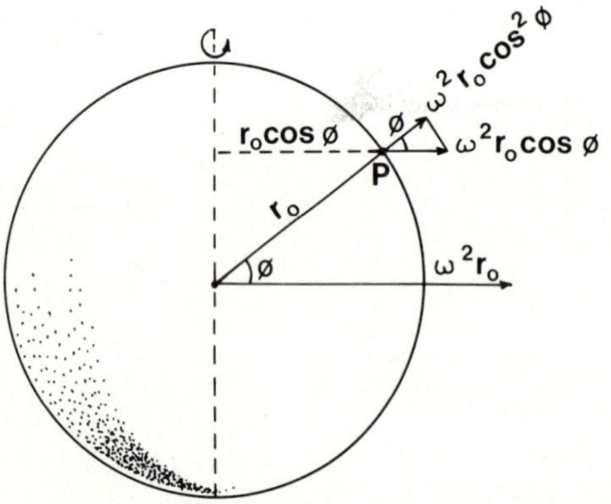

Fig. 2.6. Components of centrifugal acceleration at a point P (latitude ϕ) on the surface of the earth rotating with angular velocity ω.

Because potentials are scalar quantities, the geopotential is the sum of the attraction and rotational potentials, even though the work done by attraction moves a unit mass from infinity and the work done by rotation moves a mass from the rotation axis. The geopotential W is, therefore:

$$W = G\int_M \frac{\mathrm{d}M}{q} + \tfrac{1}{2}\,\omega^2 r_0^2 \cos^2\phi \tag{2.14}$$

which is constant (on any one geopotential).

Work which the earth's field does in moving a unit mass radially from geopotential W to $W + \mathrm{d}W$ is, similar to eq. 2.7, $\mathrm{d}W = -g_r dr$. The component of gravity in a radial direction is therefore:

$$g_r = -\frac{\partial W}{\partial r} \tag{2.15a}$$

Similarly, the components of gravity in the z direction (earth's rotational axis) and in the x and y directions (equatorial axes) are:

$$g_z = -\frac{\partial W}{\partial z} \qquad g_x = -\frac{\partial W}{\partial x} \qquad g_y = -\frac{\partial W}{\partial y} \tag{2.15b}$$

The total gravity vector at P (Fig. 2.7) is:

$$g = -\left[\left(\frac{\partial W}{\partial x}\right)^2 + \left(\frac{\partial W}{\partial y}\right)^2 + \left(\frac{\partial W}{\partial z}\right)^2\right]^{1/2} \tag{2.16a}$$

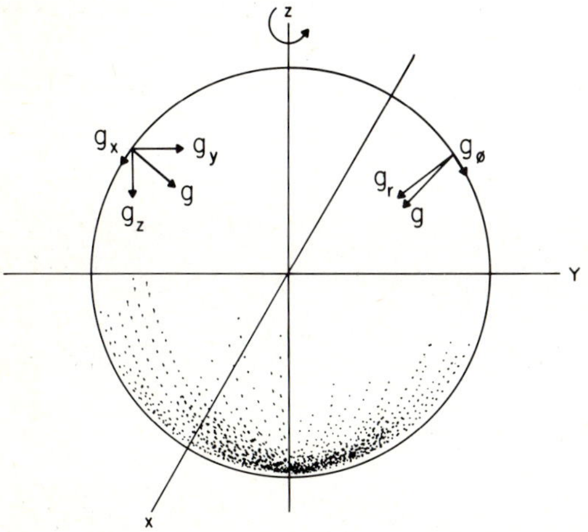

Fig. 2.7. Direction of the acceleration of gravity g and its components in rectangular and spherical coordinates (g is assumed independent of longitude).

which can be written, for g independent of longitude:

$$g = -\left[\left(\frac{\partial W}{\partial r}\right)^2 + \left(\frac{\partial W}{r\partial\phi}\right)^2\right]^{1/2} \tag{2.16b}$$

In vector notation:

$$g = -\text{ gradient } W = -\nabla W \tag{2.16c}$$

in which the operator ∇ ("del") indicates a gradient.

Eqs. 2.16 show that the gravity vector is normal to geopotential surfaces. No component of gravity thus acts along an equipotential surface, which also follows from the fact that gravity does work only from one equipotential surface to another.

Geopotentials represent surfaces of hydrodynamic equilibrium on a rotating earth, as would exist if the earth were covered by water. At sea the geoid is the geopotential surface representing mean sea level. On land the geoid can be envisaged by the heights to which water would rise in canals dissecting the continents. However, where surrounding elevations extend above the canals, water would rise above the geoid because of adjacent upward attractions.

The shape of the geoid, or any geopotential surface, results from mass distributions in the earth. If these distributions were known, the potential (eq. 2.14) could be obtained directly by integration. But mass distributions in the earth are not known, so an inverse procedure is used: mass anomalies are derived from an indirectly determined shape of the geoid.

Components of gravity on the geoid can be obtained by differentiation with respect to corresponding directions. We write eq. 2.14 in the form:

$$W = G\int_\tau \frac{\rho d\tau}{[(x_0 - x)^2 + (y_0 - y)^2 + (z_0 - z)^2]^{1/2}} + \tfrac{1}{2}\,\omega^2(x_0^2 + y_0^2) \tag{2.17}$$

The components of acceleration at P are:

$$g_z = -\frac{\partial W}{\partial z_0} = G\int_\tau \frac{\rho(z_0 - z)d\tau}{q^3} \tag{2.18a}$$

$$g_x = -\frac{\partial W}{\partial x_0} = G\int_\tau \frac{\rho(x_0 - x)d\tau}{q^3} - \omega^2 x_0 \tag{2.18b}$$

$$g_y = -\frac{\partial W}{\partial y_0} = G\int_\tau \frac{\rho(y_0 - y)d\tau}{q^3} - \omega^2 y_0 \tag{2.18c}$$

and, similarly:

$$g_r = G\int_\tau \frac{\rho(r_0 - r\cos\psi)d\tau}{q^3} - \omega^2 r_0 \cos^2\phi \tag{2.18d}$$

in which ψ is the included angle between r and r_0 (Fig. 2.5).

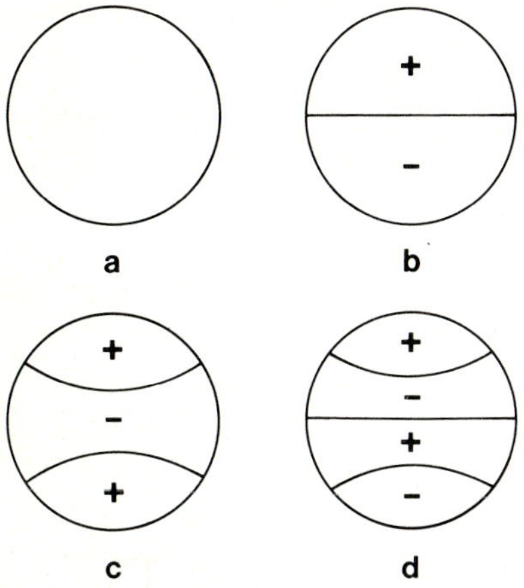

Fig. 2.8. Positive and negative regions of zonal harmonics or Legendre polynomials, $P_n(\cos\theta)$ of degree n in colatitude θ; a. $P_0(\cos\theta)$; b. $P_1(\cos\theta)$; c. $P_2(\cos\theta)$; d. $P_3(\cos\theta)$.

Because eq. 2.14 cannot be integrated, the usual procedure for describing the geoid is to define a reference spheroid on which geoidal undulations are superimposed. The potential on a spheroid can be represented by a series of spherical harmonics, which express a nonanalytic function on a sphere in a manner analogous to the way one-dimensional Fourier-series terms can represent a nonanalytic function on a circle. The zero-order term in either series represents the mean value of the function over the sphere or circle. Successively higher-order terms represent deviations from the mean value, with the higher order the term, the smaller its horizontal dimensions. Fig. 2.8 illustrates zonal harmonics, showing nodal lines where the harmonics change sign.

At points on or outside the earth, the potentials V and W satisfy Laplace's equation, and at interior points the potentials satisfy Poisson's equation. This is shown in the following section.

Laplace's and Poisson's equations

At $P(x_0, y_0, z_0)$ on or outside the earth, $\rho = 0$ and $q \neq 0$. Second derivatives of W with respect to the coordinates at P are not indeterminate. Thus:

$$
\begin{aligned}
\frac{\partial^2 W}{\partial x_0^2} &= 3G\int_\tau \frac{\rho(x_0 - x)^2\, d\tau}{q^5} - G\int_\tau \frac{\rho d\tau}{q^3} + \omega^2 \\
\frac{\partial^2 W}{\partial y_0^2} &= 3G\int_\tau \frac{\rho(y_0 - y)^2\, d\tau}{q^5} - G\int_\tau \frac{\rho d\tau}{q^3} + \omega^2 \qquad (2.19) \\
\frac{\partial^2 W}{\partial z_0^2} &= 3G\int_\tau \frac{\rho(z_0 - z)^2\, d\tau}{q^5} - G\int_\tau \frac{\rho d\tau}{q^3}
\end{aligned}
$$

Adding terms gives Laplace's equation:

$$\nabla^2 W = \frac{\partial^2 W}{\partial x_0^2} + \frac{\partial^2 W}{\partial y_0^2} + \frac{\partial^2 W}{\partial z_0^2} = 2\omega^2 \sim 0 \qquad (2.20a)$$

in which $\omega^2 = 5 \cdot 10^{-9}$ rad^2 sec^{-2} and the operator:

$$\nabla^2 = \frac{\partial^2}{\partial x^2} + \frac{\partial^2}{\partial y^2} + \frac{\partial^2}{\partial z^2}$$

is the Laplacian. For the nonrotating potential V:

$$\nabla^2 V = 0 \qquad (2.20b)$$

Inside the earth $q = 0$ at point P. The integrals forming Laplace's equation then become indeterminate. The terms must therefore be evaluated in a different way. This can be done by letting S' be the surface of a sphere surrounding a point mass m, as seen in Fig. 2.9. The outward flux of lines of gravitational acceleration across S' can be equated as:

$$\int_{S'} g_n \, dS' = -G \int_{S'} \frac{m dS'}{q^2} = -Gm \int_0^{4\pi} d\Omega = -4\pi Gm \qquad (2.21)$$

in which g_n is the outward normal of g across S'; $d\Omega = dS'/q^2$ is the solid angle ($\Omega = 4\pi$ for a sphere). Let S be an irregular closed surface and dS' the projection of dS upon S', centered about mass m. Then:

$$\frac{dS'}{q'^2} = \frac{\cos \zeta \, dS}{q^2}$$

(Fig. 2.9) where ζ is the angle between the radius vector from m and the

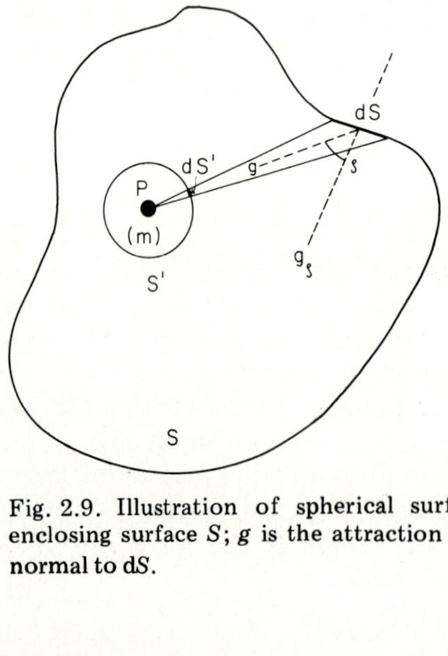

Fig. 2.9. Illustration of spherical surface S' surrounding mass m within an irregular enclosing surface S; g is the attraction produced by m on surface S and g_ζ its component normal to dS.

normal to dS, and:

$$\int_S g_\xi \, dS = \int_{S'} g_n \, dS' = -4\pi G m \qquad (2.22a)$$

By including all point masses within S, $\Sigma m = M$:

$$\int_S g_\xi \, dS = -4\pi G M = -4\pi G \int_\tau \rho \, d\tau \qquad (2.22b)$$

indicating that the total gravitational flux through any closed surface equals $-4\pi G$ times the mass enclosed by the surface; this is Gauss's law. We now apply Gauss's theorem (the surface integral of a vector g taken over a closed surface is equal to the volume integral of the divergence of g over any volume τ within the closed surface):

$$\int_S g_\xi \, dS = \int_\tau \left(\frac{\partial g_x}{\partial x} + \frac{\partial g_y}{\partial y} + \frac{\partial g_z}{\partial z} \right) d\tau \qquad (2.23)$$

in which the divergence of g is $\nabla \cdot g$, given by:

$$\nabla \cdot g = \frac{\partial g_x}{\partial x} + \frac{\partial g_y}{\partial y} + \frac{\partial g_z}{\partial z} = \frac{\partial^2 V}{\partial x^2} + \frac{\partial^2 V}{\partial y^2} + \frac{\partial^2 V}{\partial z^2} = \nabla^2 V \qquad (2.24)$$

Substituting the right side of eq. 2.24 into the right side of 2.23 and equating with 2.22b produces Poisson's equation:

$$\begin{aligned} \nabla^2 V &= -4\pi G\rho \\ \nabla^2 W &= -4\pi G\rho + 2\omega^2 \end{aligned} \qquad (2.25)$$

in which ρ is the density at interior point P with potential V or W.

Poisson's equation clearly reduces to Laplace's equation where $\rho = 0$, that is, at the surface and at exterior points. A discontinuity in ρ, as at the earth's surface or at the core-mantle boundary, produces a discontinuity in $\nabla^2 V$ and in $\nabla^2 W$

EARTH SPHEROID AND ELLIPSOID

The reference spheroid, described mathematically in terms of radius and latitude, approximates the surface of the geoid, enclosing the total mass and volume within the sea-level surface. The spheroid represents an ideal earth consisting of uniform-density layers. Spheroidal potential surfaces U are smooth, in contrast to the more irregularly shaped geoidal surface. Gravity on the spheroid also varies smoothly, in contrast to that on the geoid.

The earth was established to be an oblate spheroid from measurements of the length of a degree of latitude at different latitudes (illustrated in Fig. 2.10). The form of the spheroid is usually taken to be an ellipsoid of

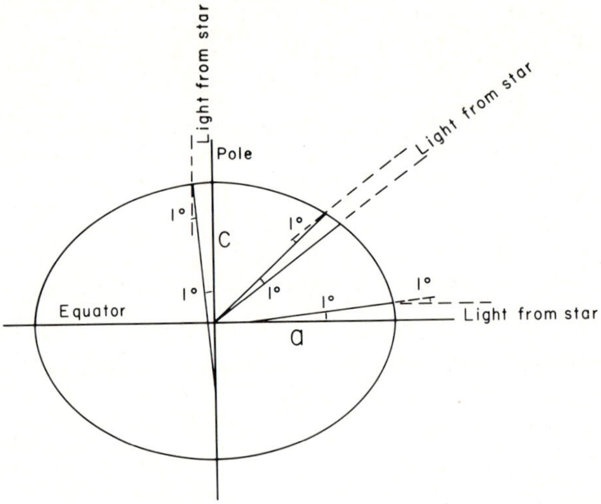

Fig. 2.10. Illustration showing differences in radial lengths corresponding to curvatures of one-degree arcs on an oblate spheroid.

revolution whose geometric ellipticity or flattening has been determined accurately from observed orbital paths of artificial satellites.

The gravitational potential on the spheroid satisfies Laplace's equation. For the attraction potential V, Laplace's equation is, in spherical polar coordinates:

$$\nabla^2 V = \frac{\partial^2 V}{\partial r^2} + \frac{2}{r}\frac{\partial V}{\partial r} + \frac{1}{r^2 \sin\theta}\frac{\partial}{\partial\theta}\left(\sin\theta \frac{\partial V}{\partial\theta}\right) + \frac{1}{r^2 \sin^2\theta}\left(\frac{\partial^2 V}{\partial\lambda^2}\right) = 0 \qquad (2.26)$$

in which r is the radius, θ the co-latitude and λ the longitude. On a spheroid of revolution, which is independent of longitude, the last term in eq. 2.26 drops out. Solutions of V then reduce to sums of zonal harmonics (Fig. 2.8), that is, sums of Legendre polynomials multiplied by coefficients which include powers of radial distance. On a spheroidal earth with equatorial radius a and no sources outside the spheroid, V has solutions of the form (e.g., Stacey, 1977, p. 320):

$$V = \frac{1}{r}\sum_{l=0}^{\infty} C_l \left(\frac{a}{r}\right)^l P_l(\cos\theta) \qquad (2.27)$$

in which C_l are constant coefficients representing internal mass sources, and $P_l(\cos\theta)$ the Legendre polynomials. The polynomials are given by Rodrigues' formula (e.g., Arfken, 1970, p. 554):

$$P_l(\cos\theta) = \frac{1}{2^l l!}\frac{\mathrm{d}^l(\cos^2\theta - 1)^l}{\mathrm{d}(\cos\theta)^l} \qquad (2.28)$$

which have known values. For l = 0, 1, 2, and 3 the values are:

$$
\begin{aligned}
P_0(\cos\theta) &= 1\\
P_1(\cos\theta) &= \cos\theta\\
P_2(\cos\theta) &= \tfrac{1}{2}(3\cos^2\theta - 1)\\
P_3(\cos\theta) &= \tfrac{1}{2}(5\cos^3\theta - 3\cos\theta)
\end{aligned}
\qquad (2.29)
$$

Potential on the spheroid

The potential at a point outside a nonrotating earth (eq. 2.27) can be written (e.g., Stacey, 1977, p. 51):

$$V = \frac{GM}{r}\left[J_0P_0(\cos\theta) - \left(\frac{a}{r}\right) J_1P_1(\cos\theta) - \left(\frac{a}{r}\right)^2 J_2P_2(\cos\theta) + \ldots\right] \qquad (2.30)$$

in which J's are dimensionless coefficients representing mass distribution within the spheroid. By inserting values of $P_l(\cos\theta)$ to terms r^{-3}, we obtain:

$$V = \frac{GM}{r}\left[J_0 - \left(\frac{a}{r}\right) J_1 \cos\theta - \frac{1}{2}\left(\frac{a}{r}\right)^2 J_2(3\cos^2\phi - 1)\right] \qquad (2.31)$$

The term J_0 must be unity, because the potential reduces to that for a point mass at very large distances, namely:

$$V = \frac{GM}{r} \qquad (2.32)$$

J_1 is zero because the origin of coordinates is taken at the center of gravity (see also eq. 2.37). J_2 remains to be determined; it describes the oblate spheroidal shape of the earth. Eq. 2.31, after substituting latitude $\phi = \pi/2 - \theta$, then becomes:

$$V = \frac{GM}{r}\left[1 - \frac{1}{2}\left(\frac{a}{r}\right)^2 J_2(3\sin^2\phi - 1)\right] \qquad (2.33)$$

The potential on a spheroid can similarly be written:

$$U = \frac{GM}{r}\left[1 - \frac{1}{2}\left(\frac{a}{r}\right)^2 J_2(3\sin^2\phi - 1) + \frac{1}{2}\left(\frac{\omega^2 r^3 \cos^2\phi}{GM}\right)\right] \qquad (2.34)$$

The J_2 term will be shown to represent ratios of moments of inertia.

The potential at an exterior point P (eq. 2.11) due to all interior masses in the earth (Fig. 2.5) can be written, applying the law of cosines:

$$V = G\int_M \frac{dM}{q} = G\int_M \frac{dM}{r_0\left[1 + \left(\frac{r}{r_0}\right)^2 - 2\left(\frac{r}{r_0}\right)\cos\psi\right]^{1/2}} \qquad (2.35)$$

The quantity q^{-1} can be expanded in powers of r^{-1}; we then rewrite eq. 2.35, to terms including r^{-3}:

$$V = G \int_M \frac{1}{r_0} \left[1 + \left(\frac{r}{r_0}\right) \cos\psi + \left(\frac{r}{r_0}\right)^2 - \frac{3}{2}\left(\frac{r}{r_0}\right)^2 \sin^2\psi \right] \mathrm{d}M \qquad (2.36a)$$

which can be rewritten:

$$V = \frac{G}{r_0} \left[\int_M \mathrm{d}M + \frac{1}{r_0} \int_M r\cos\psi\, \mathrm{d}M + \frac{1}{r_0^2} \int_M r^2\, \mathrm{d}M - \frac{3}{2}\frac{1}{r_0^2} \int_M r^2 \sin^2\psi\, \mathrm{d}M \right] \qquad (2.36b)$$

The first term on the right is the potential if the entire earth mass were concentrated at its center. The second term is zero when the origin of coordinates is at the center of gravity. This is seen when orthogonal coordinates are placed at the center with the u axis (Fig. 2.11) extending through P, such that:

$$\int_M r\cos\psi\, \mathrm{d}M = \int_M u\, \mathrm{d}M$$

The center of gravity or centroid of mass is:

$$\bar{u} = \frac{\int_M u\, \mathrm{d}M}{\int_M \mathrm{d}M}$$

which is zero when the origin of coordinates is at the center of gravity. Hence:

$$\int_M r\cos\psi\, \mathrm{d}M = 0 \qquad (2.37)$$

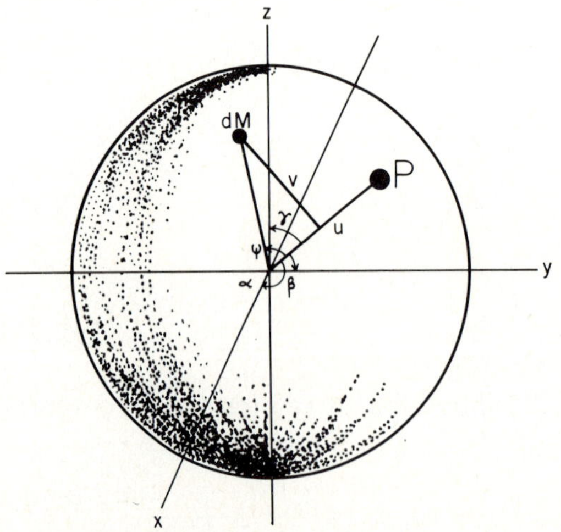

Fig. 2.11. Coordinates of dM used to obtain its moment of inertia $\int v^2 \mathrm{d}M$ about the radius vector to point P.

The third term in eq. 2.36b can be written in terms of moments of inertia:

$$\left.\begin{aligned} \frac{G}{2r_0^3}\int_M 2(x^2+y^2+z^2)\,\mathrm{d}M &= \frac{G}{2r_0^3}\left[\int_M (x^2+y^2)\,\mathrm{d}M \right. \\ &\quad \left. + \int_M (y^2+z^2)\,\mathrm{d}M + \int_M (z^2+x^2)\,\mathrm{d}M\right] \\ &= \frac{G}{2r_0^3}(C+A+B) \end{aligned}\right| \tag{2.38}$$

in which C is the moment of inertia about the earth's polar (z) axis and A and B the moments about equatorial (x and y) axes.

The fourth term in eq. 2.36b is the moment of inertia I about an axis u through P (Fig. 2.11). Thus:

$$\frac{3}{2}\frac{G}{r_0^3}\int_M r^2\sin^2\psi\,\mathrm{d}M = \frac{3}{2}\frac{G}{r_0^3}\int_M v^2\,\mathrm{d}M = \frac{3}{2}\frac{G}{r_0^3}(I) \tag{2.39}$$

Eq. 2.36b can be written, where subscripts are now dropped and B is taken equal to A:

$$V = \frac{GM}{r} - \frac{G}{2r^3}(C+2A-3I) \tag{2.40}$$

which is MacCullagh's formula.

The moment of inertia about u can be expressed as components about the x, y, and z axes. The components can be written in terms of direction cosines l, m, n (Fig. 2.11), where the moment of dM about u, $\int_M v^2\,\mathrm{d}M$, has components $\int_M (x^2+y^2)n^2\,\mathrm{d}M$ about the z axis, $\int_M (y^2+z^2)l^2\,\mathrm{d}M$ about the x axis, and $\int_M (z^2+x^2)m^2\,\mathrm{d}M$ about y. Hence:

$$\int_M v^2\,\mathrm{d}M = \int_M [(x^2+y^2)n^2 + (y^2+z^2)l^2 + (z^2+x^2)m^2]\,\mathrm{d}M \tag{2.41}$$

We substitute $n^2 = \cos^2\theta = \sin^2\phi = 1-(l^2+m^2)$ to obtain:

$$I = Cn^2 + A(l^2+m^2) = C\sin^2\phi + A(1-\sin^2\phi) \tag{2.42}$$

Equation 2.40 can now be written:

$$V = \frac{GM}{r}\left[1 - \frac{(C-A)}{2Mr^2}(3\sin^2\phi - 1)\right] \tag{2.43}$$

Comparing this equation with eq. 2.33 shows that the J_2 term in the potential on the spheroid is:

$$J_2 = \frac{C-A}{Ma^2} \tag{2.44}$$

The total geopotential on a rotating spheroid is given by:

$$U = \frac{GM}{r} \left[1 - \frac{1}{2r^2} \left(\frac{C-A}{M}\right) (3 \sin^2 \phi - 1) + \frac{1}{2} \left(\frac{\omega^2 r^3 \cos^2 \phi}{GM}\right)\right] \tag{2.45}$$

Shape of the reference spheroid (ellipsoid of revolution)

The equation for the spheroid can be obtained by using eq. 2.45, in which $U = U_0$. The spheroid has the radius:

$$r_0 = \frac{GM}{U_0} \left[1 - \frac{1}{2} \left(\frac{C-A}{Ma^2}\right) (3 \sin^2 \phi - 1) + \frac{1}{2} \left(\frac{\omega^2 a}{GM/a^2}\right) \cos^2 \phi\right] \tag{2.46}$$

in which a is substituted for r in the last two terms, both of which are small quantities. The term GM/a^2 is within 2% of the value of gravity at the equator. We define the parameter:

$$p = \frac{\omega^2 a}{GM/a^2} = \frac{\omega^2 a}{g_{eq}} = \left(\frac{\text{centrifugal acceleration at equator}}{\text{gravitational acceleration at equator}}\right) \tag{2.47}$$

The radius of the spheroid can be written:

$$r_0 = \frac{GM}{U_0} [1 - \tfrac{1}{2} J_2 (3 \sin^2 \phi - 1) + \tfrac{1}{2} p(1 - \sin^2 \phi)] \tag{2.48}$$

Rearranging terms, we obtain to first-order terms in J_2 and p:

$$r_0 = \frac{GM}{U_0} [1 + \tfrac{1}{2}(J_2 + p)] [1 - \tfrac{1}{2}(3J_2 + p) \sin^2 \phi] \tag{2.49}$$

which is of the form:

$$r = a(1 - e \sin^2 \phi) \tag{2.50}$$

an ellipse to terms of the order of e, the geometric ellipticity. For an ellipse:

$$a = \frac{GM}{U_0} [1 + \tfrac{1}{2}(J_2 + p)] \tag{2.51}$$

and:

$$e = \tfrac{1}{2}(3J_2 + p) \tag{2.52}$$

The equation for the reference spheroid (ellipsoid), including second-order terms in the theory and expressed in terms of geographic instead of geocentric latitude, is given by eqs. 2.67, 68, 70, and 72.

Gravity on the reference spheroid (normal or theoretical gravity)

Gravity along the radius vector at a point outside the spheroid can be obtained by differentiating the potential with respect to r. Thus:

$$g_r = -\frac{\partial U}{\partial r} = \frac{GM}{r^2} - \frac{3}{2}\frac{Ma^2 J_2}{r^4}(3\sin^2\phi - 1) - \omega^2 r(1-\sin^2\phi) \tag{2.53}$$

Callandreau (1889) showed that this value applies at the surface of the spheroid as well as at outside points.

Gravity on the spheroid, called normal gravity γ_ϕ at latitude ϕ is obtained by substituting eq. 2.50, where r^{-n} is given by:

$$r^{-n} = a^{-n}(1 + ne\sin^2\phi)$$

into eq. 2.53. Retaining terms of the order of e (J_2 and p are of the order e), we obtain for normal gravity:

$$\gamma_\phi = \frac{GM}{a^2}(1 + \tfrac{3}{2}J_2 - p)[1 - (\tfrac{9}{2}J_2 + p - 2e)\sin^2\phi] \tag{2.54}$$

At the equator:

$$\gamma_{eq} = \frac{GM}{a^2}(1 + \tfrac{3}{2}J_2 - p) \tag{2.55}$$

Eq. 2.54 can be written, substituting eqs. 2.52 and 55:

$$\gamma_\phi = \gamma_{eq}[1 + (2p - \tfrac{3}{2}J_2)\sin^2\phi] \tag{2.56a}$$

or:

$$\gamma_\phi = \gamma_{eq}[1 + (\tfrac{5}{2}p - e)\sin^2\phi] \tag{2.56b}$$

Eqs. 2.56 are of the form:

$$\gamma_\phi = \gamma_{eq}(1 + \beta\sin^2\phi) \tag{2.57}$$

from which it is seen that normal gravity at the pole γ_p is:

$$\gamma_p = \gamma_{eq}(1 + \beta) \tag{2.58}$$

showing that β, the gravitational ellipticity or flattening, is:

$$\beta = \left(\frac{\gamma_p - \gamma_{eq}}{\gamma_{eq}}\right) \tag{2.59}$$

Eqs. 2.56b and 2.57 give Clairaut's theorem to a first approximation, that is:

$$e + \beta = \tfrac{5}{2}p \tag{2.60}$$

which states that the sum of the geometric and gravitational flattening equals five-halves of the ratio of centrifugal to gravitational acceleration at the equator. Clairaut's theorem provides a formula for estimating the ellipticity of the spheroid.

If we include second-order terms, Clairaut's theorem takes the form (e.g., Heiskanen and Vening Meinesz, 1958, p. 52):

$$e(1 + \tfrac{17}{14}p) + \beta = \tfrac{5}{2}p \qquad (2.61)$$

and normal gravity becomes, in terms of geographic latitude Φ:

$$\gamma_\Phi = \gamma_{eq} \left\{1 + [\tfrac{5}{2}p - e(1 + \tfrac{17}{14}p)] \sin^2 \Phi - \frac{e}{8}(5p - e) \sin^2 2\Phi\right\} \qquad (2.62)$$

Geographic rather than geocentric latitude is used in this formula because the geographic latitude, the angle between the normal to the spheroid and the equatorial plane, is the latitude obtained in geodesy when sighting on a star.

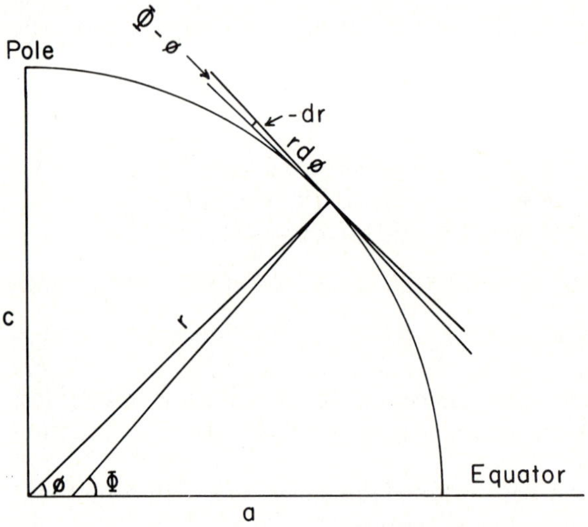

Fig. 2.12. Illustration showing the angular difference between geographic latitude Φ and geocentric latitude ϕ on a spheroidal earth.

Fig. 2.12 shows that the difference between geographic and geocentric latitude, a small angle, is:

$$\tan(\Phi - \phi) = -\frac{dr}{r\,d\phi} \doteq (\Phi - \phi) \qquad (2.63)$$

We differentiate eq. 2.50, obtaining to first order terms:

$$\frac{dr}{d\phi} = -ae \sin 2\phi \doteq -re \sin 2\phi \qquad (2.64)$$

Eq. 2.63 can now be written:

$$\Phi = \phi + e \sin 2\phi \tag{2.65}$$

We take the sine of both sides and expand, obtaining:

$$\sin \Phi = \sin \phi \cos(e \sin 2\phi) + \cos \phi \sin(e \sin 2\phi)$$

$$= \sin \phi \cos(\Phi - \phi) + \cos \phi \sin(\Phi - \phi)$$

where $\cos(\Phi - \phi) \doteq 1$ because the angle is very small. Then:

$$\sin \Phi = \sin \phi + \cos \phi \sin(e \sin 2\phi)$$

and, to first-order terms in e:

$$\sin^2 \Phi = \sin^2 \phi + 2 \sin \phi \cos \phi \sin(e \sin 2\phi)$$

We substitute $e \sin 2\phi$ for $\sin(e \sin 2\phi)$ to obtain:

$$\sin^2 \Phi = \sin^2 \phi + e \sin^2 2\phi \tag{2.66}$$

The equation of the spheroid (2.50) becomes, including second-order terms in e and expressed in geographic latitude:

$$r = a(1 - e \sin^2 \Phi + \tfrac{5}{8} e^2 \sin^2 2\Phi) \tag{2.67}$$

This is the surface on which γ_Φ (eq. 2.62) is normal gravity.

The International Ellipsoid, based on $e = 1/297.0$, was adopted by the International Geodetic Association in 1924, and is expressed by:

$$r = 6{,}378{,}388\ (1 - 0.0033670 \sin^2\Phi + 0.0000071 \sin^2 2\Phi)\,\mathrm{m} \tag{2.68}$$

Heiskanen and Vening-Meinesz (1958, pp. 54, 55) indicate that this reference surface is an ellipsoid near sea level, although at higher elevations it becomes a spheroid with a slight depression, which reaches its maximum value at 45° latitudes; the depression becomes larger with increasing elevations.

Normal gravity on an ellipsoid with $e = 1/297.0$ is given by the International Gravity Formula of 1930, the year of its adoption:

$$\gamma_{1930} = 978{,}049\ (1 + 0.0052884 \sin^2\Phi - 0.0000059 \sin^2 2\Phi)\,\mathrm{mgal} \tag{2.69}$$

In 1967 the International Geodesy Association adopted new values of earth parameters, based on $e = 1/298.247$, and a new formula for normal gravity which can be written:

$$\gamma_{1967} = 978{,}031.85\ (1 + 0.0053024 \sin^2\Phi - 0.00000587 \sin^2 2\Phi)\,\mathrm{mgal} \tag{2.70}$$

Rapp (1974) obtained improved values of earth parameters (see Appendix B), based on more recent geodetic measurements; $e = 1/298.256$. The corresponding gravity formula is:

$$\gamma_{298.256} = 978{,}031.69\,(1 + 0.0053269 \sin^2\Phi - 0.00000586 \sin^2 2\Phi)\ \text{mgal} \tag{2.71}$$

which applies to an ellipsoidal shape given by:

$$r_{298.256} = 6{,}378{,}139.00\,(1 - 0.0033528 \sin^2\Phi + 0.0000070 \sin^2 2\Phi)\ \text{m} \tag{2.72}$$

The difference between the 1967 (or 1974) and the 1930 normal gravity formulas is shown in Fig. 2.14.

The world gravity base station is Potsdam, Germany. The value of absolute gravity obtained there (in 1906) was 981.274 gal, which is now considered to be 14 mgal too high (based on free-motion measurements). Accordingly, the Potsdam Gravity System is now based on a revised Potsdam value of 981.260 gal.

Gravity anomalies computed for the 1930 gravity formula can be readily corrected to the 1967 (or 1974) formula. If the original Potsdam value is

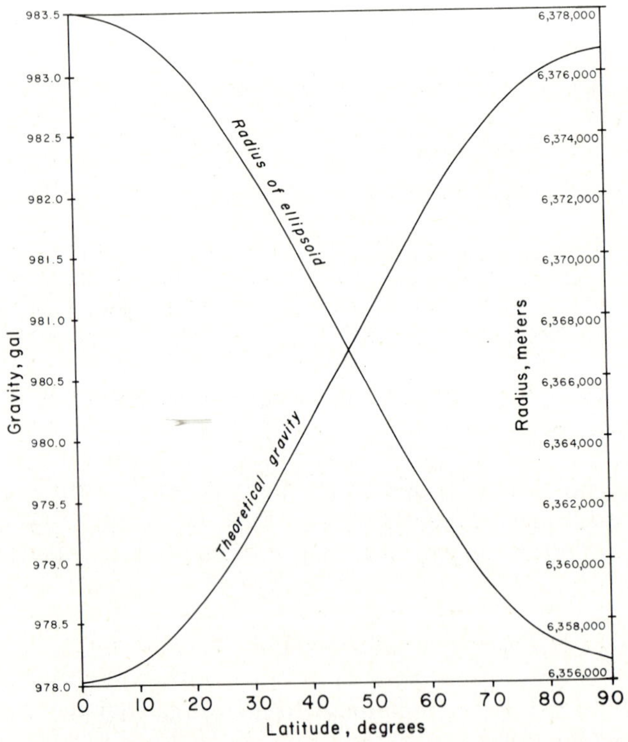

Fig. 2.13. Variations of gravity and earth radius. Values shown apply to the 1967 reference ellipsoid.

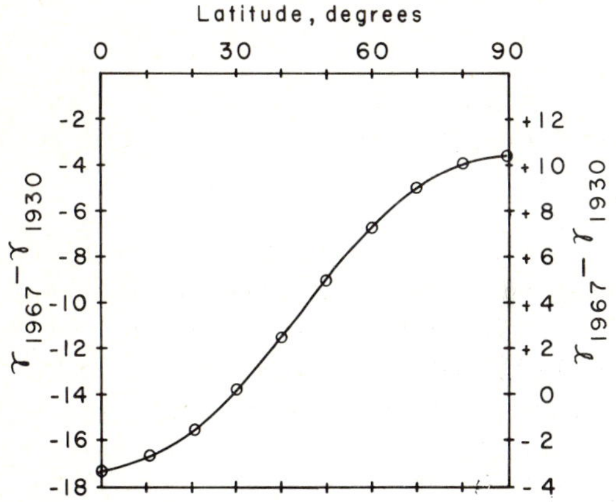

Fig. 2.14. Difference in normal or theoretical gravity between the 1967 and the 1930 gravity formulas; values on left ordinate axis assume old value of gravity at Potsdam; values on right are for the corrected gravity at Potsdam.

retained as the base, the correction applied to the anomalies is:

$$\delta g_{corr} = +(17.2 - 13.6 \sin^2 \Phi) \quad \text{mgal} \tag{2.73a}$$

and if the Potsdam datum has been corrected by −14.0 mgal, the correction is:

$$\delta g_{corr} = +(3.2 - 13.6 \sin^2 \Phi) \quad \text{mgal} \tag{2.73b}$$

GEOID DETERMINED FROM GRAVITY MEASUREMENTS

The geoid undulates about the reference spheroid. The shape of the geoid can thus be described by the height of its undulations from the spheroid. The undulations can be calculated from gravity anomaly values obtained out to large distances (actually for the entire earth), using Stokes's theorem (eq. 2.78). The undulations can also be determined from spherical harmonic expansions of the earth's gravitational potential, where tesseral harmonics (see Fig. 2.21 for illustrations of tesseral harmonic terms) to about degree and order 8 are determined from observed perturbations in the paths of artificial satellites.

Undulations of the geoid based on gravity measurements

The work gravity does in moving a unit mass from the geoidal to the spheroidal surface can be approximated by (Fig. 2.15):

$$W_0 - U_0 = g_0 N_0 \tag{2.74}$$

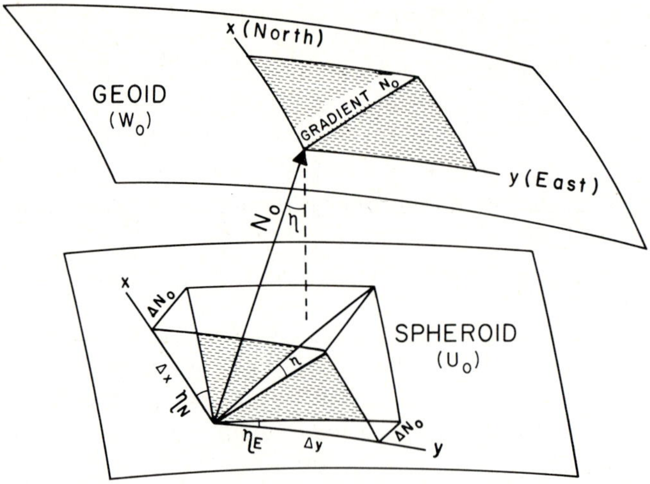

Fig. 2.15. Illustration of geoidal surface of height N_0 above the corresponding spheroid; η is the deflection of the vertical and η_N and η_E the north and east components.

in which W_0 and U_0 are the potentials on the geoid and spheroid, respectively, g_0 the value of gravity on the geoid, and N_0 the height of the geoid above the spheroid (usually small).

The gravity anomaly $\Delta g = g_0 - \gamma_\Phi$ is the difference between gravity on the geoid and the spheroid (along the normal to the geoid). The anomaly produced by the disturbing masses also includes the effect of the difference in radius to the geoid and spheroid (N_0). For N_0 outside the attracting masses, the outward gravity gradient or free-air effect at the geoid is, to sufficient accuracy:

$$\frac{\partial g_0}{\partial r} \sim \frac{\partial}{\partial r}\left(\frac{GM}{r^2}\right) = -\frac{2GM}{r^3} = -\frac{2g_0}{r} \tag{2.75}$$

The corresponding change in gravity on the geoid with respect to the underlying spheroid is $\delta g = -2g_0N_0/r$. The complete gravity anomaly is then:

$$\Delta g = -\frac{\partial}{\partial r}(W_0 - U_0) - \frac{2\bar{g}N_0}{\bar{r}} \tag{2.76}$$

in which $\bar{g}$ and $\bar{r}$ are average values for the earth, giving:

$$-\frac{2\bar{g}}{\bar{r}} = -0.3086 \text{ mgal m}^{-1} \tag{2.77}$$

Conventional anomaly determinations do not include the second term in eq. 2.76, the effect of the displacement from the geoid to the spheroid. Because anomaly calculations omit this effect, large disturbing masses calculated for corresponding anomalies are slightly underestimated. However, as most disturbing masses are relatively small, this term (the second term in eq. 2.76) is usually unimportant.

Stokes (1849) obtained expressions for geoidal undulations when values of Δg are known for the entire earth and no masses exist outside the geoid. This means that observed anomalies must be reduced to sea level (i.e., the geoid) before determinations of N_0 are made. Stokes's theorem can be written as (e.g., Heiskanen and Vening Meinesz, 1958, p. 65):

$$N_0 = \frac{1}{4\pi \bar{g} \bar{r}} \int_0^S \Delta g F(\psi)\, \mathrm{d}S \tag{2.78}$$

Integration is over the earth. $F(\psi)$ is a function of the interior angle ψ between the radius vectors to the point where N_0 is determined and the elements on the geoid with anomaly Δg. $F(\psi)$ is given by (Heiskanen and Vening Meinesz, 1958 p. 65):

$$F(\psi) = \left\{ \operatorname{cosec}\left(\frac{\psi}{2}\right) + 1 - 5\cos\psi - 6\cos\left(\frac{\psi}{2}\right) - 3\cos\psi \ln \left[\sin\left(\frac{\psi}{2}\right)\left(1 + \sin\left(\frac{\psi}{2}\right)\right)\right]\right\} \tag{2.79}$$

The value of $F(\psi)$ is small at large distances from where N_0 is determined; accordingly, distant anomalies have only slight effects on values of N_0. At

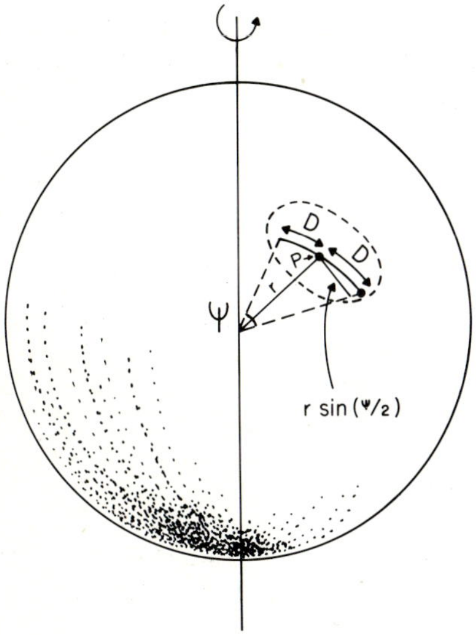

Fig. 2.16. Illustration showing portion of the earth's surface (dashed circle) and horizontal distance D from station point P; ψ is the solid angle of cone with diameter $2D$ at the surface.

short distances, where ψ is small, the first term in eq. 2.79 predominates; then (Fig. 2.16):

$$F(\psi \sim \operatorname{cosec}\left(\frac{\psi}{2}\right) \sim \frac{r}{D/2} = \frac{2r}{D} \tag{2.80}$$

where D is the horizontal distance from the point where N_0 is determined. A rough approximation for N_0 is then:

$$N_0 \sim \frac{1}{2\pi\bar{g}} \int_0^S \frac{\overline{\Delta g}\, \mathrm{d}S}{D} \tag{2.81}$$

in which $\overline{\Delta g}$ is the mean anomaly over the area of a ring of breadth $\mathrm{d}\psi$. Thus $\mathrm{d}S = 2\pi r^2 \sin\psi\, \mathrm{d}\psi$, where $\sin\psi \sim D/r$ and $\mathrm{d}\psi \sim \mathrm{d}D/r$. For distances less than 2,500 km the undulation roughly reduces to:

$$N_0 \sim \frac{1}{\bar{g}} \int_0^D \Delta g_{\mathrm{m}}\, \mathrm{d}D \tag{2.82}$$

Values of N_0 so obtained are usually too large, by as much as 20%.

Deflections of the vertical

Geoidal undulations result in variations in the direction of the vertical, that is, normals to the geoid. The deflection of the vertical is the angle between the normals to the geoid and spheroid at a point on the geoid (Fig. 2.15). The deflection is the gradient of N_0, usually expressed by its north and east components. Let η_{N} be the component of deflection in a vertical plane in the north direction. For N_0 positive, that is, the geoid outside the spheroid, η_{N} is positive to the north when:

$$\eta_{\mathrm{N}} = -\frac{\partial N_0}{\partial x} \tag{2.83a}$$

and, similarly, in the east direction, η_{E} is positive eastward when:

$$\eta_{\mathrm{E}} = -\frac{\partial N_0}{\partial y} \tag{2.83b}$$

GEOID DETERMINED FROM SATELLITE ORBITS

The earth's equatorial bulge exerts torques (moments of force) on artificial earth satellites just as it does on the moon; these torques perturb the satellite orbits as they do the lunar orbit.

Since satellites have essentially no mass, they do not affect the motion of the earth, in contrast to the moon which produces a precession of the equinoxes (a westward migration of the intersection of the ecliptic and the equator) and an 18.6-year wobble in the earth's axis (due to the lunar orbit being about 5.1° out of the ecliptic). As a consequence of the earth's bulge, and also other mass anomalies in the earth, satellite orbital paths have provided the most precise available measurements regarding the oblate spheroidal shape of the earth. Satellite paths, in conjunction with surface gravity anomalies, provide a detailed figure of the earth, the satellite geoid.

Regression of the nodes and determination of the J_2 term

Torques produced by the equatorial bulge cause all satellite orbits to precess slowly along the earth's equator (Fig. 2.17). The angular distance at which successive orbits intersect the equatorial plane, the nodes, is called the regression of the node. The node migrates westward, the direction of migration being opposite to that of the satellite.

Satellite precession can be analyzed by using Euler's angles i, Ω, and τ (Fig. 2.18), which relate the earth "inertial" axes with the rotating principal axes of inertia of the orbit; i is the orbital inclination from the equator, Ω the longitude of the node u (the angle which the nodal line makes with a fixed direction in the equatorial plane), and τ is the angle which the radius to the satellite makes with respect to the node. We take Z to be the earth's axis of rotation and X and Y to be in the equatorial plane; z is the principal axis perpendicular to the orbit and x and y are principal axes in the orbital plane.

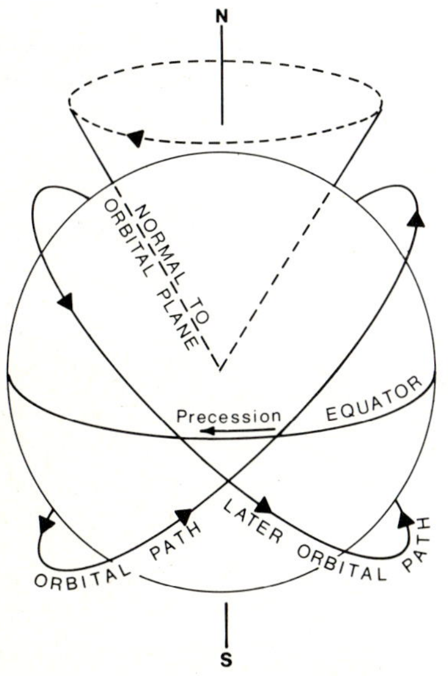

Fig. 2.17. Illustration showing satellite orbital paths, their perpendicular axes, precession of the orbital path, and the cone of orbital wobble.

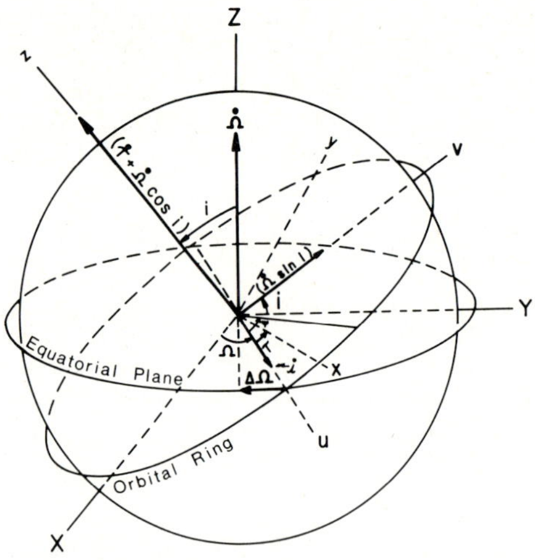

Fig. 2.18. Illustration showing components used to describe orbital paths (see text for explanations).

Gauss showed that the changes in satellite motion over many revolutions can be analyzed by replacing the satellite with a ring of radius equal to the orbital path, a mass equal to that of the satellite, and a spin velocity equal to the mean angular velocity of the satellite.

The rotational motion of the ring (satellite orbit) can be obtained by equating the moments of forces about the principal axes of inertia of the ring with the corresponding components of moments of force resulting from the attraction between the ring and the earth's bulge. Euler's equations describe the rotational motion of the ring; the equations can be written:

$$A\dot{\omega}_x + (C-B)\omega_z\omega_y = F_x r_s \tag{2.84a}$$

$$B\dot{\omega}_y + (A-C)\omega_x\omega_z = F_y r_s \tag{2.84b}$$

$$C\dot{\omega}_z + (B-A)\omega_y\omega_x = F_z r_s \tag{2.84c}$$

in which A, B, and C are moments of inertia of the ring about principal axes x, y, z; m is the mass of the ring of radius r_s; ω_x, ω_y, and ω_z are components of the angular spin velocity of the ring, where the components are about the x, y, and z directions; $\dot{\omega}_x$, $\dot{\omega}_y$, and $\dot{\omega}_z$ (dots indicate differentiation with respect to time) are components of angular acceleration of the ring about the respective directions; F_x, F_y and F_z are components of the force of attraction between the ring and the earth's bulge perpendicular to the respective axes.

The moments of inertia of a ring about its principal axes are:

$$C = mr_s^2 \tag{2.85a}$$

$$A = B = \tfrac{1}{2}mr_s^2 \tag{2.85b}$$

To obtain the torques on the left side of eqs. 2.84, we evaluate the components of angular velocity and angular acceleration. The component ω_z, about z, consists of $\dot{\tau}$ (i.e., $d\tau/dt$) plus the component of $\dot{\Omega}$ about z, which is $\dot{\Omega}\cos i$. Similarly, ω_x consists of the components of $\dot{\Omega}$ and $\dot{i}$ that act about x, and ω_y consists of the components of $\dot{\Omega}$ and $\dot{i}$ that act about y. We write (also in, e.g., Thompson, 1963, p. 37):

$$\omega_z = \dot{\Omega}\cos i + \dot{\tau} \tag{2.86a}$$

$$\omega_x = \dot{\Omega}\sin i \sin\tau + \dot{i}\cos\tau \tag{2.86b}$$

$$\omega_y = \dot{\Omega}\sin i \cos\tau - \dot{i}\sin\tau \tag{2.86c}$$

The time-derivatives of eqs. 2.86 are the angular accelerations about the respective axes, which are:

$$\dot{\omega}_z = \ddot{\Omega}\cos i - \dot{\Omega}\dot{i}\sin i + \ddot{\tau} \tag{2.87a}$$

$$\dot{\omega}_x = \ddot{\Omega}\sin i \sin\tau + \dot{\Omega}\dot{i}\cos i \sin\tau + \dot{\Omega}\dot{\tau}\sin i\cos\tau + \ddot{i}\cos\tau - \dot{i}\dot{\tau}\sin\tau \tag{2.87b}$$

$$\dot{\omega}_y = \ddot{\Omega}\sin i \sin\tau + \dot{\Omega}\dot{i}\cos i \cos\tau - \dot{\Omega}\dot{\tau}\sin i\sin\tau - \ddot{i}\sin\tau - \dot{i}\dot{\tau}\cos\tau \tag{2.87c}$$

Since the x axis can be taken arbitrarily on the ring, we place it along the nodal line; then $\tau = 0$, $\cos\tau = 1$, $\sin\tau = 0$, and the above equations reduce to:

$$\omega_x = \dot{i} \tag{2.88a}$$

$$\omega_y = \dot{\Omega}\sin i \tag{2.88b}$$

and:

$$\dot{\omega}_x = \dot{\Omega}\dot{\tau}\sin i + \ddot{i} \tag{2.89a}$$

$$\dot{\omega}_y = \dot{\Omega}\dot{i}\cos i - \dot{i}\dot{\tau} \tag{2.89b}$$

We take the average ring velocity $\bar{\omega}_s$ equal to ω_z. Then:

$$\bar{\omega}_s = \omega_z = \dot{\Omega}\cos i + \dot{\tau} \tag{2.90}$$

The orbital velocity is very much greater than the rate of change of inclination ($\bar{\omega}_s \gg \dot{i}$) and the rate of change of longitude of the node ($\bar{\omega}_s \gg \dot{\Omega}$). Hence, eq. 2.90 reduces to $\bar{\omega}_s \doteq \dot{\tau}$, and:

$$\dot{\omega}_x = \bar{\omega}_s\dot{\Omega}\sin i \tag{2.91a}$$

$$\dot{\omega}_y = -\bar{\omega}_s\dot{i} \tag{2.91b}$$

where the terms $\ddot{i}$ and $\dot{\Omega}\dot{i}$ are dropped because they are negligible compared with the terms retained.

The components of mutual attraction about the axial directions are obtained as gradients of the mutual potentials between the ring and the bulge. The gravitational potential produced by the bulge was seen to be (from eq. 2.33):

$$\Delta V = -\frac{GMa^2J_2}{2r_s^3}(3\sin^2\phi - 1) \tag{2.92}$$

The mutual potential between the ring and the bulge is $m\Delta V$. The component of force in the direction of the angle of inclination is:

$$F_i = m\left(\frac{1}{r_s}\frac{\partial V}{\partial i}\right) \tag{2.93a}$$

The component in the direction Ω is:

$$F_\Omega = m\left(\frac{1}{r_s \sin i}\frac{\partial V}{\partial \Omega}\right) \tag{2.93b}$$

and the component in the direction τ is:

$$F_\tau = m\left(\frac{1}{r_s}\frac{\partial V}{\partial \tau}\right) \tag{2.93c}$$

The torque components about the x, y, z axes produced by these force components can now be written:

$$F_x r_s = m(\partial V/\partial i) \tag{2.94a}$$

$$F_y r_s = \frac{m}{\sin i}(\partial V/\partial \Omega) \tag{2.94b}$$

$$F_z r_s = m(\partial V/\partial \tau) \tag{2.94c}$$

In evaluating $\partial V/\partial \tau$ we note that V is independent of longitude, and therefore of τ. Hence $\partial V/\partial \tau = 0$ and eq. 2.84c reduces to (substituting eq. 2.85a and noting that $B = A$):

$$mr_s^2\dot{\omega}_z = m(\partial V/\partial \tau) = 0 \tag{2.95a}$$

and therefore:

$$\dot{\omega}_z = 0 \quad \text{and} \quad \omega_z = \bar{\omega}_s \tag{2.95b}$$

Euler's equation of torques about the x axis (nodal line), produced by the attraction of the bulge, can be written (substituting eqs. 2.85 into 2.84a):

$$\tfrac{1}{2}mr_s^2(\dot{\omega}_x + \omega_y\omega_z) = m(\partial V/\partial i) \tag{2.96}$$

We substitute for $\dot{\omega}_x$, ω_y, and ω_z from eqs. 2.88b, 2.90, and 2.95b and use eq. 2.92 to obtain:

$$\tfrac{1}{2}mr_s^2\bar{\omega}_s\dot{\Omega}\sin i = m\frac{\partial}{\partial i}\left[-\frac{GMa^2J_2}{2r_s^3}(3\sin^2 i\sin^2\psi - 1)\right] \tag{2.97}$$

where the latitude in eq. 2.92 is related to the orbital inclination by:

$$\sin\phi = \sin i \sin\psi \tag{2.98}$$

and ψ is the angle on the ring measured from u.

We differentiate the right side of eq. 2.97 and integrate over the ring to obtain its mass (σ = mass/length). Then:

$$\tfrac{1}{2} m r_s^2 \bar{\omega}_s \dot{\Omega} \sin i = -\frac{GMa^2\sigma J_2}{2r_s^3}\int_0^{2\pi} 3 \sin i \cos i \sin^2\psi \, d\psi \tag{2.99a}$$

$$= -\frac{3}{2}\left(\frac{GMa^2\pi\sigma J_2}{r_s^3}\right) \sin i \cos i \tag{2.99b}$$

We substitute $\omega_s^2 r_s^3$ for GM (Kepler's third law) and simplify terms to obtain for the rate of change of the longitude of the node:

$$\dot{\Omega} = -\tfrac{3}{2}\bar{\omega}_s J_2 \left(\frac{a}{r_s}\right)^2 \cos i \tag{2.100}$$

The regression per orbital revolution $\Delta\Omega$ is obtained from the ratio:

$$\frac{\Delta\Omega}{2\pi} = \frac{\dot{\Omega}}{\bar{\omega}_s} = -\frac{3}{2}\left(\frac{a}{r_s}\right)^2 J_2 \cos i \tag{2.101a}$$

or:

$$J_2 = \frac{-\Delta\Omega}{3\pi \cos i}\left(\frac{a}{r_s}\right)^2 \tag{2.101b}$$

Since J_2 is a constant, it follows that along polar orbits, where $i = \pi/2$, $\Delta\Omega = 0$; for equatorial orbits $i = 0$ and $\Delta\Omega$ takes on its maximum value.

Fig. 2.19 shows values of $\Delta\Omega$ and $\dot{\Omega}$ as functions of inclination for circular orbits 300 km above the earth (r_s = 6,670 km) with a period of 90 min. Fig. 2.20 illustrates the relation between orbital height and period of revolution. Heights of more than 4,000 satellites now orbiting the earth range from a few hundred to about 30,000 km.

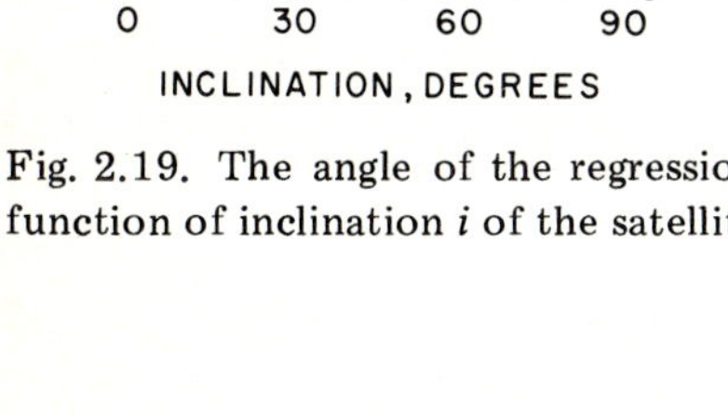

Fig. 2.19. The angle of the regression of the node $\Delta\Omega$ and the regression velocity $\dot{\Omega}$ as function of inclination i of the satellite orbit.

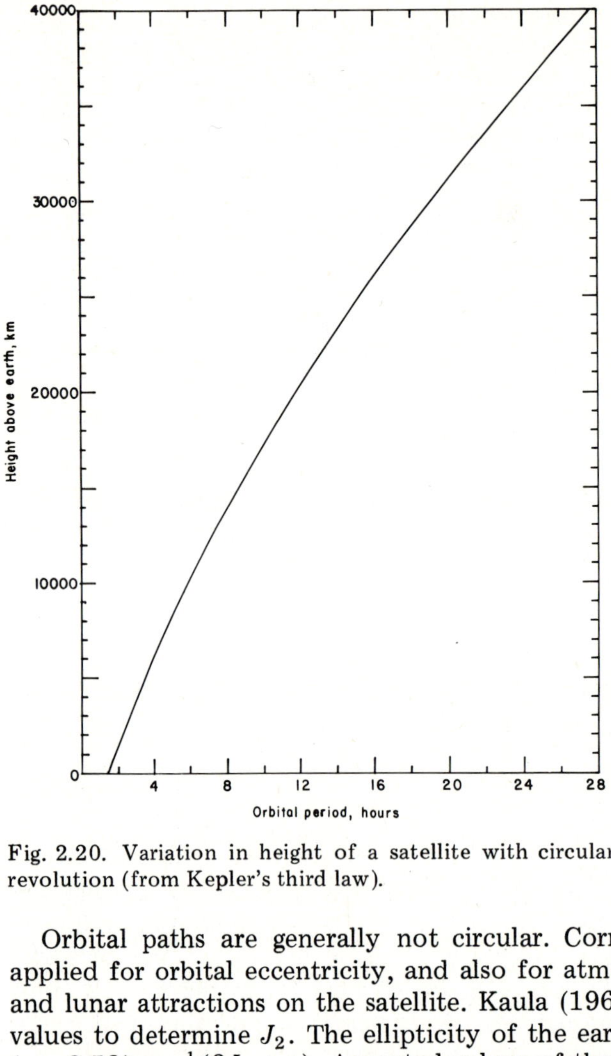

Fig. 2.20. Variation in height of a satellite with circular orbit as function of period of revolution (from Kepler's third law).

Orbital paths are generally not circular. Corrections must therefore be applied for orbital eccentricity, and also for atmospheric drag and for solar and lunar attractions on the satellite. Kaula (1966) obtained such corrected values to determine J_2. The ellipticity of the earth spheroid was seen to be (eq. 2.52) $e = \frac{1}{2}(3J_2 + p)$. Accepted values of the spheroid (Appendix B) are $J_2 = 0.0010827$, $e = 0.0033528 = 1/298.256$, and $p = 0.0034678$.

Geoidal potentials and spherical harmonic coefficients

Satellite data provide not only the J_2 term in the geopotential equation, but also higher-order J terms, the zonal harmonics, which describe the spheroid as a function of latitude alone. Similar tesseral harmonic terms (Fig. 2.21) can be obtained, which describe the spheroid as functions of both latitude and longitude. The satellite geoid can be determined from the tesseral harmonic series.

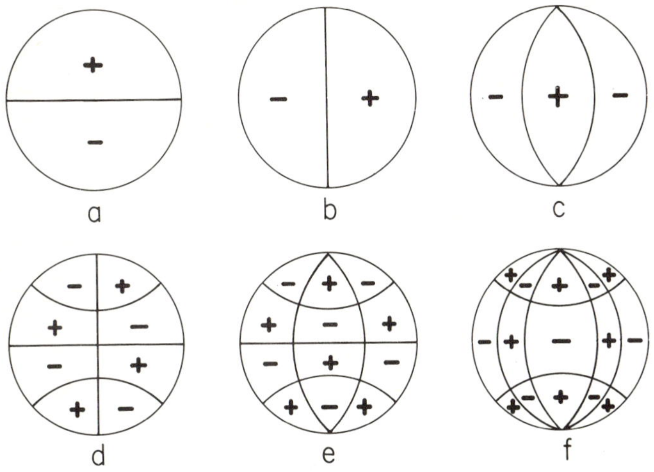

Fig. 2.21. Illustration of positive and negative areas of spherical harmonic components of Legendre (zonal) and associated Legendre polynomials (sectorial, dependent upon longitude only, and tesseral harmonics): a. $P_1(\sin\phi)$; b. $P_1^1(\sin\phi)$; c. $P_2^2(\sin\phi)$; d. $P_4^1(\sin\phi)$; e. $P_5^2(\sin\phi)$; f. $P_6^4(\sin\phi)$. (c, d, and f are modified after Kaula, 1965.)

Spheroid potentials as functions of latitude only are obtained from potentials containing a series of zonal harmonics. The potential at the satellite can be written (e.g., Merson, 1961, p. 20):

$$V = \frac{GM}{r_s}\left[1 - \sum_{n=2}^{\infty}\left(\frac{a}{r_s}\right)^n J_n P_n(\sin\phi)\right] \tag{2.102}$$

where the J terms include J_2 and higher-order terms and $P_n(\sin\phi)$ are the Legendre polynomials in eq. 2.28, except that latitude ϕ has replaced co-latitude θ. The first two terms in eq. 2.102 essentially determine the orbital path, the first describing an elliptical path and the second a perturbation due to the equatorial bulge. Third- and higher-order J terms represent small orbital perturbations owing to mass anomalies in the earth which cause torques to act on the angular momentum of the satellite.

Rates of orbital precession are proportional to all J terms. A series of zonal harmonic coefficients in the gravitational potential can thus be determined from observed orbital paths. Lower-order J terms have been determined for satellites with various inclinations. Kozai (1969) obtained J values up to 21 terms, listed in Table 2.I. The geopotential U containing these J terms defines the Standard Earth Model 1 as published by the Smithsonian Astrophysical Observatory (e.g., Gaposchkin, 1974). Gravity along this spheroid is obtained from $-\partial U/\mathrm{d}r$, as in eq. 2.53, and the shape of the spheroid is described in a manner similar to eq. 2.49.

Geoid potentials as functions of both latitude and longitude can be

TABLE 2.I

Values in J terms in zonal harmonics obtained from satellite data (Kozai, 1969)

$J_2 = 1082.62800 \cdot 10^{-6}$	$J_{12} = -0.4200 \cdot 10^{-6}$
$J_3 = -2.53800 \cdot 10^{-6}$	$J_{13} = -0.12300 \cdot 10^{-6}$
$J_4 = -1.59300 \cdot 10^{-6}$	$J_{14} = -0.07300 \cdot 10^{-6}$
$J_5 = -0.23000 \cdot 10^{-6}$	$J_{15} = -0.17400 \cdot 10^{-6}$
$J_6 = +0.50200 \cdot 10^{-6}$	$J_{16} = +0.18700 \cdot 10^{-6}$
$J_7 = -0.36200 \cdot 10^{-6}$	$J_{17} = +0.08500 \cdot 10^{-6}$
$J_8 = -0.11800 \cdot 10^{-6}$	$J_{18} = -0.23100 \cdot 10^{-6}$
$J_9 = -0.10000 \cdot 10^{-6}$	$J_{19} = -0.21600 \cdot 10^{-6}$
$J_{10} = -0.35400 \cdot 10^{-6}$	$J_{20} = +0.00500 \cdot 10^{-6}$
$J_{11} = +0.20200 \cdot 10^{-6}$	$J_{21} = +0.14400 \cdot 10^{-6}$

obtained from a series of tesseral harmonics. The potential at the satellite has the form (e.g., Gaposchkin and Lambeck, 1970):

$$V = \frac{GM}{r_s}\left\{1 - \sum_{l=2}^{\infty}\left(\frac{a}{r_s}\right)^l \sum_{m=0}^{l}\left[C_l^m \cos m\lambda + S_l^m \sin m\lambda\right] P_l^m(\sin\phi)\right\} \qquad (2.103)$$

where C_l^m and S_l^m are tesseral harmonics arising from masses within the earth, λ is longitude and P_l^m (sin ϕ) are associated Legendre polynomials with solutions of the form (e.g., Stacey, 1977, p. 321):

$$P_l^m(\sin\phi) = \cos^m\phi \frac{\mathrm{d}^m P_l(\sin\phi)}{\mathrm{d}(\sin\phi)^m} \qquad (2.104)$$

where $P_l(\sin\phi)$ is the Legendre polynomial in eq. 2.102.

Gaposchkin (1974) determined Legendre coefficients C_l^m and S_l^m to degree and order 18 (l and m = 18) and a number of higher order terms from the equation:

$$V = \frac{GM}{r_s}\left\{1 - \sum_{n=2}^{\infty}\left(\frac{a}{r_s}\right)^n J_n P_n(\sin\phi) + \sum_{l=2}^{\infty}\sum_{m=1}^{l}\left(\frac{a}{r_s}\right)^l \left[C_l^m \cos m\lambda + S_l^m \sin m\lambda\right] P_l^m(\sin\phi)\right\} \qquad (2.105)$$

which defines Standard Earth Model 3. The model represents the satellite geoid, illustrated in Fig. 2.22; the corresponding free-air anomaly is shown in Fig. 2.23. The J_n (zonal harmonic) terms are obtained from satellite orbital perturbations; the C_l^m and S_l^m (tesseral harmonic) terms are obtained from satellite perturbations for values of l and m up to degree and order 8 (corresponding to anomaly wavelengths greater than about 4,000 km), and from surface gravity anomalies for values of l and m greater than 8 (corresponding to anomaly wavelengths shorter than about 4,000 km). Satellites thus provide information on long-wavelength features on the geoid and gravity anomalies the shorter features.

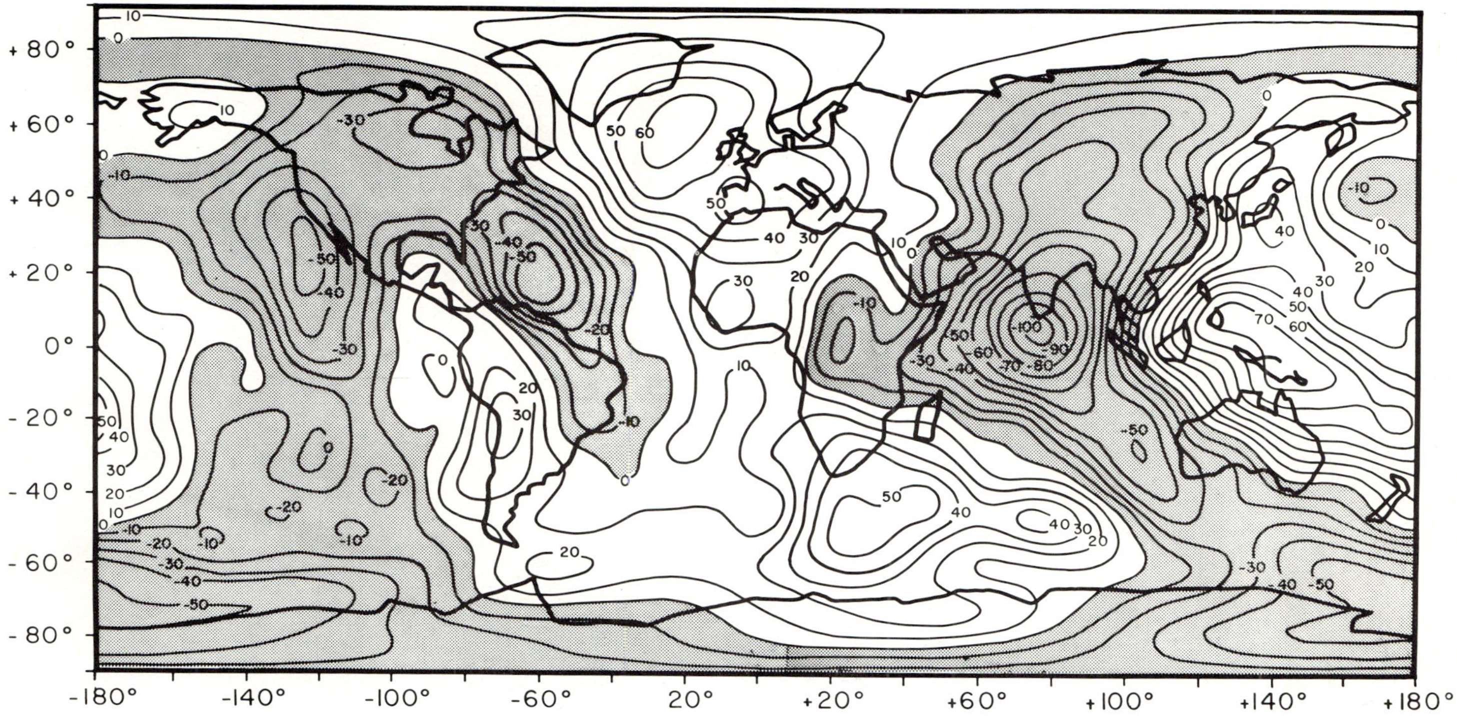

Fig. 2.22. SE 3 geoid height, in meters, calculated for an ellipsoid with $e = 1/298.256$. (Reproduced from Gaposchkin, 1974, with permission.)

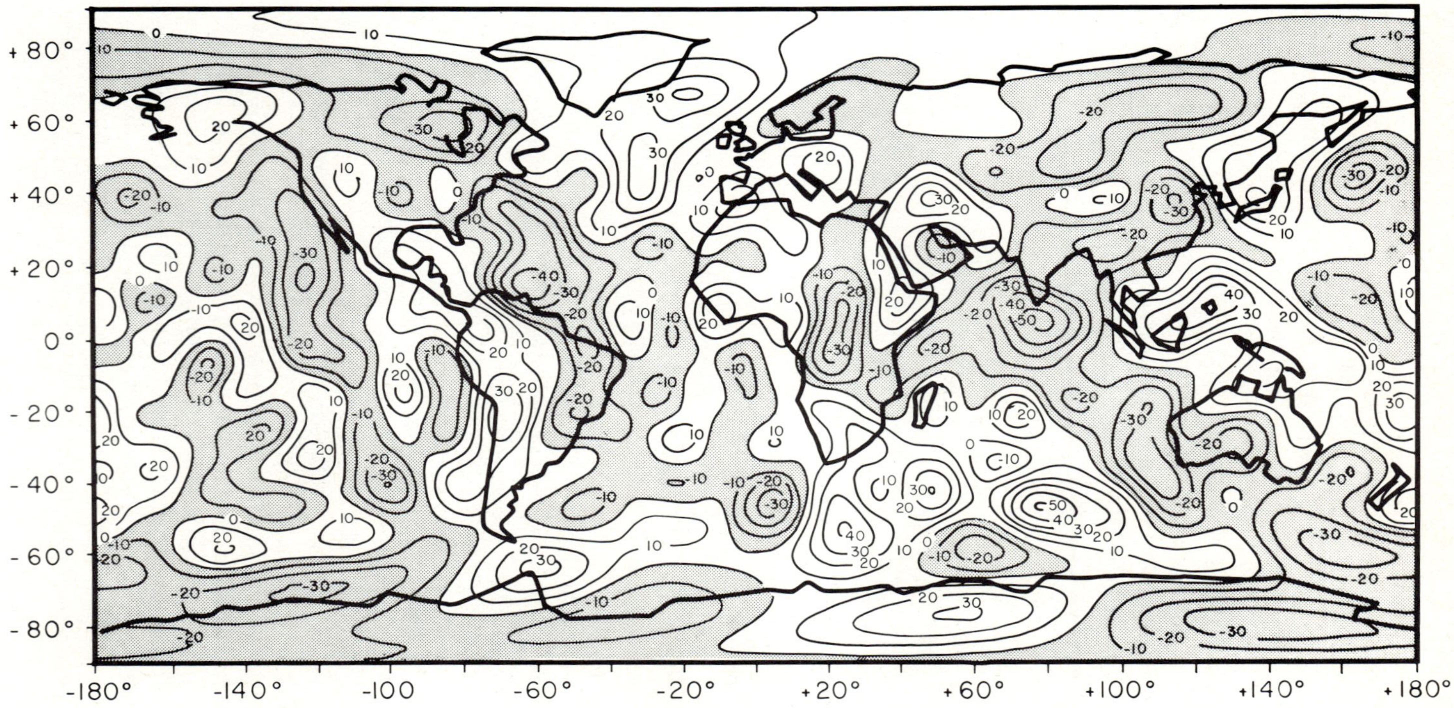

Fig. 2.23. SE 3 gravity anomaly in milligals calculated for an ellipsoid with e = 298.256. (Reproduced from Gaposchkin, 1974, with permission.)

The recent Goddard Space Flight Center Goddard Earth Model 8 (GEM-8) provides geopotentials derived from satellite and gravity data complete to Legendre degree and order 25, and includes additional terms up to degree 30 (Wagoner et al., 1976; included are geoidal height and free-air anomaly maps).

FORM OF A HYDROSTATIC EARTH

Once the reference ellipsoid is determined, we wish to know whether it represents equilibrium conditions on a rotating fluid earth, and thus an earth in hydrostatic equilibrium. If the earth is not in hydrostatic equilibrium, the inference is that its interior must be capable of supporting long-term stress differences.

Darwin (1898) showed that for a rotating fluid earth, the external shape could be an ellipsoid if the fluid is of uniform density. He and also De Sitter (1924) analyzed the external shape of a fluid body for the case of a density increase toward the center. Such a density variation requires the inclusion of the next-higher harmonic coefficients, which results in a small departure from the reference ellipsoid for realistic increases of density with depth.

Whether the earth is in hydrostatic equilibrium can be investigated by comparing the ellipticities of the fluid and actual earth. The J_2 term and ellipticity of the reference spheroid are known from the regression of the nodes of satellites; however, the J_2 term corresponding to a hydrostatic earth is not well known. Caputo (1965) showed that for hydrostatic equilibrium, there is a linear relationship between J_2 and the earth's mechanical ellipticity H, which equals $(C - A)/C$, where C and A are the moments of inertia included in the J_2 term (eq. 2.44). H is determined from observed precessions of the equinoxes. The corresponding J_2 and H values for the solid earth, based on observed satellite and precessional data, do not exhibit the same linear relationship. This difference is taken to indicate that the earth is not in hydrostatic equilibrium.

By assuming a ratio of J_2/H in the hydrostatic earth equal to the observed ratio in the solid earth, an ellipticity of $e = 1/299.8$ is obtained, which is 0.4% smaller than the observed ellipticity of $e = 1/298.256$. The respective ellipticity values are considered to be reliable, so that the difference cannot be attributed to uncertainties in measurements.

The ellipticity of the solid earth is thus seen to be greater than that of the fluid earth, which means that the actual equatorial bulge is larger than the bulge corresponding to hydrostatic equilibrium. The implication is that the earth contains laterally asymmetrical density (mass) variations, suggesting either a tri-axial ellipsoidal shape or that the earth previously had rotated more rapidly and that it has retained a portion of its "fossil bulge". The former explanation is more likely. In either event, the earth appears to have sufficient strength to sustain long-term shearing stresses.

Chapter 3

EARTH TIDES AND TIDAL DEFORMATIONS

Earth tides are deformations of the solid earth produced by the gravitational attractions of the sun and moon, attractions which are periodic because of the earth's rotation in the solar and lunar gravity fields. The lunar—solar attractions also produce elastic deformations in the solid earth, thereby changing the shape of the earth and distorting the direction of the vertical.

The periodic tidal attractions are added vectorially to the very much larger secular accelerations produced by the earth's mass attraction and rotation. The total gravity vector undergoes periodic variations in both magnitude and direction. Variations in direction are observed as small deviations of the vertical, ranging up to 40 msec of arc; variations in gravity range up to 0.3 mgal; variations in radius of the deformable earth range up to 56 cm. The variations in the gravity vector produce dynamic ocean tides, which are manifestations of the ocean continually adapting its sea surface in a direction perpendicular to the instantaneous vertical.

Earth deformations can be measured, but they are affected by displacements which the measuring instruments encounter over tidal cycles. For example, oceanic tidal measurements at coastal stations include oscillations in the height of the station. Similarly, gravity measurements are affected not only by the lunar—solar attraction, but, because of the resultant earth deformation, they are also affected by a change in the earth's radius. The elastic deformations can be analyzed in terms of Love numbers h, k, and also l, which are related to rigidity and density in the solid earth. The deformations which result in tidal variations in gravity and in ocean heights can be expressed by parameters defining combinations of Love numbers (eqs. 3.40 and 3.47). The parameters, and therefore the Love numbers, are affected by the amplitude and phase of the open-ocean tides. Thus, while lunar—solar forces can be calculated with great accuracy, the resultant Love numbers must be corrected for ocean-loading effects. An inversion technique has recently been developed (Kuo and Jachens, 1975) using tidal gravity measurements at numerous locations to map open-ocean tides.

Clearly, routine gravity measurements include the effects of lunar—solar attractions and resultant earth deformations. From known tidal equations (eqs. 3.30), tidally generated variations in gravity can be computed for given times and positions. The resultant tidal corrections can be readily applied to observed gravity anomalies.

In the solid earth, the highly rigid interconnections between adjacent molecules result in small displacements of particles, such that equilibrium is established with little phase lag. This contrasts with oceanic tides, where liquid particles may be moved large distances and, because of inertia, undergo oscillations.

Tides in the solid earth can therefore be analyzed in terms of static theory, particularly as the shortest tidal periods of 12 h are an order of magnitude greater than the longest periods of free oscillations in a liquid earth, about 1½ h. Free oscillations do not produce resonances at tidal periods.

Since the earth—moon radius equals many (about 60) earth radii, the lunar and solar gravitational attraction in the different parts of the earth are nearly the same as the average lunar attraction on the earth as a whole. Consequently, attractions between the earth and moon and the earth and sun can be calculated on the basis of point masses, and tides treated as small deviations from these attractions.

According to static tidal theory, the instantaneous gravitational acceleration at a point P (Fig. 3.1) consists of the attraction due to the moon and sun superimposed on that of the earth. Total gravity can be expressed by its vertical and horizontal components (where the horizontal component is in the plane containing P and the earth—moon axis). Writing the components as $(g_r)_P$ and $(g_\psi)_P$ at P, $(g_r)_0$ and $(g_\psi)_0$ at the center of the earth, and g as secular gravity due to the whole earth (positive inward), the attraction at P due to the moon of mass m is:

$$(g_r)_P = -Gm\left[\frac{\cos \psi'_m}{(R'_m)^2}\right] \qquad (g_\psi)_P = Gm\left[\frac{\sin \psi'_m}{(R'_m)^2}\right] \tag{3.1}$$

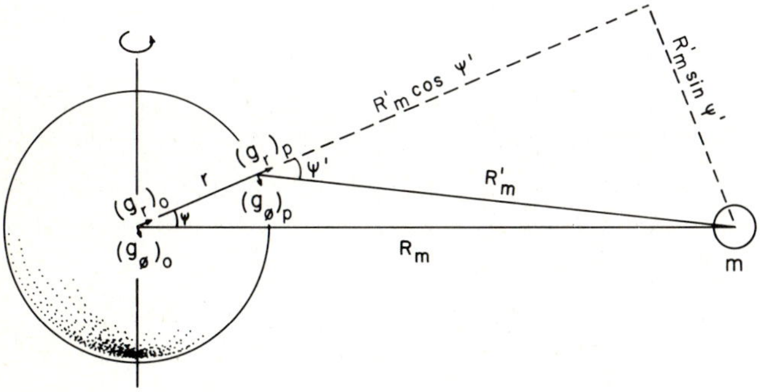

Fig. 3.1. Illustration of the components of lunar attraction acting at the earth's surface and parallel components acting at its center. Angles ψ and ψ' are in a plane containing the surface point and the earth—moon centers.

in which ψ'_m and R'_m are the zenith angle and distance from P to the center of the moon. The vertical component at P, produced by the earth and the moon, can be written:

$$g + (g_r)_P = g - Gm\left[\frac{\cos\psi'_m}{(R'_m)^2}\right] \tag{3.2}$$

Components in these directions at the center of the earth can be written:

$$(g_r)_0 = -Gm\left(\frac{\cos\psi_m}{R_m^2}\right) \qquad (g_\psi)_0 = Gm\left(\frac{\sin\psi_m}{R_m^2}\right) \tag{3.3}$$

in which ψ_m is the zenith angle and R_m the distance between the centers of the earth and moon (Fig. 3.1). Components of tidal accelerations are defined by:

$$(\Delta g_r)_m = (g_r)_P - (g_r)_0 = -Gm\left[\frac{\cos\psi'_m}{(R'_m)^2} - \frac{\cos\psi_m}{R^2}\right] \tag{3.4a}$$

$$(\Delta g_\psi)_m = (g_\psi)_P - (g_\psi)_0 = Gm\left[\frac{\sin\psi'_m}{(R'_m)^2} - \frac{\sin\psi_m}{R_m^2}\right] \tag{3.4b}$$

We can eliminate the terms R' and ψ' by substituting:

$$\left.\begin{aligned} (R')^2 &= R^2 + r^2 - 2rR\cos\psi \\ R'\sin\psi' &= R\sin\psi \\ R'\cos\psi' &= R\cos\psi - r \end{aligned}\right| \tag{3.5}$$

into eqs. 3.4 to obtain:

$$(\Delta g_r)_m = -Gm\left[\frac{R_m\cos\psi_m - r}{(R_m^2 + r^2 - 2rR_m\cos\psi_m)^{3/2}} - \frac{\cos\psi_m}{R_m^2}\right] \tag{3.6a}$$

$$(\Delta g_\psi)_m = Gm\left[\frac{R_m\sin\psi_m}{(R_m^2 + r^2 - 2rR_m\cos\psi_m)^{3/2}} - \frac{\sin\psi_m}{R_m^2}\right] \tag{3.6b}$$

These equations can be simplified by expanding terms as $(1+x)^{-3/2} = 1 - \frac{3}{2}x + \ldots$. Retaining terms to R_m^{-3} (where $r/R_m = 1/60$), we obtain:

$$(\Delta g_r)_m = -Gm\left[\left(\frac{\cos\psi_m}{R_m^2} - \frac{r}{R_m^3}\right)\left(1 + \frac{3r\cos\psi_m}{R_m}\right) - \frac{\cos\psi_m}{R_m^2}\right] \tag{3.7a}$$

$$(\Delta g_\psi)_m = Gm\left[\frac{\sin\psi_m}{R_m^2}\left(\frac{3r\cos\psi_m}{R_m}\right)\right] \tag{3.7b}$$

Simplifying terms produces:

$$(\Delta g_r)_m = -\frac{Gmr}{R_m^3}(3\cos^2\psi_m - 1) \tag{3.8a}$$

$$(\Delta g_\psi)_m = \frac{3Gmr}{R_m^3} (\sin \psi_m \cos \psi_m) \tag{3.8b}$$

By substituting:

$$\cos^2 \psi = \tfrac{1}{2}(1 + \cos 2\psi) \tag{3.9a}$$

$$\sin 2\psi = 2 \sin \psi \cos \psi \tag{3.9b}$$

$$g = \frac{GM}{r^2} \tag{3.9c}$$

we obtain formulas for the components of lunar attraction:

$$(\Delta g_r)_m = -\tfrac{3}{2} g \left(\frac{m}{M}\right)\left(\frac{r}{R_m}\right)^3 (\cos 2\psi_m + \tfrac{1}{3}) \tag{3.10a}$$

$$(\Delta g_\psi)_m = \tfrac{3}{2} g \left(\frac{m}{M}\right)\left(\frac{r}{R_m}\right)^3 (\sin 2\psi_m) \tag{3.10b}$$

Corresponding equations apply for solar attractions.

Vectors $[(\Delta g_r)^2 + (\Delta g_\psi)^2]^{1/2}$ are illustrated for the earth's surface in Fig. 3.2, where the outward direction of Δg_r in the equatorial regions results from the negative sign in eq. 3.10a (from eq. 3.2, it is seen that Δg_r acts in the opposite direction to g at low latitudes).

The deviations of the vertical, α_m, owing to the lunar attraction, can be written:

$$\tan \alpha_m = \frac{(\Delta g_\psi)_m}{g + (\Delta g_r)_m} = \frac{\frac{3}{2}\left(\frac{m}{M}\right)\left(\frac{r}{R_m}\right)^3 \sin 2\psi_m}{\left[1 - \frac{3}{2}\left(\frac{m}{M}\right)\left(\frac{r}{R_m}\right)^3 (\cos 2\psi_m + \frac{1}{3})\right]} \tag{3.11}$$

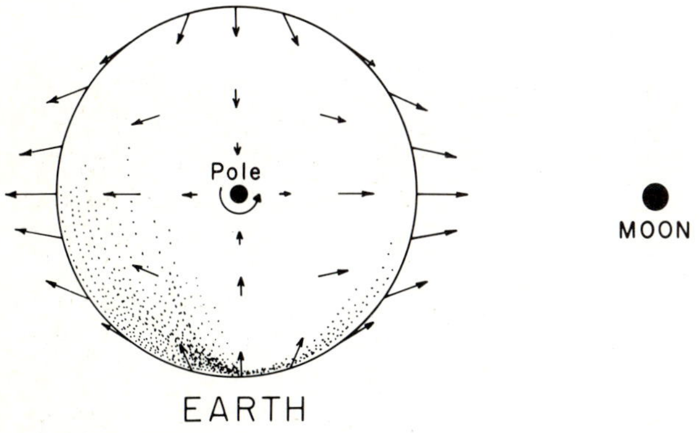

Fig. 3.2. Illustration of tidal forces produced by lunar attraction.

Since α_m is very small:

$$\alpha_m = \frac{3}{2}\left(\frac{m}{M}\right)\left(\frac{r}{R_m}\right)^3 \sin 2\psi_m \left[1 + \frac{3}{2}\left(\frac{m}{M}\right)\left(\frac{r}{R_m}\right)^3 (\cos 2\psi_m + \tfrac{1}{3})\right] \qquad (3.12)$$

If we neglect terms smaller than R_m^{-3}, the deviation becomes:

$$\alpha_m = \frac{3}{2}\left(\frac{m}{M}\right)\left(\frac{r}{R_m}\right)^3 \sin 2\,\psi_m \qquad (3.13)$$

The deviation can be expressed in seconds of arc (1 rad = 57.30°, and sin 1″ = 1/206,265):

$$\alpha_m = \tfrac{3}{2}(206{,}265)\left(\frac{m}{M}\right)\left(\frac{r}{R_m}\right)^3 \sin 2\psi_m \quad \text{arc sec} \qquad (3.14)$$

Expressions for the corresponding solar attractions are:

$$(\Delta g_r)_s = -\tfrac{3}{2}g\left(\frac{S}{M}\right)\left(\frac{r}{R_s}\right)^3 (\cos 2\psi_s + \tfrac{1}{3})$$

$$(\Delta g_\psi)_s = \tfrac{3}{2}g\left(\frac{S}{M}\right)\left(\frac{r}{R_s}\right)^3 \sin 2\psi_s \qquad (3.15)$$

$$\alpha_s = \tfrac{3}{2}(206{,}265)\left(\frac{S}{M}\right)\left(\frac{r}{R_s}\right)^3 \sin 2\psi_s$$

By inserting numerical values in Appendix B into the previous equations, we obtain:

$$\left.\begin{aligned}
(\Delta g_r)_m &= -0.08226\ (\cos 2\psi_m + \tfrac{1}{3})\ \text{mgal}\\
(\Delta g_r)_s &= -0.03788\ (\cos 2\psi_s + \tfrac{1}{3})\ \text{mgal}\\
(\Delta g_\psi)_m &= +0.08226 \sin 2\psi_m\ \text{mgal}\\
(\Delta g_\psi)_s &= +0.03788 \sin 2\psi_s\ \text{mgal}\\
\alpha_m &= 0.017317 \sin 2\psi_m\ \text{arc sec}\\
\alpha_s &= 0.0077647 \sin 2\psi_s\ \text{arc sec}
\end{aligned}\right| \qquad (3.16)$$

Fig. 3.2 is obtained from eqs. 3.16, showing that:

(1) Vertical components of tidal attractions have amplitudes which are a maximum where $\psi = 0, \pi$, for which $(\cos 2\psi + \frac{1}{3}) = \frac{4}{3}$; a minimum along the ring where $\psi = \pi/2$, for which $(\cos 2\psi + \frac{1}{3}) = -\frac{2}{3}$, $(\Delta g_r)_{min}$ having half the magnitude in the opposite direction of $(\Delta g_r)_{max}$; and are zero along rings where $(\cos 2\psi + \frac{1}{3}) = 0$, that is, where $\psi = \pm 54^\circ.7$.

(2) Horizontal components of tidal attractions have amplitudes which are a maximum along rings where $\psi = \pm\pi/4$; zero where $\psi = 0, \pi$, and along the ring where $\psi = \pm\pi/2$.

(3) Deviations of the vertical are a maximum along rings where $\psi = \pm\,\pi/4$; zero for $\psi = 0, \pi$, and along the ring where $\psi = \pm\pi/2$.

The maximum range in tidal gravity variations on a rigid earth are thus:

$$\left.\begin{aligned}
[(\Delta g_r)_m]_{\max} &= -0.08226\ (\tfrac{4}{3}) = -109.7\ \mu\text{gal}\\
[(\Delta g_r)_m]_{\min} &= +0.08226\ (\tfrac{2}{3}) = +\ \ 54.8\ \mu\text{gal}\\
&\qquad \text{Total lunar range} \quad 164.5\ \mu\text{gal}\\
[(\Delta g_r)_s]_{\max} &= -0.03788\ (\tfrac{4}{3}) = -\ \ 50.5\ \mu\text{gal}\\
[(\Delta g_r)_s]_{\min} &= +0.03788\ (\tfrac{2}{3}) = +\ \ 25.3\ \mu\text{gal}\\
&\qquad \text{Total solar range} \quad 75.8\ \mu\text{gal}
\end{aligned}\right| \qquad (3.17)$$

The attraction is a maximum when the sun and moon are in line with the earth; it is then 0.165 + 0.076 = 0.24 mgal. However, this attraction applies to a rigid earth. It will be shown (eqs. 3.42) that the change in gravity owing to the earth's deformation ranges up to 0.04 mgal. Maximum variations in gravity occurring over a tidal cycle thus are 0.24 + 0.04 = 0.3 mgal.

Deviations of the vertical are greatest along rings where $\psi = \pm 45^\circ$ and have the value $\alpha = 0''.017317 + 0''.007765 = 0''.025082$ when the sun and moon are in line with the earth. Horizontal components of gravity are similarly greatest when the sun and moon are aligned with the earth, the maximum occurring along rings where $\psi \pm 45^\circ$; Δg_ψ ranges in value up to 0.12 mgal.

The ratio of lunar to solar tidal attractions can be determined from pertinent values included in eqs. 3.16 and 3.17; these include:

$$\frac{110}{51} \sim \frac{165}{76} \sim \frac{173}{78} \sim \frac{822}{379} \sim 2.2 \qquad (3.18)$$

The lunar attraction is therefore about 2.2 times greater than the solar attraction. When the sun and moon are aligned with the earth, tides are a maximum (spring tides) and are 3.2 times greater than the solar tide. When the sun and moon are perpendicular with respect to the earth, tides are minimal (neap tides) and are 1.2 times the solar tide. Consequently, spring tides are 2.7 (3.2/1.2) times larger than neap tides.

TIDAL POTENTIALS

The tidal potential W_2, due to both the sun and moon, represents the work of the tides. On a fluid earth this work raises the tides a height Δr and, similar to eq. 2.7, can be expressed by $W_2 = (\Delta r)g$.

On a rigid earth tidal accelerations at P (Fig. 3.1) are obtained from the tidal potential, similar to eq. 2.10. Thus,

$$\Delta g_r = -\frac{\partial W_2}{\partial r} \qquad \Delta g_\psi = \frac{\partial W_2}{r\,\partial \psi} \tag{3.19}$$

Tidal potentials produced by the moon and the sun, which satisfy eqs. 3.8 and 3.19, are written:

$$(W_2)_m = \tfrac{1}{2} Gm(r^2/R_m^3)(3\cos^2\psi_m - 1) \tag{3.20}$$

$$(W_2)_s = \tfrac{1}{2} GS(r^2/R_s^3)(3\cos^2\psi_s - 1) \tag{3.21}$$

We see from eq. 2.29 that these tidal potentials are second-degree spherical harmonics, that is:

$$\tfrac{1}{2}(3\cos^2\psi - 1) = P_2(\cos\psi)$$

Laplace's tidal equation

Zenith angles ψ_m and ψ_s in eqs. 3.20 and 3.21 are usually awkward to determine. They can be described (Fig. 3.3) in terms of the astronomic (geocentric) latitude ϕ, the declination δ of the moon and sun with respect to the equatorial plane (the declination of the sun is zero at the time of the

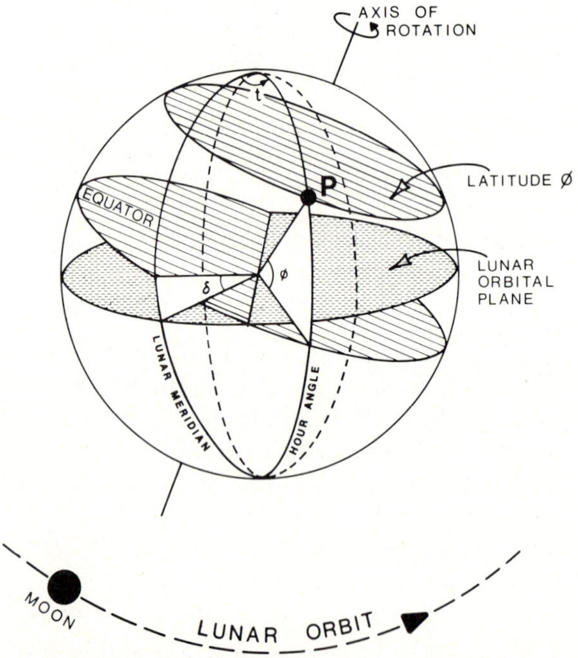

Fig. 3.3. Illustration of moon's declination δ and hour angle t at point P (latitude ϕ).

equinoxes and ±23°.47 during the solstices; the declination of the moon ranges between ±28°.6), and the hour angle t of the moon or sun (1 h of siderial time equals 15° longitude). From spherical trigonometry:

$$\cos \psi = \sin \phi \sin \delta + \cos \phi \cos \delta \cos t \tag{3.22}$$

We substitute eq. 3.22 into eqs. 3.20 and 3.21 to obtain the coordinates of P in terms of its known latitude, declination, and hour angle. Thus:

$$(3 \cos^2 \psi - 1) = 3 \sin^2 \phi \sin^2 \delta - 1 + \tfrac{3}{2} \cos^2 \phi \cos^2 \delta (\cos 2t + 1) + \tfrac{3}{2}(\sin 2\phi \sin 2\delta \cos t) \tag{3.23}$$

which simplifies to:

$$(3 \cos^2 \psi - 1) = \tfrac{3}{2}[(\cos^2 \phi \cos^2 \delta \cos 2t) + (\sin 2\phi \sin 2\delta \cos t) + (3 \sin^2 \phi \sin^2 \delta - \sin^2 \phi - \sin^2 \delta + \tfrac{1}{3})] \tag{3.24}$$

The lunar potential $(W_2)_m$ then can be written:

$$(W_2)_m = \tfrac{3}{4} Gm \left(\frac{r^2}{R_m^3}\right)[(\cos^2 \phi \cos^2 \delta \cos 2t) + (\sin 2\phi \sin 2\delta \cos t) + 3(\sin^2 \phi - \tfrac{1}{3})(\sin^2 \delta - \tfrac{1}{3})] \tag{3.25}$$

We substitute $G = gr^2/M$ (eq. 3.9c) to obtain Laplace's tidal equation for the moon:

$$(W_2)_m = \tfrac{3}{4} g \left(\frac{m}{M}\right) \left(\frac{r^4}{R_m^3}\right)[(\cos^2 \phi \cos^2 \delta \cos 2t) + (\sin 2\phi \sin 2\delta \cos t) + 3(\sin^2 \phi - \tfrac{1}{3})(\sin^2 \delta - \tfrac{1}{3})] \tag{3.26a}$$

For the sun:

$$(W_2)_s = \tfrac{3}{4} g \left(\frac{S}{M}\right) \left(\frac{r^4}{R_s^3}\right)[(\cos^2 \phi \cos^2 \delta \cos 2t) + (\sin 2\phi \sin 2\delta \cos t) + 3(\sin^2 \phi - \tfrac{1}{3})(\sin^2 \delta - \tfrac{1}{3})] \tag{3.26b}$$

Clearly the bracketed terms in eqs. 3.26 are second-degree spherical harmonic functions. The first term represents a sectorial harmonic, varying as a function of longitude; the second term a tesseral harmonic, varying with both latitude and longitude; the third term a zonal harmonic, varying with latitude. The equations have far-ranging implications in describing the various tidal components which the earth experiences.

The sectorial harmonic term has nodal lines where:

$$\cos^2 \phi \cos^2 \delta \cos 2t = 0 \tag{3.27}$$

Nodal lines thus occur along the meridians $t = \pm \pi/4$ and nodal points at the poles ($\phi = \pm\pi/2$), such that the meridional lines divide the earth into four

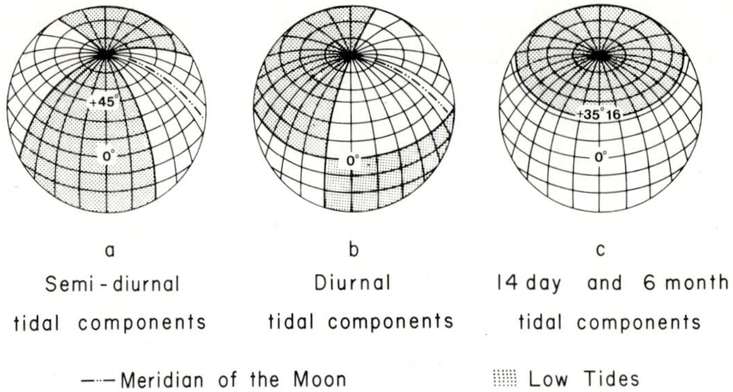

Fig. 3.4. Aerial distributions of high and low lunar-tidal components with respect to the lunar meridian. (Reproduced from Melchior, 1966, with permission.)

quadrants (Fig. 3.4a), two of which have positive and two negative potentials. The corresponding tides are high on the quarters facing and opposite the moon $[(W_2)_m > 0]$, and low on the two transverse quarters $[(W_2)_m < 0]$. Two high and two low tides occur per earth revolution; these harmonics therefore describe semidiurnal tides. We also see from this harmonic term that the semi-diurnal tides are highest along the equator, where $\phi = 0$, when the declination is zero (at the equator during the equinoxes), and smallest along the equator at the times of the solstices. The tides remain zero at the poles.

The tesseral harmonic term in eqs. 3.26 has nodal lines where:

$$\sin 2\phi \sin 2\delta \cos t = 0 \tag{3.28}$$

The nodal lines are along the equator, where $\phi = 0$, and along the meridian $t = \pm \pi/2$, dividing the earth into four quadrants (Fig. 3.4b). The tides exhibit one maximum and one minimum per earth revolution and are, accordingly, diurnal tides. The tidal amplitudes change sign in accordance with the sign of the declination of the moon or sun. Largest amplitudes occur where $t = 0, \pi$ at latitude 45°N, and smallest at 45°S, when δ is a maximum in the northern hemisphere, and vice versa when δ is a maximum in the southern hemisphere. The tidal components are zero at all points along the equator, along the ±90° meridian, at the poles, and everywhere when $\delta = 0$ (as the time of the equinoxes for the solar component).

The zonal harmonic, the third term in eqs. 3.26, indicates that nodal lines occur where:

$$(\sin^2 \phi - \tfrac{1}{3})(\sin^2 \delta - \tfrac{1}{3}) = 0 \tag{3.29}$$

which is along latitudes 35°.16N and 36°.16S. The tidal amplitudes are permanently high in latitudes lower than 35° and permanently low in higher latitudes (Fig. 3.4c), producing essentially constant deformations which very slightly increase the earth's ellipticity. As eq. 3.29 varies with both lunar and solar declination, the tides are greatest when the declination reaches its maximum both above and below the equator. There are, consequently, two tidal cycles during the moon's orbit around the earth and two during the earth's orbit around the sun; the zonal harmonic term represents the 14-day and 6-month tides.

Tidal components

The moon and the sun each produce three tidal components on the earth. Considering also the fact that the lunar distance from the earth varies during each orbital revolution, the resultant phase relationships between the tidal components become complex; a large number of combinations of components or tidal waves exist. Of 30 of the more important components, 12 are essentially semi-diurnal, 12 approximately diurnal, and 6 are long-period (e.g., Melchior, 1966, p. 33). Tidal components having the largest amplitudes are listed in Table 3.I

TABLE 3.I

Tidal components which exhibit largest amplitudes

Symbol	Tidal component	Period
M_2	Principal lunar	12 h 25 min, semi-diurnal
S_2	Principal solar	12 h 00 min, semi-diurnal
N_2	Lunar ellipticity (due to monthly variation in the moon's distance)	12 h 39 min, semi-diurnal
K_2	Lunar—solar declination	11 h 58 min, semi-diurnal
O_1	Principal lunar	25 h 49 min, diurnal
P_1	Principal solar	24 h 04 min, diurnal
K_1	Lunar—solar declination	23 h 56 min, diurnal
M_0	Lunar flattening	13.66 days, 14-day period
S_0	Solar flattening	182.5 days, 6 months

Tidal corrections to gravity measurements

Since solid-earth tidal components range up to 0.3 mgal, land-based gravity measurements are usually corrected for tidal attractions. Sea-based measurements are usually not accurate to more than 0.5—1.0 mgal, hence they are generally not so corrected.

Formulas for computing tidal accelerations produced by both the sun and the moon, as they act at any point on the earth as a function of time (in the future as well as in the past), have been published by Longman (1959) and others (e.g., Bartels, 1957). The Longman formulas are programable on digital computers, so that tidal corrections can be readily applied to observed measurements.

Longman (1959) obtained formulas for vertical attractions produced by the moon and sun, similar to eq. 3.8a, which are:

$$(\Delta g_r)_m = -\frac{3\,Gmr}{R_m^3}\left[(3\cos^2\psi_m - 1) + \frac{3}{2}\left(\frac{r}{R_m}\right)(5\cos^3\psi_m - 3\cos\psi_m)\right] \tag{3.30a}$$

$$(\Delta g_r)_s = -\frac{3\,GSr}{R_s^3}(3\cos^2\psi_s - 1) \tag{3.30b}$$

Point P is described by its geographic latitude and longitude and its earth radius (eq. 2.72) and elevation. The zenith angles ψ_m and ψ_s and the distances between centers of the earth and moon, R_m, and earth and sun, R_s, are expressed as variable functions of orbital parameters.

Expansions of eqs. 3.30a and b, amenable to computer programing, appear in Appendix C.

Theoretical amplitudes of tidal components on a rigid earth are computed by differentiating Laplace's tidal equation, obtaining values of $-\partial W_2/\partial r$ in eqs. 3.26. Fig. 3.5 illustrates variations in gravity with latitude for the main tidal components.

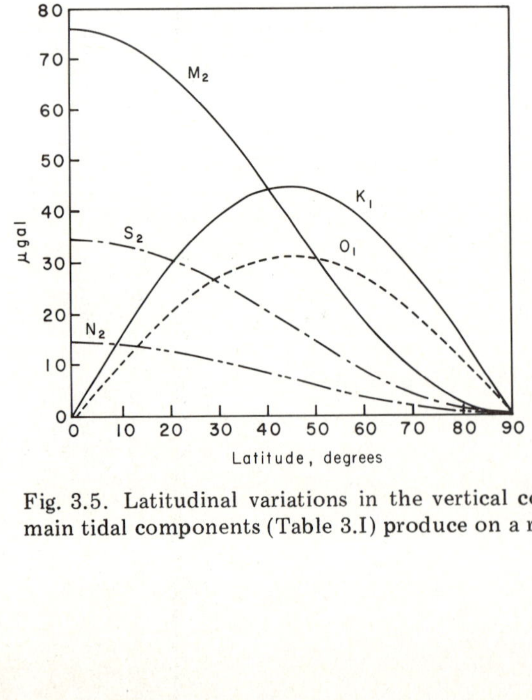

Fig. 3.5. Latitudinal variations in the vertical component of tidal attractions which the main tidal components (Table 3.I) produce on a rigid earth.

Because the tidal potential W_2 can be expressed by spherical harmonic functions, all deformation in the earth produced by W_2 can be expressed by the same functions when multiplied by appropriate numerical coefficients. These coefficients are the Love numbers h and k. In 1912 T. Shida in Japan showed that a third coefficient l must be included to characterize completely the effects of the tides in the solid earth.

Love numbers h, k, and also l are defined as follows: h is the ratio of the height of the body tide to the height of the equilibrium (static) ocean tide; k is the ratio of the additional potential (W_2') produced by the deformation of the earth to the deforming potential (W_2); l is the ratio of horizontal displacement in the crust to the corresponding displacement of the equilibrium (static) ocean tide. A fourth number, f, has been added, where f is the ratio of the cubic expansion (relative change of volume in rocks) to the height of the equilibrium tide at the surface.

It follows that for an equilibrium ocean tidal height of W_2/g, the corresponding body tidal height is hW_2/g. The potential produced by tidal deformation is related to the tidal potential by $W_2' = kW_2$.

Variations in gravity due to tides (gravimetric factor δ)

The total gravitational potential W at a point on the earth consists of the potential W_0 due to the earth itself, the lunar—solar potential W_2, the earth-deforming potential W_2', and the potential produced by the change in the height of the geoid. The potential resulting from the last term is $-(\Delta r)g_0$. As $g_0 = -\partial W_0/\partial r$ (to sufficient accuracy), we write for the total potential:

$$W = W_0 + W_2 + W_2' + (\Delta r)\frac{\partial W_0}{\partial r} \tag{3.31}$$

The corresponding change in gravity on the earth is:

$$\Delta g_r = g - g_0 = -\left(\frac{\partial W_0}{\partial r} + \frac{\partial W_2}{\partial r} + \frac{\partial W_2'}{\partial r} + \Delta r\,\frac{\partial^2 W_0}{\partial r^2}\right) + \left(\frac{\partial W_0}{\partial r}\right) \tag{3.32}$$

which simplifies to:

$$\Delta g_r = -\left(\frac{\partial W_2}{\partial r} + \frac{\partial W_2'}{\partial r} + \Delta r\,\frac{\partial^2 W_0}{\partial r^2}\right) \tag{3.33}$$

Differentiation of eqs. 3.20 and 3.21 shows that:

$$\frac{\partial W_2}{\partial r} = 2\left(\frac{W_2}{r}\right) \tag{3.34}$$

which equals $-\Delta g_r$ on a rigid earth.

W'_2 can be treated as a function of the lunar—solar zonal potential terms in Laplace's tidal equation (eqs. 3.26), and can be written in the form (e.g., Melchior, 1966, p. 110):

$$W'_2 = \frac{3D}{r^3}(\sin^2\phi - \tfrac{1}{3})(\sin^2\delta - \tfrac{1}{3}) \tag{3.35}$$

in which D is the Doodson tidal contant (D= 26.206 cm^2 sec^{-2}; refer to Melchior, 1966, p. 19). By differentiating eq. 3.35 we obtain:

$$\frac{\partial W'_2}{\partial r} = -3\left(\frac{W'_2}{r}\right) \tag{3.36}$$

To sufficient accuracy:

$$\frac{\partial^2 W_0}{\partial r^2} = -\frac{\partial g_0}{\partial r} = -\frac{\partial}{\partial r}\left(\frac{GM}{r^2}\right) = 2\left(\frac{g}{r}\right) \tag{3.37}$$

which is the same as eq. 2.75.

The total gravity variation at P becomes:

$$\Delta g_r = -\left[2\left(\frac{W_2}{r}\right) - 3\left(\frac{W'_2}{r}\right) + 2\Delta r\left(\frac{g}{r}\right)\right] \tag{3.38}$$

which can be simplified, since the tidal height on a deformable earth is $h(W_2/g)$, and $W'_2 = kW_2$. The tidal gravity variation on a deformable earth is thus given by:

$$\Delta g_r = -\left(\frac{2W_2}{r}\right)(1 + h - \tfrac{3}{2}k) = (1 + h - \tfrac{3}{2}k)\left(-\frac{\partial W_2}{\partial r}\right) \tag{3.39}$$

The multiplying factor δ is the gravimetric factor:

$$\delta = (1 + h - \tfrac{3}{2}k) \tag{3.40}$$

which represents the magnification in tidal gravity on a deformable relative to a rigid earth. The value of δ is obtained by dividing the observed tidal anomaly by the theoretical value on a rigid earth ($-\partial W_2/\partial r$). Uncertainties in the true value of δ are introduced by distortions in the tidal anomaly measurements, produced primarily by ocean loading. Ocean-loading effects are large near coastal regions, and even significant in the interiors of continents. Observed tidal anomalies should be corrected for ocean-loading effects, where possible.

Observed values of δ range from 1.13 to 1.24, a representative value being 1.15. The elastic yielding of the earth thus appears to cause tidal gravity variations of about 15% (and as much as 24%) greater than they would be on a rigid earth. Hence:

$$\Delta g_r \sim 1.15\left(-\frac{\partial W_2}{\partial r}\right) \tag{3.41}$$

Maximum variations in gravity which occur over tidal cycles can be estimated, similar to eqs. 3.17; they are:

$$\left.\begin{aligned}(\Delta g_r)_{max} &= -1.15[0.082(\tfrac{4}{3}) + 0.038(\tfrac{4}{3})] &&= -0.190 \text{ mgal}\\ (\Delta g_r)_{min} &= -1.15[-0.082(\tfrac{2}{3}) - 0.038(\tfrac{2}{3})] &&= +0.085 \text{ mgal}\end{aligned}\right| \quad (3.42)$$

$$\text{Maximum variation in } \Delta g_r = 0.028 \text{ mgal}$$

The maximum variation in Δg_r on a rigid earth (eq. 3.17) was seen to be 0.24 mgal; the corresponding multiplying factor is unity. The difference between the gravity variations on a deformable and a rigid earth is thus approximately 0.04 mgal. Approximately 1/7 of the observed tidal gravity variations appear to be attributable to elastic yielding of the earth.

Height of ocean tide above tide gage (parameter γ)

In the previous section, measurements of gravity variations over tidal cycles were shown to provide one combination of the Love numbers, the gravimetric factor δ. Other measurements can provide values of additional combinations. An important combination is the parameter concerned with water-tide measurements along coastal areas.

In a fluid earth, the height of the equilibrium oceanic tide Δr_1 was seen to be the lunar—solar potential W_2 divided by the attracting force per unit mass, that is:

$$\Delta r_1 = \frac{W_2}{g} \quad (3.43)$$

In a deformable earth, the effective height of the equilibrium ocean tide Δr_2 becomes:

$$\Delta r_2 = \frac{W_2}{g} + \frac{W_2'}{g} = (1 + k)\left(\frac{W_2}{g}\right) = (1 + k)\Delta r_1 \quad (3.44)$$

The height of the body tide Δr_3 was seen to be:

$$\Delta r_3 = h\left(\frac{W_2}{g}\right) = h\Delta r_1 \quad (3.45)$$

The height of the tide above the tide gage (excluding effects of tidal currents) is the difference between the height of the ocean tide and the increased height of the crust, that is, $\Delta r_2 - \Delta r_3$. Thus:

$$\Delta r_2 - \Delta r_3 = (1 + k - h)\left(\frac{W_2}{g}\right) = \gamma\left(\frac{W_2}{g}\right) = \gamma\Delta r_1 \quad (3.46)$$

where γ is the multiplying factor:

$$\gamma = (1 + k - h) \quad (3.47)$$

which is the reduction in the theoretical ocean tide that gives the ocean height above the tide gage. As W_2 is known, W_2/g can be readily computed, and as $(\Delta r_2 - \Delta r_3)$ is measured by the tide gage, γ can be obtained (as accurately as the true tidal height is measured). In this way a second combination of h and k is obtained. Heights of equilibrium ocean tides in a fluid earth can be derived from eqs. 3.20 and 3.21; thus:

$$\left.\begin{aligned} \frac{(W_2)_m}{g} &= \frac{3}{2}\left(\frac{Gm}{g}\right)\left(\frac{r^2}{R_m^3}\right)(\cos 2\psi_m + \tfrac{1}{3}) \\ \frac{(W_2)_s}{g} &= \frac{3}{2}\left(\frac{GS}{g}\right)\left(\frac{r^2}{R_s^3}\right)(\cos 2\psi_s + \tfrac{1}{3}) \end{aligned}\right| \quad (3.48)$$

which, when we substitute values for the appropriate constants (see Appendix B) reduce to:

$$\left.\begin{aligned} \frac{(W_2)_m}{g} &= 26.7(\cos 2\psi_m + \tfrac{1}{3})\text{ cm} \\ \frac{(W_2)_s}{g} &= 12.3(\cos 2\psi_s + \tfrac{1}{3})\text{ cm} \end{aligned}\right| \quad (3.49)$$

Heights of the static ocean tide above the tide gage on the deformable earth can then be written:

$$\left.\begin{aligned} \gamma(\Delta r_1)_m &= \gamma(26.7)(\cos 2\psi_m + \tfrac{1}{3})\text{ cm} \\ \gamma(\Delta r_1)_s &= \gamma(12.3)(\cos 2\psi_s + \tfrac{1}{3})\text{ cm} \end{aligned}\right| \quad (3.50)$$

Tidal and other measurements have shown that values of γ range between 0.64 and 0.84; an average value for the earth is 0.71. Eqs. 3.50 take on average values of:

$$\left.\begin{aligned} \overline{\gamma(\Delta r_1)_m} &= 19.0(\cos 2\psi_m + \tfrac{1}{3})\text{ cm} \\ \overline{\gamma(\Delta r_1)_s} &= 8.7(\cos 2\psi_s + \tfrac{1}{3})\text{ cm} \end{aligned}\right| \quad (3.51)$$

The maximum variation in observed equilibrium ocean tidal heights is the sum of the maximum height, which according to eqs. 3.51 occurs where $\psi = 0, \pi$, and is approximately $25 + 12 = 37$ cm; the minimum height, which occurs at $\psi = \pi/2$, is approximately $-13 - 6 = -19$ cm; the total range is 56 cm.

Variations of the vertical with respect to the crust

Tidal deformations cause the direction normal to the crust to vary with respect to the vertical. A horizontal pendulum (e.g., long-period horizontal seismograph), or a level seeks an equilibrium position with respect to the vertical. As the vertical undergoes deviations over tidal cycles, the seismograph position or the level follows the changing vertical. Vertical deviations corresponding to a rigid earth can be readily calculated; observed deviations pertaining to the deformable earth are somewhat smaller. An analysis shows (e.g., Melchior, 1966, p. 112) that the amplitude of the observed deviation is reduced by the same parameter γ used to determine the height of the tide above the tide gage. Thus, the observed deviation equals γ times the theoretical deviation. It is not surprising that the same γ applies to crustal deformation as well as to ocean tidal heights above bottom because ocean tides are affected by variations in the vertical of the ocean bottom.

Numerical values of h and k

Values of δ and γ obtained from physical measurements permit calculations of h and k from the relationships in eqs. 3.40 and 3.47. Representative values are $h = 0.59$ and $k = 0.27$. Average heights of earth tides thus appear to be about 59% of the height of the equilibrium ocean tide. The average value of the potential produced by tidal deformation appears to be about 27% of the tidal potential.

Values obtained for δ, and therefore h and k, are affected by ocean loading even at considerable distances from the ocean, and also by both small and large ground fractures at the station site. In so far as corrections can be applied to compensate for these factors, measurements of δ and γ can be used to investigate variations in the earth's elastic properties.

Values of k can also be determined from variations in the potential perturbing paths of artificial satellites, giving $k \sim 0.39$ (see, e.g., Cook, 1973, p. 100) and 0.25 (Kolenkiewicz et al., 1973), from fortnightly variations in the length of day giving $k \sim 0.34$, and from the Chandler period giving $k \sim 0.28$.

Measurement and value of l

The measurement of l can be obtained with extensometers or with horizontal pendulums designed to record differential crustal movements. A parameter obtained from geophysical measurements (e.g., Melchior, 1966, p. 300) is:

$$\Lambda = (1 + k - l) \tag{3.52}$$

Established values of Λ and l vary considerably. Those based on observations of different tidal components indicate that l ranges between 0.04 and 0.08, a representative value being 0.05.

TIDAL FRICTION

The moon and sun raise tidal bulges on the earth, the amplitude being about one third of a meter in the solid earth. The tidal bulge would be in phase with the tidal stresses if the solid earth consisted of perfectly elastic materials and the liquid oceans and core exhibited zero viscosity, that is, if there were no dissipative reactions.

The solid earth is not perfectly elastic, however, and the viscosity of the oceans is substantial. This results in dissipative reactions which make the tidal deformations (strain) lag behind the tidal attractions (stress). In the rotating earth the tidal bulge is thus carried forward (Fig. 3.6), which creates an asymmetrical mass that acts on the moon (and to a negligible extent on the sun).

The mass asymmetry produces a couple, in which two equal but opposite forces act about the earth—moon center of mass. The attraction of the bulge produces a torque which causes the lunar orbital velocity to increase; an equal but opposite torque causes the earth's rotational velocity to decrease.

There has been continued speculation regarding the magnitude of the lag in the tidal bulge. Darwin (1898) showed that a phase lag of one degree could produce the estimated slowdown. The lag could be derived from tidal gravity measurements (see, e.g., MacDonald, 1964; Kuo et al., 1970) if the measurements can be adequately corrected for the effects of ocean loading. Since ocean loading is significant, even in the interior of continents, the corrections include considerable uncertainties.

The tidal potential W_2 (eqs. 3.20 and 3.21) has the form $\frac{1}{2}$ $(3 \cos^2 \psi - 1)$, where the angle ψ is shown in Fig. 3.1. This is equal to $P_2(\cos \psi)$ in eq. 2.29 and the potential is thus a second-degree zonal harmonic, which means that it is ellipsoidal in shape. The most direct estimates of the deformation of the earth's gravitational field produced by tidal attractions are obtained from perturbations of satellite orbital paths. Newton (1969) and later Kolenkiewicz et al. (1973) showed that the tidal ellipticity is represented by a second-degree Love number and phase lag. Kolenkiewicz et al. used

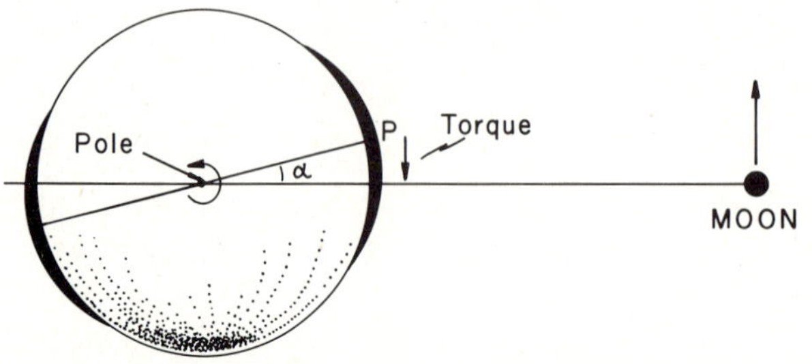

Fig. 3.6. Plan view of the earth and moon, showing the tidal bulge carried forward when tidal deformation lags behind the deforming stresses.

laser-range measurements to determine perturbations of the orbital inclination of a satellite to compute the amplitude and phase of earth tides, obtaining values for the second-degree harmonic Love number $k_2 = 0.245$ and phase lag $\chi_2 = 3.2°$. This value of the lag is similar to that which Smith and Jungels (1970) obtained from linear earth-strain-meter measurements. The earth slow-down rate obtained by Kolenkiewicz et al. is greater than that obtained by Newton, but smaller than that based on astronomical observations of lunar orbital acceleration.

The potential produced by the earth tidal deformation, $k_2 W_2$, decreases outward from the earth as the inverse cube of the distance. At the moon, the potential V_m due to earth is (e.g., Stacey, 1977, p. 98), if we use eq. 3.20 for W_2:

$$V_m = k_2 W_2 \left(\frac{a}{R}\right)^3 = \frac{k_2 G m a^5}{2R^6} (3 \cos^2 \psi - 1) \qquad (3.53)$$

The tidal torque produces an orbital acceleration of the earth and the moon about the common center of mass. The torque acting on the moon is:

$$L = m \left(\frac{1}{R} \frac{\partial V}{\partial \psi}\right)_{\psi = \chi} (R) = \frac{3}{2} \frac{k_2 G m^2 a^5}{R^6} \sin 2\chi_2 \qquad (3.54)$$

We substitute 0.245 for k_2, 3.2° for χ_2, and other constants as given in Appendix B to obtain for the torque:

$$L = 4.86 \cdot 10^{23} \text{ dyne cm} \qquad (3.55)$$

The moon exerts an equal but opposite torque on the tidal bulge of the earth.

The torque due to the bulge is equal to the rate of change of the moon's orbital angular momentum, H_m, about the center of mass. Thus:

$$L = \frac{dH_m}{dt} = \frac{d}{dt} \left[\left(\frac{M}{M+m}\right) m R^2 \omega_L \right] \qquad (3.56)$$

in which ω_L is the angular velocity of the moon. R and ω_L are related by Kepler's third law of planetary motion, which is derived from the equality of centripetal force in the lunar attraction about the earth—moon center of mass with the gravitational force of attraction. For a circular orbit:

$$\left(\frac{M}{M+m}\right) m \omega_L^2 R = G\left(\frac{Mm}{R^2}\right) \qquad (3.57)$$

which reduces to Kepler's third law:

$$\omega_L^2 R^3 = G(M+m) \qquad (3.58)$$

If we substitute eq. 3.58 into 3.56, the torque can be obtained in terms of

R or ω_L. Thus:

$$L = \frac{d}{dt}\left[Mm\left(\frac{GR}{M+m}\right)^{1/2}\right] = \frac{d}{dt}\left\{Mm\left[\frac{G^2}{\omega_L(M+m)}\right]^{1/3}\right\} \tag{3.59}$$

or

$$L = -\left(\frac{H_m}{3\omega_L}\right)\frac{d\omega_L}{dt} = \left(\frac{H_m}{2R}\right)\frac{dR}{dt} \tag{3.60}$$

in which $d\omega_L/dt$ is the acceleration of the moon in its orbit and dR/dt the rate of lunar recession. Applying values in Appendix B to eq. 3.56, the value for H_m is $2.85 \cdot 10^{43}$ g cm^2 sec^{-1}. We substitute this value for H_m and that in eq. 3.55 for L to obtain:

$$\frac{dR}{dt} = \frac{2RL}{H_m} = 1.3 \cdot 10^{-7} \text{ cm/sec} \sim 4 \text{ cm/year} \tag{3.61a}$$

for the rate of lunar recession and

$$\frac{d\omega_L}{dt} = -\frac{3\omega_L L}{H_m} = -1.38 \cdot 10^{-23} \text{ rad/sec}^2 \sim -28 \text{ arc sec/century}^2 \tag{3.61b}$$

for the decrease in the moon's orbital velocity.

Newton (1969) used ancient eclipses to determine the value of $d\omega_L/dt$, obtaining $-42''$ century^{-2}. Oesterwinter and Cohen (1972) used a half century of instrument observations on the moon, obtaining a value of $-38''$ century^{-2}. Van Flandern (1975) considered a best value to be $-38''$ century^{-2}.

The slowdown of the earth's rotation, $d\omega/dt$, can be calculated from the torque which the moon exerts on the bulge, which is the negative of the above calculated torque. Thus:

$$-L = \frac{d}{dt}\left(C\omega\right) = C\frac{d\omega}{dt} \tag{3.62}$$

The slowdown rate is then:

$$\frac{d\omega}{dt} = -\frac{L}{C} = -6.0 \cdot 10^{-22} \text{ rad/sec}^2 \tag{3.63}$$

where the value of C, the earth's moment of inertia, is given in Appendix B. This slowdown rate is equal to 2.2 msec day^{-1} century^{-1}.

The slowdown has been calculated by numerous investigators using astronomically determined values for $d\omega_L/dt$. Munk and MacDonald (1960, p. 202) used $d\omega_L/dt = 1.09 \cdot 10^{-22}$ rad sec^{-2} ($-22''4$ century^{-2}). The value of $-38''$ century^{-2} results in a lunar recession of 5.6 cm year^{-1} and a slowdown rate of 3 msec day^{-1} century^{-1}.

It is readily seen that the earth's slowdown rate is about 45 times greater than the rate of change in the moon's orbital velocity:

$$\frac{(d\omega/dt)}{(d\omega_L/dt)} = \frac{-6.0 \cdot 10^{-22}}{-1.38 \cdot 10^{-23}} \sim 45 \qquad (3.64)$$

The tidal friction produced by the sun, which also slows down the earth's rate of rotation, was omitted in the previous discussion. It is seen in eq. 3.18 that tidal amplitudes generated by the sun are approximately 45% as large as those generated by the moon. If the energy dissipation is assumed to be proportional to the square of the tidal amplitude, which applies to linear-type dissipative processes (possibly not a good assumption), tidal dissipation produced by the sun is about 20% of that produced by the moon.

Dissipation of energy in the earth as a result of tidal friction has been treated by numerous investigators. Munk and MacDonald (1960, p. 203) estimated the dissipation rate to be $2.74 \cdot 10^{19}$ ergs sec^{-1}. If the value for L in eq. 3.55 is used (e.g., Stacey, 1977, p. 100), the dissipation rate is $3.4 \cdot 10^{19}$ ergs sec^{-1}. If the astronomic-determined value for $d\omega_L/dt$ of $-38''$ $century^{-2}$ is used, the dissipation rate becomes $4.6 \cdot 10^{19}$ ergs sec^{-1}.

The sink for at least a part of the energy dissipation is attributable to turbulence of ocean tidal currents. Taylor (1922) showed that dissipation varies with the cube of current velocity, and since the larger velocities are in shallow seas, the oceanic sink occurs essentially in the shallow seas. Miller (1966) calculated the tidal-energy dissipation produced by all shallow seas and obtained a value of approximately $1.5 \cdot 10^{19}$ ergs sec^{-1}. If Miller's value is correct, and if we take the total dissipation rate to be approximately $4 \cdot 10^{19}$ ergs sec^{-1} (i.e., between 3.4 and 4.6), then dissipation will be substantial in the solid earth. Such dissipation is unlikely to occur throughout the mantle because of its high rigidity and high average value of Q (see eq. 4.1). A possibility is that the low-velocity layer of the upper mantle, which appears to have a Q as low as 70, may be the source of an important sink. On the other hand, Lambeck (1975, and personal communication, 1976) concludes from satellite data that all of the dissipated energy could occur in the oceans.

Tidal friction will cause the earth to rotate more slowly so that eventually the earth and moon will rotate synchronously, showing the same faces to one another. The length of day will then be about 48 h. This stable condition means that no tidal dissipation will occur in the earth due to the moon. However, the small amount of tidal friction due to the sun will cause the earth's rotation to continue to slow down. The effect will be to produce an instability in the lunar orbit, the lunar centripetal force decreasing with the resultant decreased orbital velocity until it is exceeded by the force of attraction, then causing the moon to approach or fall into the earth.

Chapter 4

INTERIOR STRUCTURES AND PROCESSES — GRAVITY-RELATED ASPECTS

INTERIOR OF THE EARTH

Knowledge about the composition and structure of the earth's interior is derived almost entirely from geophysical investigations. Most of this information is obtained from seismic (principally earthquake) wave propagation. The following provides an overview of the earth's interior composition.

Variations of seismic-wave velocity with depth

The basic source of information concerning the composition of the earth's interior is seismic travel-time curves These are plots of observed time of travel, as function of epicentral distance, of various direct, reflected, and refracted waves propagated through the earth. Using slopes of travel-time curves for the compressional (*P*) and shear (*S*) waves as they propagate directly from source to recording station through the crust and mantle (and, with modifications, through the core), Wiechert (1907) and Herglotz (1907) applied wave propagation theory to calculate values of seismic velocity as a function of depth. These analyses led to the famous Wiechert-Herglotz equations, which have been subsequently reproduced in many publications (see, e.g., Richter, 1958, p. 667; Bullen, 1963, p. 120; Cook, 1973, p. 295; Stacey, 1977, p. 325). With these equations, slopes of *P*- and *S*-wave travel-time curves have been used to obtain generalized *P* and *S* velocity—depth curves. These curves have provided the principal evidence for velocity discontinuities within the earth. Discontinuities are identified by an abrupt change in velocity (a first-order discontinuity) or a change in velocity gradient (a second-order discontinuity) with depth. Discontinuities identified from the depth—velocity curves have stood the test of time, the curves requiring only minor modifications over the years, although there still remain differences of opinion in the interpretation of parts of the travel-time data. Fig. 4.1 illustrates velocity—depth curves which incorporate generally accepted modifications of the earlier developed curves.

Discontinuities within the earth

Velocity—depth curves (Fig. 4.1) indicate that the earth consists of a central core, surrounded by a massive mantle, which is covered by an outer

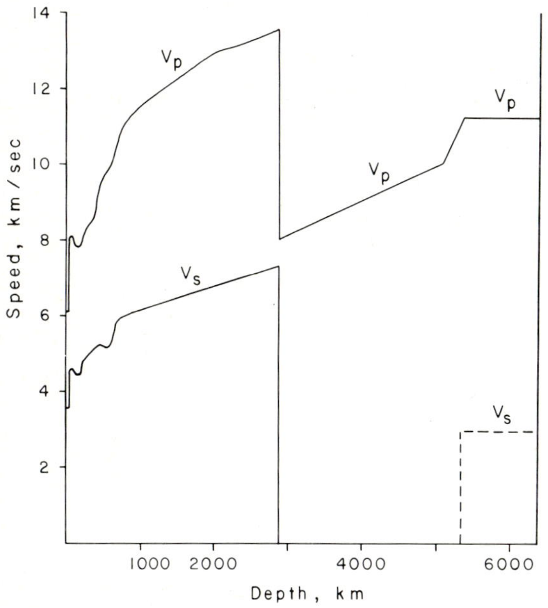

Fig. 4.1. Variations in velocities of P and S waves with depth in the earth.

crust. Velocity discontinuities clearly separate these major regions, which in turn are seen to be further separated by less pronounced discontinuities. Each of these discontinuities could represent either a chemical change or a solid—solid mineral phase change. Nevertheless, a generally accepted view regarding the composition of materials between the discontinuities has emerged.

Current usage divides the crust and subcrustal region into the lithosphere, the relatively rigid outer layer of the earth which varies in thickness from 50 to 150 km, and the asthenosphere which underlies the lithosphere. The asthenosphere is a softer more plastic part of the upper mantle which exhibits limited strength. A low-velocity layer (LVL) appears to exist in the upper mantle, characterized in Fig. 4.1 by the decrease in P and S velocities at a depth of about 150 km. The upper part of the LVL appears to be gradational, and is generally considered to comprise the top of the asthenosphere. In the theory of plate tectonics, rigid lithospheric plates move relative to one another over the softer asthenosphere.

The two major discontinuities in the earth, which are also first-order types, are the Mohorovicic (Moho) discontinuity, between the crust and mantle and characterized by a sudden increase in P and S velocities with depth, and the Wiechert-Gutenberg discontinuity, between the mantle and core and characterized by a sharp decrease in P velocity and a drop to zero for S (Fig. 4.1) in the core. Depths to the Moho vary; representative depths

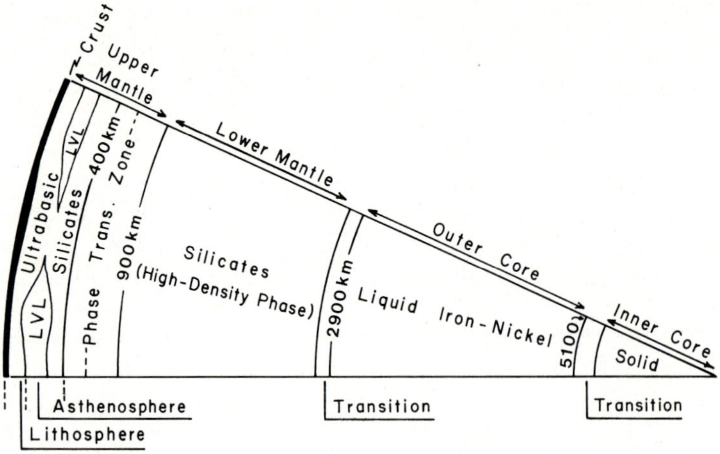

Fig. 4.2. Discontinuities, layering, and probable materials in the interior of the earth.

are 10—11 km on oceans and 30—33 km on continents. The Wiechert-Gutenberg discontinuity is at a depth of nearly 2,900 km. Both discontinuities appear to be quite sharp, on the order of a kilometer in thickness.

The mantle consists of an upper part, which extends to a depth between about 650 and 900 km, and a lower part which extends to the top of the core. The upper mantle includes the LVL, which is generally considered to be world-wide for *S* waves (Nuttli, 1969), if not also for *P*, and which appears to have ill-defined transitional upper and lower boundaries (second-order discontinuities) whose depths appear to vary widely in different types of regions. In oceanic provinces a typical depth to the top of the LVL is 60—75 km, and to the bottom about 200 km. In continental provinces the depth to top is typically 100 km or more, although in areas of recent volcanism (e.g., mid-ocean ridges) and some Tertiary tectonics the LVL appears to extend up to the Moho locally; under old continental shield areas the LVL is considerably deeper and thinner, if it exists there at all. When classifications of lithosphere and asthenosphere rather than crust and mantle are used, the lithosphere appears to extend down to the LVL.

Beneath the LVL two and possibly three discontinuities are generally recognized (e.g., Anderson et al., 1972; Johnson, 1969; Nuttli, 1969; Brian Kennett, personal communication, 1976). The upper two are at depths of approximately 400 and 650 km. The base of the asthenosphere, which is not well defined, is generally considered to be at one of these two discontinuities. A third discontinuity appears to be at an approximate depth of 900 km, forming the base of the upper mantle. In the core, a discontinuity, comprised of a transition zone, occurs between the inner and outer core, at depths from about 5,000 to 5,200 km.

The crust and mantle were once considered as regions characterized by horizontal variations in type of rock (crust) overlying a laterally homogeneous material (mantle). During the past several decades, it has become increasingly clear that the upper mantle, at least from the Moho to the base of the LVL, is also inhomogeneous laterally and that depths of isotherms in this region can vary considerably. Isotherms are shallower under volcanic regions, including mid-ocean ridges, and appear to be substantially shallower under oceans than continents (Sclater and Francheteau, 1970); they are relatively deep under shield areas, and extend sharply downward along the cooler subducting plates (e.g., McKenzie, 1969).

Compositions of layers in the earth are not known exactly, but generally accepted ones are consistent with available measurements and known mineralogy. The lower crustal layer appears to extend around the earth, although it may vary in composition and age. Beneath oceans the layer (oceanic layer 3) has a typical thickness of 5 km, and beneath continents a thickness of 10—20 km. In both types of provinces the layer is characterized by a *P*-wave velocity near 6.8 km sec^{-1}. Laboratory experiments have shown that basic igneous rocks have such velocities at these overburden pressures. The composition of this layer is generally considered to be gabbro or basalt, with a density of 2.9—3.0 g cm^{-3}.

Materials just beneath the Moho have typical *P*-wave velocities of 8.1 km sec^{-1} under both oceans and continents, ranging from about 7.5 km sec^{-1} (or less) beneath mid-ocean ridges and volcanic regions up to about 8.6 km sec^{-1}, especially under shield areas. Laboratory experiments have shown that among known rocks only ultrabasic ones (e.g., peridotite, dunite, harzburgite, pyroxinite, or eclogite), have the high sub-Moho velocities at the respective pressures and temperatures. The sub-Moho material is commonly considered to be peridotite or harzburgite, although according to Ringwood (1969), the material may likely have a composition of 3/4 peridotite and 1/4 basalt, which he calls pyrolite. The density of the ultrabasic material is approximately 3.3 g cm^{-3}.

From the Moho down to the LVL, the mineralogical composition could be the same as that beneath the Moho, with the effects of increasing pressure and temperature producing a lower seismic velocity, or the composition could change slightly, for example, to a pyrolite, or there could be phase changes. Ringwood (1969) discussed in detail the probable mineralogy of the upper mantle. In any event, the velocity of propagation in the LVL is near 7.8 km sec^{-1}, and the density likely somewhat below that of the overlying mantle, probably below 3.3 g cm^{-3}. Also, the temperature is considered to be slightly below the melting temperature for the respective depths, possibly resulting in a small amount of partial melting, and the rigidity, viscosity, and Q (e.g., Knopoff, 1964) are decreased. $1/Q$ is a measure of energy dissipation in a vibrating system:

$$\frac{1}{Q} = \frac{\Delta W}{2\pi W} \qquad (4.1)$$

in which W is the total energy stored in a vibrating system (as a rock) and ΔW the energy dissipated per cycle of vibration (as in the passage of a seismic wave). Values of Q of approximately 400 have been reported for the crystalline rocks comprising the continental crust; values under oceans are reported to be several times greater. Values in the LVL appear to be about 70. In the lower mantle Q is considered to increase to about 2,000, and possibly even 5,000 near the base of the mantle (MacDonald, 1964; Knopoff, 1964).

Viscosity, which is the proportionality factor between shearing stress and time-rate of shearing strain, generally increases in rocks with increasing values of Q. Values of viscosity in the upper lithosphere generally range from 10^{24} P (poise) to 10^{26} P, based on analyses of earthquake and laboratory data (e.g., Murrell, 1976). Values in the LVL are estimated to be between 10^{18} and 10^{20} P (Elsasser, 1969; Walcott, 1973), and may possibly be lower. The relatively low viscosity and low Q in this layer are expected to affect isostatic equilibrium and lithospheric flexuring in response to topographic loads.

Values of viscosity beneath the LVL are thought to increase with depth by some investigators but not by others. For example, MacDonald (1964) calculated that the viscosity near the base of the mantle may be as high as 10^{26} P, thereby concluding that mantle-wide convection is not possible. On the other hand, Tozer (1965, 1970a,b) believes that mantle-wide convection occurs as a principal mode of heat transfer in the earth, and investigated plausible viscosity—temperature relationships which are consistent with convection. He obtained a viscosity of 10^{20}—10^{21} P throughout the mantle (e.g., Jacobs, 1975, p. 108). Similar arguments led O'Connell (1976) to conclude the viscosity is 10^{21}—10^{22} P throughout the mantle. Cathles (1975, p. 3 and others) concludes from post-Glacial uplift analyses that the viscosity throughout the entire mantle, except possibly for the LVL, has a constant value of 10^{22} P.

Velocities of P and S waves (Fig. 4.1) between depths of 400 to about 900 km increase more rapidly than can be accounted for by pressure increases alone, suggesting that materials in this depth range undergo solid—solid phase transformations; the crystals possibly take on a spinel-type structure. Throughout the lower mantle (depths below about 900 km) the composition is considered to be mineralogically uniform because observed velocity gradients can be accounted for by pressure and temperature increases. Ringwood (1972) discussed the inferred mineralogy of the lower mantle. The entire mantle can be taken to consist of silicate minerals; the chemical composition could be either uniform, at least below the LVL, or the upper and lower mantle could consist of layers with slightly different compositions.

Because the outer core does not transmit *S* waves, it must be a fluid which means its rigidity is nearly zero. Gans (1972) calculated typical viscosities to be as low as $6 \cdot 10^{-3}$ P, and David Gubbins (personal communication, 1976) calculated that for pure iron they may be as low as 10^{-3} P. The density of the core must be relatively high because the earth's moment of inertia about its rotational axis is approximately 83% that of a sphere of uniform density; a mass concentration in the interior region is required. The composition of the core is considered to be essentially iron, a common element with the expected density which also occurs in iron meteorites. Nickel is a likely constituent because it too occurs in iron meteorites. However, a pure Fe—Ni core produces a density which is too high and a *P*-wave velocity too low to fit available geophysical data (Anderson et al., 1972, p. 57). Hence, probably about 10% of the core consists of lighter elements; the most likely constituents are S, Si, or MgO.

The inner core is considered to be solid, the principal evidence being derived from free oscillations of the earth (Gilbert et al., 1973). The composition of the inner core is probably Fe—Ni (Verhoogen, 1973) containing fewer light elements than the outer core, implying a somewhat higher density.

Jacobs (1971) and Birch (1972) conclude that temperatures in the core may be near the melting point. An interesting consequence of the same melting point and adiabatic gradients throughout the core would be that the outer region of the core will at times become solidified, thus becoming part of the lower mantle, and at others the lower part of the mantle becomes soluble in the outer core. The effects would be to form "bumps" in the core—mantle boundary, as proposed by Hide (1969).

Temperatures in the core, according to Jacobs (1975, p. 111), which are consistent with data on density and seismic-velocity distributions, heat flow data, and rates of production of magnetic energy appear to be 3,200°—4,700°C at the core—mantle boundary and 4,700°—6,200°C at the inner—outer core boundary. Higgins and Kennedy (1971) proposed a temperature of about 4,250°C at the inner—outer core boundary, based on melting point—pressure curves extrapolated to core pressures.

Variation of density with depth

Densities within the earth cannot be determined directly; hence, they are inferred from boundary conditions and fitted with all available observations, particularly seismic discontinuities. Bullen (1963, p. 231; 1975, p. 169) estimated density distributions which satisfy: (1) the mean density of the earth; (2) the calculated moment of inertia of the earth about its rotational axis; (3) the density of ultrabasic rocks beneath the Mohorovicic dis-

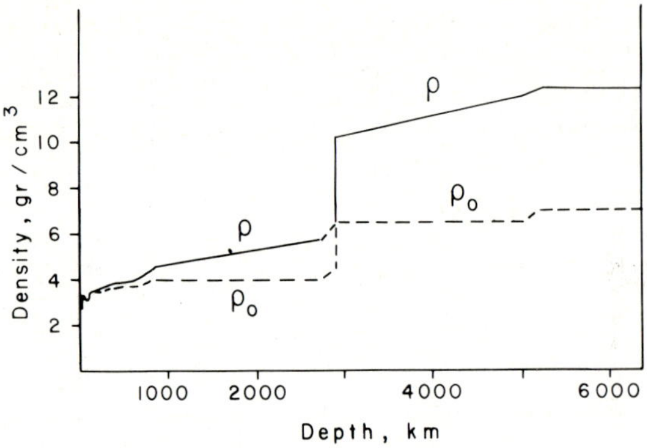

Fig. 4.3. Variations in density (ρ) and in uncompressed density (ρ_0) with depth in the earth.

continuity; (4) established *P*- and *S*-wave velocity—depth curves; (5) surface-wave data; and (6) in the latest model, free earth-oscillation data. Bullen's initial density model was published in 1936; over the years he proposed several earth models and has modified them as new information about the earth's interior became available. The current density Model A″ (Bullen and Haddon, 1967; Bullen, 1975, p. 173) is a best fit of the available data. Fig. 4.3 shows a slight modification of this model by including a small density decrease in the LVL of the upper mantle, and incorporates effects of solid—solid phase transitions in the depth range of 400—900 km, as proposed by A.E. Ringwood and co-workers (e.g., Ringwood, 1972).

Fig. 4.3 also shows estimates of uncompressed (zero pressure) as well as compressed densities for materials having the likely compositions shown in Fig. 4.2. The inferred lower density in the LVL has implications regarding source materials for the growth of lithospheric plates at ocean ridges and would play a role in isostatic adjustments and lithospheric flexures in response to topographic loads.

Variation of gravity and pressure with depth

It is shown in Chapter 7 that gravity at an interior point in a spherical earth is the same as if all mass within a sphere of that radius were concentrated at the center, and also that the value is not affected by outer spherical layers of uniform composition. Hence, gravity at interior points in the earth can be readily approximated for assumed density distributions. Fig. 4.4 shows variations of gravity with depth consistent with Bullen Model A″. We see that gravity increases with depth over parts of the mantle, reaching a maximum near the base of the mantle, as a consequence of the relatively high density in the core. Gravity is clearly zero at the earth's center of mass.

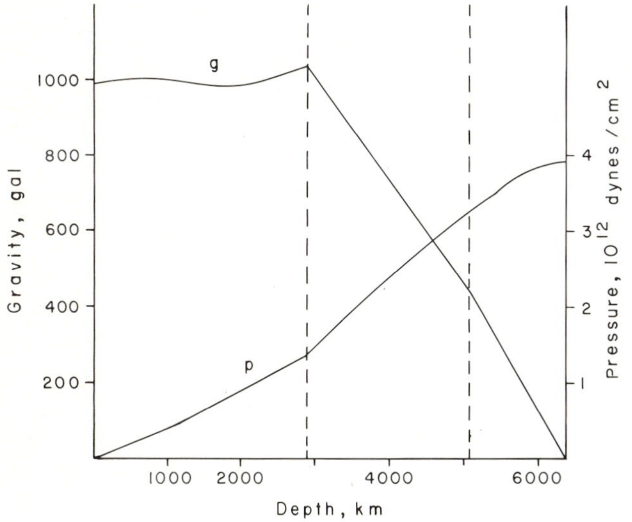

Fig. 4.4. Variations in gravity (g) and pressure (p) with depth in the earth.

Hydrostatic pressure within a spherically symmetric earth can be readily calculated when density and gravity are known as functions of depth (Figs. 4.3 and 4.4). Overburden pressure is given by $p = \bar{\rho}\bar{g}(a - r_1)$ where $\bar{\rho}$ and $\bar{g}$ are the average density and gravity, a the earth's radius, and r_1 the radius at which pressure p applies. The pressure gradient $\mathrm{d}p/\mathrm{d}r = -\rho g$, where $g = GM_1 r_1^{-2}$, M_1 being the total mass within radius r_1. Fig. 4.4 shows variations of pressure with depth (Bullen, 1975, p. 361).

PLATE TECTONICS

The theory of plate tectonics (discussed in detail in Morgan, 1968; McKenzie, 1972; Le Pichon et al., 1973; Cox, 1973) evolved from analyses of many of the geophysical and geological data obtained particularly during the 1960's. The concept provides a unified theory of tectonic processes which have occurred throughout much, if not all of geologic time.

The plates in plate tectonics consist of sections of the rigid, high-strength lithosphere which overlies the considerably less-rigid, low-strength asthenosphere. Whereas typical thicknesses of the lithosphere are estimated to be from 50 to 100 km, its strength, the ability to withstand stress without permanent deformation, appears to be highest in its upper part. The effective lithospheric thickness is therefore expected to vary with the time during which the lithosphere behaves as an elastic plate. The part which behaves as a rigid plate for durations up to several hundred million years, as required for the reconstructed past plate motions, may be relatively thin. The corresponding viscosity appears to decrease from the upper lithosphere downward to the LVL (evidenced by Q decreasing from about 400 and even considerably higher values to a minimum of about 70). The transition from the lithosphere to asthenosphere is poorly defined; it is likely variable and probably extends over a considerable depth range.

Plates are characterized by three types of seismically active borders, illustrated in Fig. 4.5. One is a mid-ocean ridge, an oceanic linear topographic high along which materials well up from the asthenosphere as plates separate away from the ridge, generating new material which is added to the trailing edge of the plate. Growth rates vary, ranging from 1 to 5 cm yr^{-1} or more on each plate. A second is the down-going or subducted leading edge of a plate, characteristic of island-arc—trench systems, which is consumed as it descends into the mantle (Fig. 4.5) at rates up to 12 cm yr^{-1} or more. A third is a transform fault, first described by Wilson (1965), a fracture zone along which plates move relative to one another without material being generated or consumed.

Seismic foci along transform faults and ocean ridges are characteristically shallow, especially along transform faults where foci are observed to be predominantly in the upper part of the lithosphere. Offsets of materials beneath the zones of these shallow foci may be produced by rock creep, where stresses do not build up sufficiently to generate earthquakes. Earthquakes along the down-going plate, the largest of which have high magnitudes, occur over nearly its entire length. In fact, the subducted plate is assumed to terminate about where its deepest earthquakes occur. Each down-going plate is characterized by its maximum focal depth. The deepest shocks observed anywhere are at depths of just over 700 km, which is approximately 1,000 km along the slab, depending upon its dip, from where it bends downward under an oceanic trench (Fig. 4.5). On the ocean side of

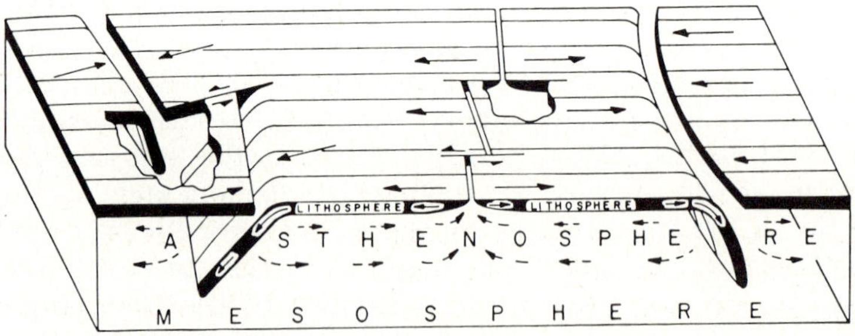

Fig. 4.5. Block diagram illustrating the lithosphere and underlying asthenosphere. Arrows on lithosphere indicate relative movements of adjacent blocks, those in the asthenosphere possible compensating flows. Materials well up beneath an ocean ridge; the lithosphere descends into the mantle along ocean trenches; lithospheric plates shear against one another along transform faults. (Reproduced from Isacks et al., 1968, with permission.)

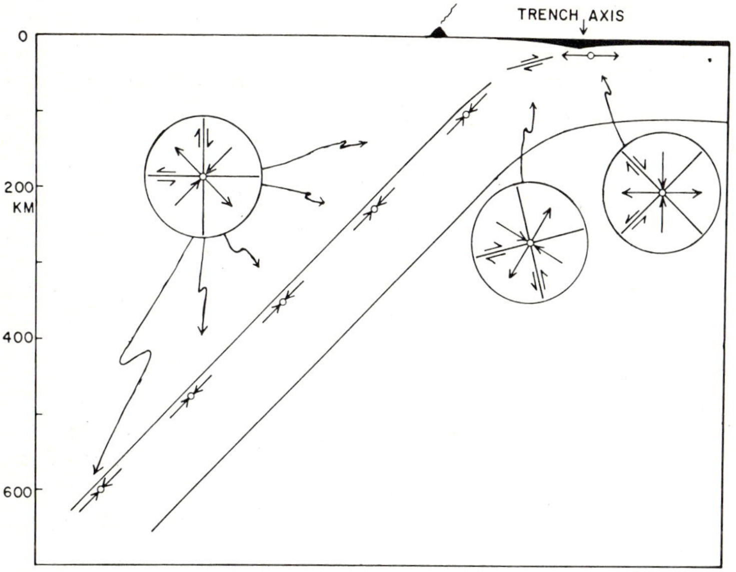

Fig. 4.6. Vertical section normal to an oceanic trench axis, showing typical orientations of principal stress or of shearing directions. The circular blowups show the sense of motion of two possible slip planes resulting from the principal stresses indicated. (Reproduced from Isacks et al., 1968, with permission.)

the trench, where the subduction zone begins to bend downward, earthquakes are shallow, being characterized by relative extension in a horizontal direction normal to some trench axes (Stauder, 1968) and by compression at others (Hanks, 1971). Landward of the trench and at depths of about 75—150 km, foci generally exhibit underthrusting (Fig. 4.6). At greater depths their maximum principal stresses are in the general direction of the down-going plate (Isacks et al., 1968); the source mechanism is less well understood, but is likely associated with the presence of fluids which have the effect of reducing stresses required to produce fractures. Isacks et al. also observed that slip rates are higher along those subducted plates which exhibit deeper maximum focal depths. This is possibly related to the time required for the cooler slab to be heated to the higher temperatures of the surrounding mantle materials, resulting in the slab eventually losing its identity. The duration required for this identity loss appears to be millions of years, because of the low thermal conductivities of the lithospheric materials.

Borders of the plates consist of any combination of the three different types of boundaries. The relative stability (or instability) and possible eventual destruction of a point of intersection of three plates, known as a triple junction, were analyzed by McKenzie and Morgan (1969). Under certain conditions triple junctions are stable over many millions of years, such that the three adjoining plates are either not in motion or all move with the same velocity. Under other conditions triple junctions are unstable, the plates then moving relative to one another, such that plates and even the triple junction can be consumed as down-going slabs.

FLEXURE OF THE LITHOSPHERE

A basic question concerning sources of gravity anomalies is the relationship between plate tectonics and classical, local-type of isostatic compensation. The rigid lithosphere indicated by plate tectonics implies that it behaves as an elastic beam overlying a weak fluid (asthenosphere). Isostasy implies regional and even local flow at possibly shallow depths. It has not been clear whether the response of the lithosphere to long-duration topographic loads is essentially elastic, viscoelastic, or viscous. Recent evidence indicates that the response can be elastic to loads for durations of several tens of millions of years; however, probably not for durations as short as tens of thousands of years because the asthenosphere then behaves less like a fluid.

Since the early 1930's investigators have emphasized that the lithosphere has long-term strength (i.e., the ability to support stresses without permanent deformation). Vening Meinesz (1931) considered regional compensation for topographic features supported on a strong lithosphere; he explained (Vening Meinesz, 1941b) gravity anomalies near Hawaii as associated with a lithosphere which supports extensive volcanic flows, that is, an elastic plate overlying a fluid. Gunn (1943, 1947, 1949) and others also considered the strength of the upper lithosphere in accounting for geophysical measurements in the vicinity of surface loads.

Recent investigators analyzed flexures of the lithosphere produced by two- and three-dimensional loads. Walcott (1970a, b, c, 1972, 1976) concluded from investigations of surface loads on the lithosphere that such loads are supported as if the lithosphere were a viscoelastic substance, that is, a non-Newtonian (Maxwellian) fluid in which viscosity decreases as stress increases. This implies that although the lithosphere supports applied loads, with time the load becomes increasingly more localized. Later investigations by Watts and co-workers (Cochran, 1973; Watts and Talwani, 1974; Watts and Cochran, 1974; Watts et al., 1975), based on analyses of gravity anomalies over various two- and three-dimensional oceanic structures, indicate that the lithosphere is capable of resisting permanent deformation over periods up to tens of millions of years. This implies that long-term responses may be more elastic than viscoelastic.

The flexural rigidity of an elastic beam supported at both ends is given by EI, the product of Young's modulus (E) and the moment of inertia (I) of the

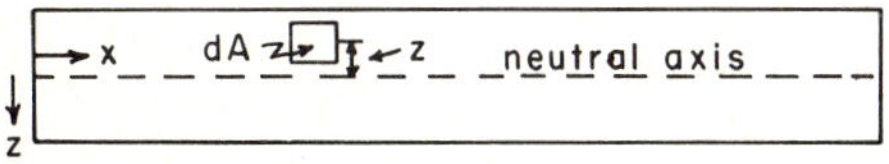

Fig. 4.7. Diagram of cross section of a beam showing the neutral axis and element of area at a distance z from the axis.

cross-sectional area (Fig. 4.7) with respect to the neutral axis of the beam (intersection of the neutral surface with the cross section). I is given by (Timoshenko, 1955, p. 95):

$$I = \int z^2 \mathrm{d}A \tag{4.2}$$

in which dA is the cross-sectional area per unit beam width. The beam thickness B is related to I by (Timoshenko, 1955, p. 417):

$$I = \frac{B^3}{12} \tag{4.3}$$

The "effective flexural rigidity" (EFR) is a term (Gunn, 1943; McKenzie, 1967, 1968; Walcott, 1970a; Lambeck, 1972; Watts and Cochran, 1974) used to indicate the value of flexural rigidity that applies to the lithosphere in a particular region. The best-fitting EFR is the value most in agreement with geophysical and geological data in the vicinity of a topographic load. Watts and Cochran conclude that the EFR in an area attains a constant value some time after the surface load has reached an equilibrium condition; thereafter the lithosphere is able to support surface loads for many tens of millions of years. Watts and Cochran also show that loads of greatly different ages can produce essentially the same value of EFR.

Two-dimensional analyses of surface loads (Cochran, 1973; Watts and Talwani, 1974; Watts and Cochran, 1974) were extended to three-dimensional loads by Watts et al. (1975), who fitted their analyses with observed free-air anomalies to obtain best-fitting values of EFR. These analyses showed that three-dimensional bodies cannot be adequately determined with two-dimensional methods. A general rule of thumb is that three-dimensional analyses are required unless the length of the body is more than four times its width.

Theory of beam and plate deflection

The lithosphere can be modeled as an elastic beam overlying a fluid because the asthenosphere has a low viscosity (low strength) for loads applied for more than about 10^5 years. The theory of loads on elastic beams overlying fluids has been treated in the engineering literature (e.g., Timoshenko, 1955; Nadai, 1963; summarized as it applies to the bending of

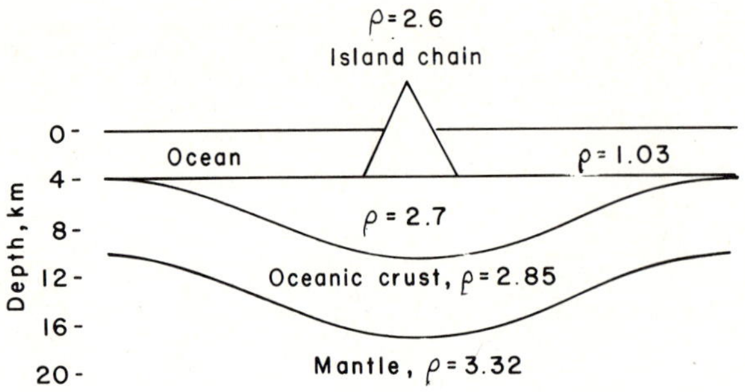

Fig. 4.8. Vertical section of a lithosphere down-warped by infilled rock and an island chain.

the lithosphere by Gunn, 1943, 1947; Le Pichon et al., 1973; and more recently by Watts and Cochran, 1974).

Let a distributed load with intensity p, as produced by materials infilling a down-warped lithosphere (Fig. 4.8), act on a cross section of the beam (Fig. 4.9). The shearing force F_s, per unit beam width, acting at cross section mn is the sum of all downward and upward forces along mn; the bending moment L, per unit beam width, at mn is the sum of the products of these forces and their respective distances from the cross section. Equilibrium in the beam requires then an increase in bending moment $\mathrm{d}L$ equal to $F_s\mathrm{d}x$, where $\mathrm{d}x$ is the distance between adjacent cross sections along the beam. Thus:

$$F_s = \frac{\mathrm{d}L}{\mathrm{d}x} \qquad (4.4)$$

Since the distributed load acts between the two cross sections, the load intensity p (per unit length and cross section) on a beam element in $p\mathrm{d}x$. The difference in shearing force between adjacent cross sections is $\mathrm{d}F_s = -p\mathrm{d}x$, the negative sign indicating a net downward load. Hence:

$$p = -\frac{\mathrm{d}F_s}{\mathrm{d}x} = -\frac{\mathrm{d}^2 L}{\mathrm{d}x^2} \qquad (4.5)$$

The bending moment of a beam clamped at both ends is (e.g., Timoshenko, 1955, p. 95):

$$L = \frac{EI}{R_c} \qquad (4.6)$$

in which L is the moment per unit length and R_c the radius of curvature of the deflected beam. For downward deflections small compared to beam

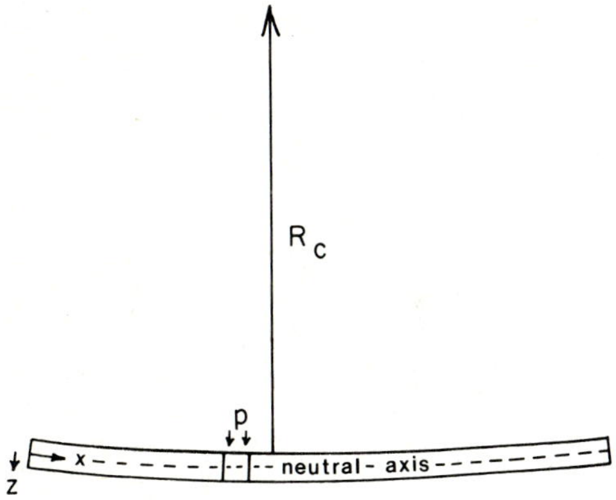

a

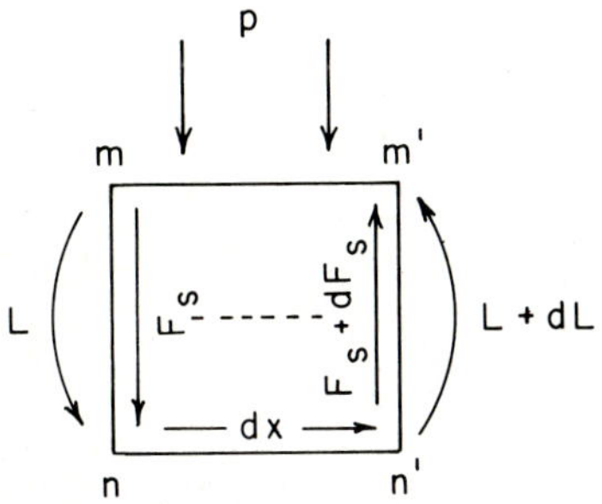

b

Fig. 4.9. a. Cross section of a deflected beam with radius of curvature R_c; p is the load intensity on the beam. b. Illustration of shearing forces F_s across section mn and F_s + dF across $m'n'$ produced by load p; L and L + dL are the applied bending moments.

length, with z positive upward:

$$\frac{1}{R_c} \sim \frac{d^2 z}{dx^2} \tag{4.7}$$

We combine eqs. 4.7 and 4.6 to obtain:

$$L = EI \frac{d^2 z}{dx^2} \tag{4.8}$$

and differentiate eq. 4.8 twice with respect to x and combine with eq. 4.5, to obtain:

$$EI \frac{d^4 z}{dx^4} + p = 0 \tag{4.9}$$

Applying beam theory to two-dimensional plates, eq. 4.9 becomes (e.g., Watts and Cochran, 1974; Walcott, 1976):

$$\frac{EI}{(1-\sigma^2)}\left(\frac{\partial^2 z}{\partial x^2}+\frac{\partial^2 z}{\partial y^2}\right)^2 + p = 0 \tag{4.10}$$

in which σ is Poisson's ratio, which is the negative relative change in diameter of a rod to its relative change in length when under uniaxial extension. The flexural rigidity of a plate is $EI/(1-\sigma^2)$.

The beam load per unit length, p, is a pressure which is the difference between the buoyancy and weight of the beam (per unit area), that is:

$$p = (\rho_u - \rho_r)gz \tag{4.11}$$

in which ρ_u is the density of the underlying asthenosphere and ρ_r that of the overlying distributed material (water or rock). Eq. 4.10 can then be written for a plate of unit width:

$$\frac{EI}{(1-\sigma^2)}\left(\frac{d^4 z}{dx^4}\right) + (\rho_u - \rho_r)gz = 0 \tag{4.12}$$

which has solutions of the form

$$z = \left[C_1 \cos\left(\frac{x}{f}\right) + C_2 \sin\left(\frac{x}{f}\right)\right] e^{x/f} + \left[C_3 \cos\left(\frac{x}{f}\right) + C_4 \sin\left(\frac{x}{f}\right)\right] e^{-x/f} \tag{4.13}$$

in which:

$$f = \left[\frac{4EI}{(1-\sigma^2)(\rho_u - \rho_r)g}\right]^{1/4} \tag{4.14}$$

When a line force F_l per unit beam width is applied, eq. 4.12 takes the form:

$$\frac{EI}{(1-\sigma^2)}\left(\frac{d^4 z}{dx^4}\right) + (\rho_u - \rho_r)gz = F_l \tag{4.15}$$

The beam deflection is a maximum where F_l is applied, that is where $dz/dx = 0$, and it is a minimum at large distances from the application of F_l. The deflection can then be written (e.g., Watts and Cochran, 1974):

$$z = \frac{F_l}{2fge^{x/f}(\rho_u - \rho_r)}\left[\cos\left(\frac{x}{f}\right) + \sin\left(\frac{x}{f}\right)\right] \tag{4.16}$$

and the deflection produced by a topographic load is calculated by integrating over the load width.

The parameter f has been variously described in the literature. Vening Meinesz (1941b) designated the radius of regionality by $2.905f$, Gunn (1943) called $1/f$ the lithospheric constant, and Walcott (1970a) described f as the flexure parameter. Watts and Cochran (1974) use the flexural rigidity of a beam, EI, and of a plate, $EI/(1-\sigma^2)$, as the significant parameters, these quantities being a property of the beam or plate, to describe the amplitude and wavelength of the lithospheric deflection.

A differential equation for a point load on a three-dimensional plate, solved by Hertz (1884), relates deflections of a plate overlying a fluid with density ρ_u and overlain by a material of density ρ_r (infilling the deflected plate), plus the effect of a surface point load F_p, can be written (Nadai, 1963), similar to eqs. 4.10 and 4.15:

$$\frac{EI}{(1-\sigma^2)}\left(\frac{\partial^2 z}{\partial x^2}+\frac{\partial^2 z}{\partial y^2}\right)^2+(\rho_u-\rho_r)gz=F_p \tag{4.17}$$

The plate thickness B is related to I by:

$$I=\frac{B^3}{12(1-\sigma^2)} \tag{4.18}$$

The solution to eq. 4.17 obtained by Hertz is (see, e.g., Gunn, 1943, or Watts et al., 1975):

$$z=\frac{F_p}{2\pi fg(\rho_u-\rho_r)}\int_0^{\infty}\frac{e^{-rv/f}\sin(rv/f)dv}{(v^2-1)^{1/2}} \tag{4.19}$$

in which r is the distance from F_p to the deflection point and v is a variable of integration. Hertz obtained values for z in terms of r in the vicinity of the point load. Watts et al. (1975) obtained the deflection for a volume source by integrating over the volume. As illustrated in Fig. 4.10, integration is over r, the distance from a load compartment to the deflection point; an element of area is $rdrd\theta$. For a compartment bounded by circles r_{n+1} and r_n and angles θ_{n+1} and θ_n, Watts et al. obtained:

$$z=\frac{F_p(\Delta\theta)}{2\pi f^2 g(\rho_u-\rho_r)}\left\{\frac{r^4}{4f^2}\left[\frac{\ln(r/f)}{4}-\frac{1}{16}\right]-\frac{r^8}{(2^2)(4^2)(6^2)(f^2)}\left[\frac{\ln(r/f)}{8}-\frac{1}{8^2}-\frac{1}{8}\left(\frac{5}{6}\right)\right]+\frac{\pi}{4}\left[\frac{r^2}{2}-\frac{r^6}{6(2^2)(4^2)(f^4)}+\frac{r^{10}}{10(2^2)(4^2)(6^2)(8^2)(f^8)}\right]-(1+\ln 2-K)\left[\frac{r^4}{(4^2)(f^2)}-\frac{r^8}{8(2^2)(4^2)(6^2)(f^6)}\right]\right\} \tag{4.20}$$

where K is a constant ($K=0.577$) and $\Delta\theta=\theta_{n+1}-\theta_n$. The condition of integration is that $\bar{r}$, the distance from the center of the load compartment

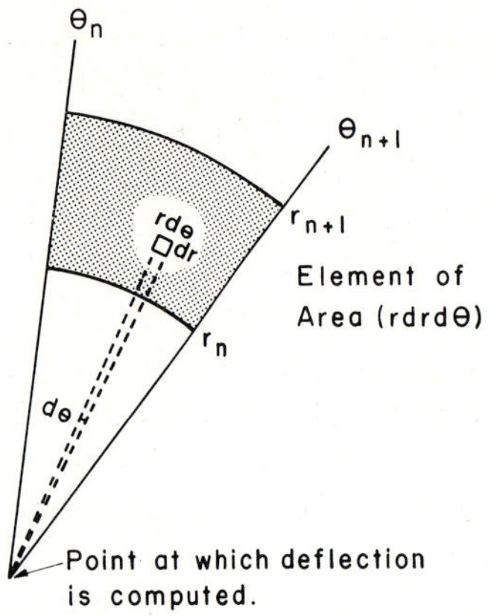

Fig. 4.10. Schematic representation of a three-dimensional load as a differential sector of two circles. (Modified after Watts et al., 1975.)

to the point of deflection, is much greater than $r_{n+1} - r_n$. The deflection is evaluated between limits:

$$r = \bar{r} + \tfrac{1}{2}(r_{n+1} - r_n) \quad \text{and} \quad r = \bar{r} - \tfrac{1}{2}(r_{n+1} - r_n)$$

Deflections near load compartments can be calculated by subdividing compartments so as to satisfy the above conditions.

Fig. 4.11 illustrates the flexure produced by a "long" load compared to that by a "square" load of the same width and height. As the long load has a mass four or more times greater than the square load, this is not surprising. The curves do show, however, the need for using three-dimensional analyses unless the load length is at least four times its width.

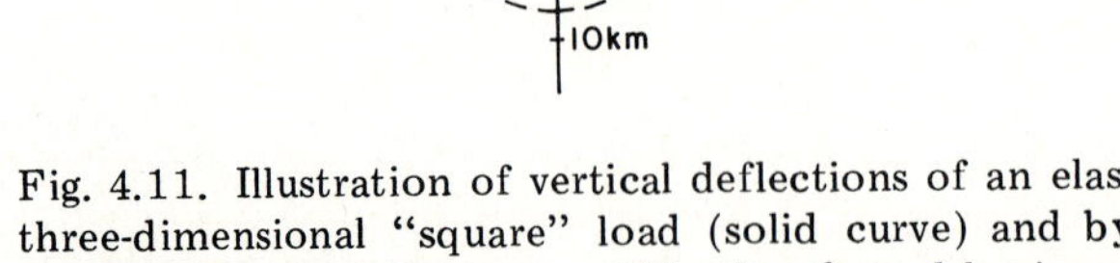

Fig. 4.11. Illustration of vertical deflections of an elastic plate or a beam produced by a three-dimensional "square" load (solid curve) and by a two-dimensional or long load (dashed curve) of the same width. Depths and horizontal distances are in km. (Modified after Watts et al., 1975.)

Determination of lithospheric flexures

Flexures of the lithosphere produced by topographic loads which have existed for more than about a million years can be estimated from comparisons of computed and observed free-air anomalies extending across the region. A range of values of flexural rigidity is assumed and the corresponding downwarps computed (eqs. 4.16 and 4.20); attractions are then computed for each assumed downwarping, including the effects of the material infilling the downwarped basin (water and/or rock) of the topographic feature. The anomaly best fitting the observed amplitude and wavelength characterize the EFR of the lithosphere. Watts and co-workers obtained EFR values of $5 \cdot 10^{29}$ to $5 \cdot 10^{30}$ dyne cm around the Hawaiian Archipelago and the Great Meteor Seamount (northeast Atlantic Ocean). From geologic evidence both structures are known to have existed for tens of millions of years. However, the lithosphere may not be entirely homogeneous within an area, as might be expected in provinces where topographic features (e.g., the Hawaiian Archipelago) are superimposed upon a regional plate.

Regarding an elastic-viscoelastic response of the lithosphere to long-duration stresses, Watts and Cochran (1974) present evidence for an elastic response or a viscoelastic (i.e., an initial elastic response followed by viscous flow) response with a very long relaxation time (the time for the flexure to decrease to 1/e of the original displacement). Walcott (1970a, b, c; 1976), on the other hand, proposes viscoelastic responses with varying relaxation times. The elastic part of a lithospheric plate is restricted to its upper more rigid part. This is evidenced by the fact that earthquake foci along transform faults occur mostly in the upper part of the plate, seldom in the middle or lower part.

Where topographic loads act for relatively shorter durations, as due to Late-Pleistocene glaciation with associated post-Glacial uplift, the lithospheric flexure has not reached steady-state conditions. This is because the asthenosphere responds as a highly viscous material to short-duration stresses, not then behaving like a fluid. The rate of plate flexing, such as in post-Glacial rebound, provides a measure of the viscosity of the asthenospheric materials; the slower the rate the lower the viscosity. By contrast, the rate of rebound provides little information on the rheology of the lithosphere.

Estimating lithospheric thicknesses (from gravity)

The thickness of the elastic part of the lithosphere can be estimated when its EFR has been determined. In an elastic beam, the thickness B is related to its moment of inertia I about a neutral axis (eq. 4.3) by $B^3 = 12I$. In an elastic plate, which has a flexural rigidity $EI/(1 - \sigma^2)$, the corresponding relationship is $B^3 = 12(1-\sigma^2)I$. By combining quantities for the elastic lithosphere, we obtain (also given in Le Pichon et al., 1973, p. 11):

$$B^3 = \frac{12(1-\sigma^2)(\text{EFR})}{E} \tag{4.21}$$

in which Poisson's ratio σ is approximately 0.27, based on *P*- and *S*-wave velocities in the lithosphere (i.e., for short-duration stresses).

If appropriate values for the Hawaiian area are used, we can determine the thickness of the lithosphere that has supported the topographic load there for up to seventy million years. Watts and Cochran (1974) obtained an EFR $\sim 2 \cdot 10^{30}$ dyne cm; using $E = 10^{12}$ dynes cm^{-2} gives $B \sim 28$ km. A lithospheric thickness based on seismic surface waves is about 100 km, and that based on thermal analyses (Sclater and Francheteau, 1970) is about 60—75 km. The thinness of the elastic plate is attributable to the upper part of the lithosphere being relatively more rigid. A second example is the lithospheric thickness beneath the Great Meteor Seamount. For an EFR $\sim 6 \cdot 10^{29}$ dyne cm (Watts et al., 1975) and $E = 10^{12}$ dynes cm^{-2}, $B \sim 19$ km.

Walcott (e.g., 1976) calculated effective thicknesses of the elastic part of the lithosphere in several regions. He obtained values of 20 km in the Hawaiian area, 29 km in the Interior Plains of North America, 10 km in the Canadian Cordillera, and 10 km in the Lake Bonneville area of the Basin and Range Province. He obtained a value of 170 km for the short-term loading by Late-Pleistocene ice sheets in Canada, for which case the asthenosphere exhibits some strength.

ISOSTASY

Large Bouguer anomalies have consistently been observed on prominent topographic features, as mountains, high plateaus, and oceans. The anomalies have large negative values on mountains and high-elevation areas in general, and even larger positive values on oceans. The source of these large values is explained if mass deficiencies (lower-density material) exist beneath mountains and mass excesses (higher-density material) beneath oceans. It is partly as a consequence of these Bouguer-anomaly characteristics, although primarily from deflections of the vertical obtained from astro-geodetic measurements, that the concept of isostatic equilibrium was developed.

Isostatic equilibrium is usually taken to mean that lateral stresses, hence overburden pressures, are equal at some depth in (or beneath) the lithosphere, independently of the composition of the overlying structures, such as mountains, oceans, or continents. Where the lithosphere is assumed to flex, the concept of equilibrium allows for lateral stress differences, so that the isostatic balance is regional rather than local.

Because rocks, particularly at confining pressures corresponding to depths of more than about 50 km, flow in response to substantial stress differences, the effect of long-duration stresses is to produce flow that reduces and even eliminates stress differences. This behavior suggests that materials at depth, particularly in the asthenosphere, will flow toward a stress-equilibrium state, provided no large tectonic forces are acting. In contrast to the asthenosphere, the lithosphere has sufficient strength that relatively little flow is expected to occur within it.

Isostatic equilibrium is achieved at the depth of compensation, characterized by uniform lateral stresses. The depth of compensation can vary from place to place, depending in part upon the extent to which an area is out of equilibrium. The calculated compensation depth depends on the assumed isostatic mechanism, particularly whether based on the Pratt or Airy density distributions.

The high strength of the crust (and lithosphere) leads us to expect that isostatic compensation is regional in nature, although small blocks (diameter less than about 50 km) may be compensated locally. Lithospheric flexuring implies regional compensation; so does post-Glacial uplift in North American and in Fennoscandia, as does extensive sedimentation in large areas such as the Black Sea or the Gulf of Mexico. Complete isostatic equilibrium would hardly be achieved because of the high strength of the crust and lithosphere.

Isostatic computations are greatly simplified if compensation is assumed to be local rather than regional. Isostatic anomaly calculations therefore are commonly developed for local compensation. Justification for this computational procedure has been generally demonstrated by the achievements in isostasy, which have been described by Heiskanen (Heiskanen and Vening Meinesz, 1958, Chapter 7, pp. 187—219).

Pratt-Hayford isostatic method

Hayford (1909, 1910) developed the Pratt concept of isostatic equilibrium into a method which can be readily applied to determine isostatic gravity anomalies and densities of compensating masses.

Pratt's explanation for compensation is illustrated in Fig. 4.12, where the higher the elevation, the lower the density of the underlying lithosphere, and vice versa for ocean depths. The Pratt-Hayford method (Hayford, 1909; Heiskanen and Vening Meinesz, 1958, p. 131) assumes that: (1) densities beneath higher land elevations are less than those beneath level land; (2) the layer of compensation or equilibrium lies at a depth D beneath a particular elevation; (3) the depth of compensation is everywhere equal when measured from the actual surface (not from sea level); and (4) the density reduction $\Delta\rho$ of a column of uniform density ρ_h, extending to elevation h, compared to a column with density ρ_n extending to sea level, is:

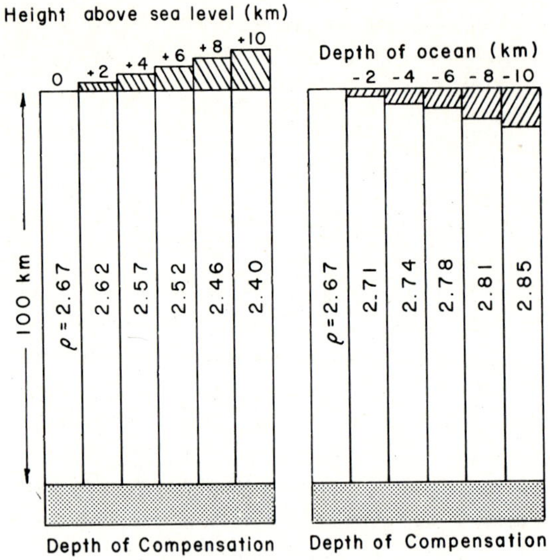

Fig. 4.12. Schematic representation of the Pratt-Hayford concept of isostatic compensation. Densities are in units of g cm^{-3}.

$$\Delta\rho = -\left(\frac{h}{D}\right)\rho_h \tag{4.22}$$

Eq. 4.22 is derived on the assumption that stresses are equal at depth D beneath the surface. Equal stresses at the base of columns of unit cross section can be expressed by:

$$p = \int_{-D}^{h} \rho(z) g \, dz = \text{constant} \tag{4.23}$$

where $\rho(z)$ is the density and g the value of gravity applicable in the column. Changes in g are very small over column heights and are neglected in the Pratt-Hayford concept. Thus, for constant density ρ_h in a unit column to elevation h and density ρ_n in a normal column to sea level:

$$\int_{-D}^{h} \rho_h \, dz = \int_{-D}^{0} \rho_n \, dz \tag{4.24}$$

which integrates into:

$$\rho_h (D + h) = \rho_n D \tag{4.25}$$

The density reduction in the two columns is $\Delta\rho = \rho_h - \rho_n$ (eq. 4.22).

The Pratt-Hayford assumption approximates equal pressures at depth D. Heiskanen and Vening Meinesz (1958, p. 133), in discussing the difference between equal-pressure and equal-mass conditions, show that the Hayford compensation lies between the two assumptions and is four times nearer the equal-pressure than the equal-mass assumption.

Airy-Heiskanen isostatic method

Heiskanen (1924, 1938a, b) developed the Airy concept of isostatic equilibrium into a method readily applicable to the determination of isostatic anomalies and the calculation of "roots" under mountains and "antiroots" under oceans. The Airy-Heiskanen method, illustrated in Fig. 4.13, assumes that: (1) isostatic compensation is complete; (2) compensation occurs directly beneath the topography (i.e., compensation is local); and (3) the density of the crust is 2.67 g cm^{-3} and of the mantle 3.27 g cm^{-3} (i.e., a contrast of $\Delta\rho = 0.6$ g cm^{-3} everywhere).

Compensation can be assumed on the basis of equal pressure, except that g is assumed to be constant throughout the column. Thus, the mass in a column of density ρ and elevation h will be compensated by a root of density contrast $\Delta\rho$ and thickness t. This requires that $\rho h = \Delta\rho t$, or a root thickness of:

$$t = \left(\frac{\rho}{\Delta\rho}\right) h = \lambda h \tag{4.26}$$

Similarly, the thickness of an antiroot beneath the ocean is:

$$t' = \left(\frac{\rho - 1.027}{\Delta\rho}\right) h' = \mu h' \tag{4.27}$$

in which 1.027 is the density of ocean water and h' the water depth.

Heiskanen constructed his final compensation tables in 1938. They are based on equal masses for the topographic and the compensating masses rather than on equal pressure, however, in response to general requests that equal-mass computations will not change the total mass of the earth. Let T be the normal thickness of the crust. The mass of a topographic cap of height h (Fig. 4.14) is:

$$\Delta M_1 = \frac{4}{3}\pi\rho \sin^2(\theta/2) \; [(r+h)^3 - r^3] \tag{4.28}$$

in which r is the earth's radius and θ the central angle subtended by the cap. The mass of the compensating column from depth T to $T + t$ is, similarly:

$$\Delta M_2 = \frac{4}{3}\pi\Delta\rho \sin^2(\theta/2) \{(r-T)^3 - [r-(T+t)]^3\} \tag{4.29}$$

Equating the topographic and compensating masses produces:

$$\rho[(r+h)^3 - r^3] = \Delta\rho\{(r-T)^3 - [r-(T+t)]^3\} \tag{4.30}$$

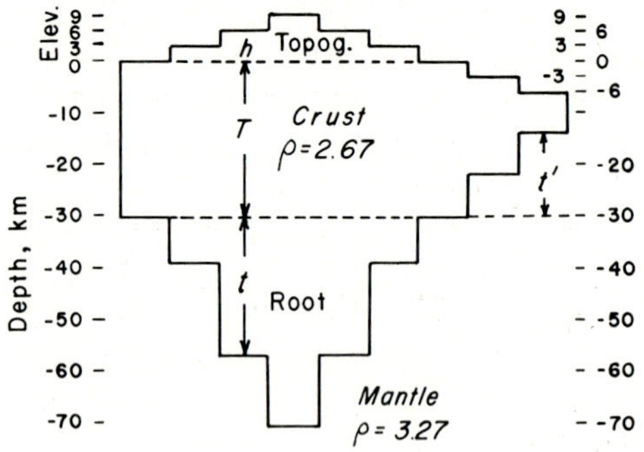

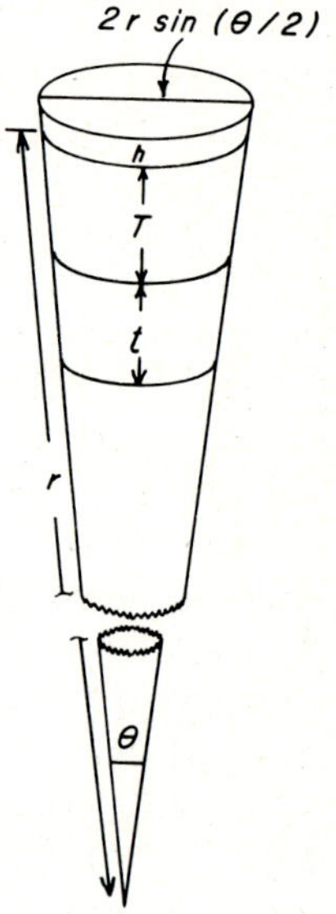

Fig. 4.13. Schematic representation of the Airy-Heiskanen concept of isostatic compensation. Densities are in units of g cm^{-3}.

Fig. 4.14. Sketch of a cone through the earth; h is elevation above sea level, T the normal crustal thickness (i.e., to depth of compensation), and t the thickness of the root.

Heiskanen expanded this equation by series, retaining up to second-order terms, and substituting λ for $\rho/\Delta\rho$, to obtain the compensation in a land compartment of:

$$t = \lambda h \left\{1 + \frac{2T + (\lambda + 1)h}{r} + \frac{(2T + \lambda h)[2T + (\lambda + 1)h]}{r^2} - \frac{T(T + \lambda h)}{r^2} - \frac{(\lambda^2 - 1)h^2}{3r^2}\right\} \quad (4.31)$$

For an ocean compartment Heiskanen obtained a corresponding equation:

$$t' = \mu h' \left\{1 + \frac{2T - (\mu + 1)h'}{r} + \frac{(2T - \mu h')[2T - (\mu + 1)h']}{r^2} - \frac{T(T - \mu h')}{r^2} - \frac{(\mu^2 - 1)(h')^2}{3r^2}\right\} \quad (4.32)$$

Eqs. 4.26, 4.27, 4.31, and 4.32 show that the smaller the density contrast between crust and mantle, the larger the root or antiroot. For a density contrast of 0.6 g cm^{-3}, a root is approximately 4.5 times larger than the land elevation, and an antiroot 2.7 times larger than the water depth.

Two-dimensional Airy-isostatic computations

J. L. Worzel, while at the Lamont-Doherty Geological Observatory, developed two-dimensional, pseudo-Airy-isostatic computational procedures for eliminating the edge effect of crustal-mantle structures from a gravity profile across a continental margin. Rabinowitz (1974; 1976) used this procedure to analyze basement structures across continental margins. The antiroot beneath the sea is determined from observed water depths, assuming constant pressue (local compensation) at some depth, such as 30 km (illustrated in Fig. 4.15). The water is filled with rock, resulting in a density contrast with water of about 1.6 g cm^{-3}, and an antiroot with a density contrast of about 0.5 g cm^{-3}. A two-dimensional computation is made for the attraction of the ocean part filled with rock, which is added to the observed free-air anomaly, producing an approximate Bouguer anomaly. A two-dimensional computation is then made of the attraction produced by the antiroot with reduced density, the attraction being subtracted from the approximate Bouguer anomaly. This computation is made for all stations along the profile, thus providing two-dimensional, pseudo-isostatic anomalies. These anomalies are based on a minimal number of assumptions, but it does not follow that the assumptions included are good approximations of actual structures. The method has been applied in the analysis of smaller structures across continental margins, where margins are in approximate isostatic equilibrium and where seismic control data are absent. The isostatic anomalies probably exclude most of the edge-effects of deep structures at the margin, such that the anomalies appear to be fairly representative of smaller structural effects across the margin.

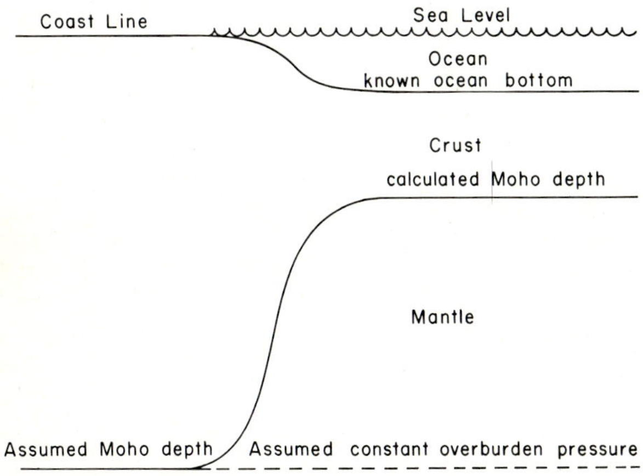

Fig. 4.15. Representation of a vertical section as used to obtain computed values of two-dimensional, pseudo-isostatic anomalies across a continental margin.

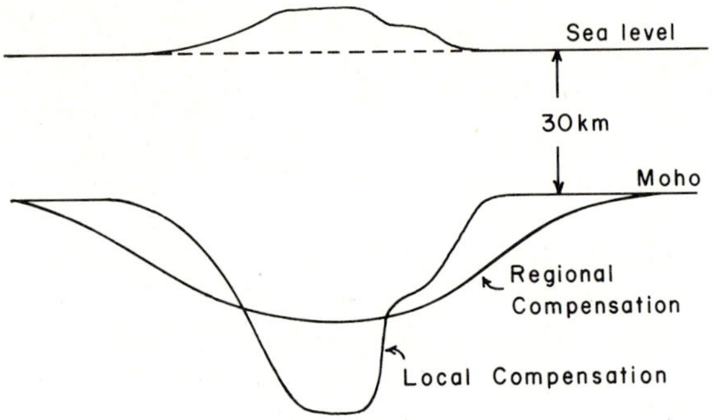

Fig. 4.16. Schematic representation of Vening Meinesz local and regional curves of compensation corresponding to a topographic feature.

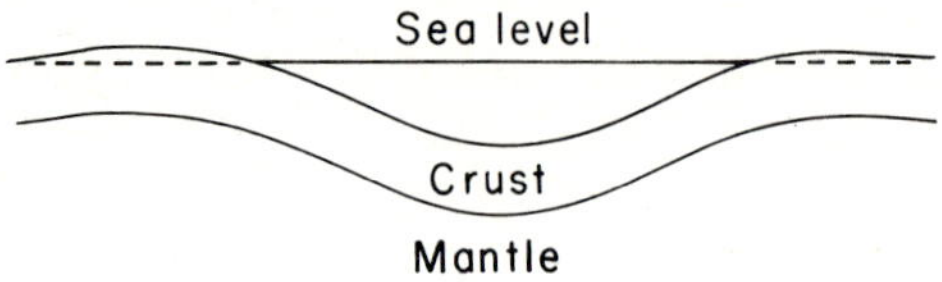

Fig. 4.17. Illustrations of Vening Meinesz crustal bending produced by a concentrated load, based on Hertz's formula (eq. 4.19). (After Heiskanen and Vening Meinesz, 1958, p. 322.)

Vening Meinesz regional isostatic method

Vening Meinesz (1931) modified the Airy concept of compensation on the assumption that the crust responds as an elastic plate capable of resisting shear stresses produced by a topographic load. The amount of down-bending at each point on the plate is a measure of isostatic compensation at the point. Fig. 4.16 illustrates the difference between the Vening Meinesz regional and conventional local compensation; Fig. 4.17 shows the Vening Meinesz bending curve obtained from Hertz's formula for the bending of a plate caused by a concentrated load (see eq. 4.16).

Vening Meinesz's radius of regionality, 2.905 f, where f is given in eq. 4.14, is a measure of lithospheric down-bending. This downbending led to the analyses of lithospheric flexures, discussed in the previous section. More detailed descriptions of the Vening Meinesz Regional Method are given in Heiskanen and Vening Meinesz (1958, pp. 137—142).

Discussion of isostatic concepts

Pratt-Hayford and Airy-Heiskanen isostatic-anomaly determinations produce astonishingly similar values, considering the radically different assumed crustal and subcrustal structures and density distributions. It is known from seismic refraction profiling that the Pratt concept of uniform-density columns to a depth D (which happens to be near the base of the lithosphere) is at variance with observed structures. Similarly, the Airy concept of roots and antiroots, which, according to Heiskanen, involves a single density contrast of 0.6 g cm^{-3} between crust and mantle, is too simplistic to conform with known structures. Fortunately, calculated isostatic anomalies are not sensitive to assumed structures.

Extensive geophysical measurements have been obtained which provide reasonable pictures of crustal and subcrustal structures in different types of provinces. Gross features can include substantially different types of compensating structures. For example, roots of mountains appear to exist according to the Airy concept beneath the Sierra Nevada Mountains of California and the Alps of Europe, but are not characteristic of all mountains. In areas of Late-Tertiary orogeny, for example the Basin and Range Province of western North America, the Moho is considerably shallower than predicted by the root concept (e.g., Pakiser and Robinson, 1966); yet the area is in isostatic equilibrium, which requires that the sub-Moho material has a lower-than-typical sub-Moho density (it exhibits a low regional seismic velocity). Possibly the lower-density LVL extends up to the Moho in such regions. Similarly, beneath the most extensive mountain ranges on earth, the mid-ocean ridges, the Moho is relatively shallow, a typical depth appears to be 7 km below sea level. Instead of a root, there is a lack of root; yet the ridges are essentially in isostatic equilibrium. This appears to be achieved by low-density, low-velocity mantle material extending up to the Moho (see, e.g., Fowler, 1976, and Fig. 8.21). In continental shield areas, by contrast, the Moho is usually deeper than predicted by the Airy-Heiskanen concept, and the sub-Moho materials exhibit relatively high seismic velocities, implying relatively high mantle densities. These areas are characterized by a thinner LVL, if one exists there at all.

Crustal and mantle structures determined from seismic methods in different areas of the world have been described by many investigators, notably Mueller (1974), Herrin and Taggart (1962), Pakiser and Steinhart (1966), Cleary and Hales (1966), Healy and Warren (1969), J. Ewing (1969), Shor et al. (1970), McCamy and Meyer (1966), Woollard (1966), and Keen et al. (1970).

Since deep seismic-refraction profiles have provided extensive information about the crustal and subcrustal structures in various continental and oceanic provinces, isostatic anomalies are now computed, if at all, to determine deflections of the vertical, and not commonly to determine thicknesses of roots or antiroots.

Analyses of deep structures from gravity anomalies should utilize all structural information available for an area, particularly seismic-refraction data at nearby locations. Effects of known layers can be computed, such as the ocean, sediment layers (as obtained from bore-hole drilling or seismic profiling), or crustal thickness, and these effects are subtracted from the observed anomaly, analogous to the stripping technique (Hammer, 1963) where the effects of known structures are removed from observed anomalies. The remaining anomaly can be analyzed in terms of reasonable density contrasts in the underlying materials. Free-air anomalies are most appropriate for this type of analysis when configurations of the upper layer include topographic and bathymetric variations.

RELATION BETWEEN GRAVITY ANOMALIES AND ELEVATION

Gravity anomalies are clearly produced by density contrasts of materials extending from the surface to considerable depth. The largest contrasts occur at the earth's surface and at the ocean bottom (or at the base of sediments in the ocean). Anomalies are therefore expected to be greatly affected by ground elevation and by water depth; however, anomalies are also dependent upon deeper structures. We wish to consider the relative effect which elevation and terrain have on observed anomalies. It is desirable to know how well and under what conditions gravity anomalies can be predicted from elevations or bathymetry.

Local anomalies, characterized by short wavelengths, are produced by uncompensated features and are therefore directly dependent upon terrain or bathymetry. This dependence is seen in the difference between local free-air and Bouguer anomalies, where the effects of elevation and terrain variations have been essentially removed from the Bouguer but not from the free-air values. Talwani et al. (1961, 1971) and McKenzie and Bowin (1976) showed that free-air anomalies with wavelengths not exceeding about 100 km correlate well with bathymetry.

Bouguer anomalies have been known to correlate approximately with elevation, as reported by Heiskanen and Vening Meinesz (1958), Woollard (1962, 1968, 1969), Wilcox (1976) and others. There is a general inverse relationship between increasing Bouguer anomaly values and increasing elevation (Fig. 4.18), although within a large province anomaly values may range by as much as ±50 mgal because of variations in the amounts of topographic compensation. Predictions of Bouguer anomalies from known elevations require corrections for isostatic compensation; it is not often possible to compute such corrections reliably.

Relationships between free-air anomaly and bathymetry have recently been investigated by Cochran and Talwani (1977) in nearly all oceans; they correlated amplitudes and wavelengths of a large number of free-air gravity anomaly profiles with corresponding bathymetry. McKenzie and Bowin (1976) investigated the relationship by applying filters of the Wiener type to

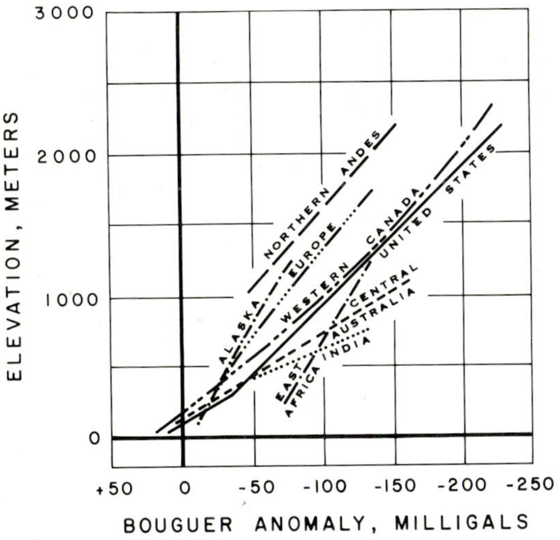

Fig. 4.18. Regional 3° x 3° mean simple Bouguer anomalies as functions of average elevation for different geographic provinces. (Reproduced from Woollard, 1969, with permission.)

several gravity profiles in the Atlantic Ocean, in order to predict gravity from bathymetry.

Cochran and Talwani found good correlations with bathymetry under certain conditions, questionable ones under others, and no correlations in some instances. Anomalies with wavelengths of less than 50 km are essentially uncompensated for topography, and because the largest density contrasts encountered are commonly associated with topographic relief (or the ocean sediment-basement contact), the short wavelength anomalies correlate well with bathymetry.

Anomalies with wavelengths from 50 to several hundred kilometers are produced by features essentially uncompensated for topography, but they are affected by density contrasts in the lower crust and uppermost mantle. These anomalies sometimes correlate fairly well with bathymetry, but not always. Anomalies across the Hawaiian Ridge and adjacent moat, or across a large seamount, where a down-warped lithosphere provides regional compensation, are examples of such wavelengths.

Anomalies with wavelengths from about 1,000 to 4,000 km correlate with some types of topographic features, such as mid-ocean ridges or the entire Hawaiian swell, but not with other types of features. Anomalies with wavelengths exceeding 4,000 km, such as those across continental and oceanic provinces or across plate boundaries, appear to show no relationship with elevation.

McKenzie and Bowin (1976) conclude from their filter analyses that isostatic compensation becomes significant where wavelengths exceed about

100 km, becoming more pronounced as the wavelength increases. Gravity anomalies predicted by their analysis account for about three-quarters of typically observed anomalies, although some large anomalies clearly do not correlate with bathymetry. Subtracting their predicted from observed anomalies produces residuals which are a measure of buried structures. They conclude that the predicted anomalies could be used to fill short gaps in anomaly profiles.

To sum up, it appears that gravity anomalies are related to topographic features and regional elevation in varying degrees; considerable care must be exercised in attempting to predict anomalies reliably from bathymetry.

Chapter 5

GRAVITY-MEASURING INSTRUMENTS

Gravity instruments measure either gravity differences or absolute gravity. Gravimeters measure gravity differences on land and, with modifications, at sea. Pendulums were formerly used to measure gravity differences on land and, with special design, in a submerged submarine. Falling-body apparatuses are used to measure absolute gravity in the laboratory; reversible pendulums (see section on absolute-gravity pendulums) were formerly used to obtain these measurements.

Modern instruments measure gravity or gravity differences with remarkable sensitivity. Land gravimeters can routinely measure differences of 0.1 and even 0.01 mgal. Tidal gravimeters can measure gravity variations of 1 μgal (0.001 mgal). Sea gravimeters can measure differences of 1 mgal and some even to 0.1 mgal. Falling-body apparatuses can measure gravity to nearly 1 μgal.

Other types of measuring instruments include the torsion balance, which measures the horizontal gravity gradient and the curvature of the geopotential field, and gradiometers, which measure the gravity gradient only (vertical or horizontal). These instruments cannot provide gravity values with the same precision as gravimeters, however, and are therefore not commonly used.

A gravimeter is an instrument which measures extremely small changes in weight. The weight of a mass varies with changes in the gravitational field. To detect a change in weight (force of gravity) of 0.1 mgal, a gravimeter must have a sensitivity of 10^{-7}g. If the mass in a meter hangs on a 50-cm-long spring, a 0.1 mgal change produces a $0.5 \cdot 10^{-5}$cm change in length. Such a small distance cannot be measured by even as sensitive an instrument as an interferometer, which does not measure distances smaller than a wavelength (i.e., about $5 \cdot 10^{-5}$cm). The resolution of a sensitive gravimeter must be more than 10 times that of an interferometer.

Gravimeters consist of a mass attached to either a coiled spring, a torsion fiber, or a vibrating string (a strip of metal). Two basic types of gravimeters measure these very small changes in spring length, in torsion angle, or in frequency of vibration. The first is a stable or static gravimeter, in which high optical or mechanical magnification provides the sensitivity required to measure change in displacement of the weight (or in torsion angle or in resonant frequency). Two forces act on the mass: gravity and spring tension (or its counterpart). The second type is an unstable or astatic gravimeter, in

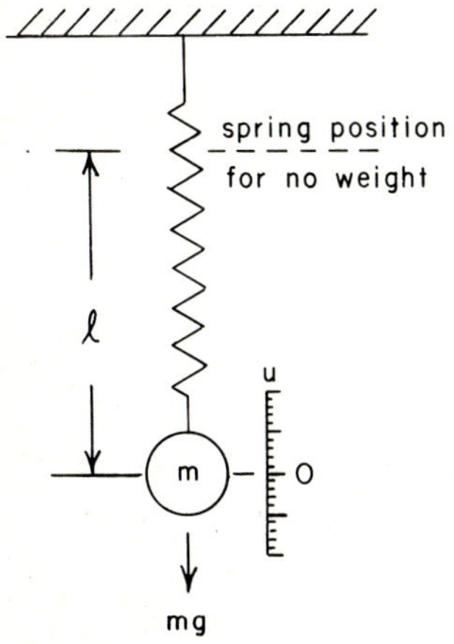

Fig. 5.1. A simple spring; weight mg increases the length of the spring a distance l over its no-weight length.

which a third or balancing force acts to produce instabilities so that small changes in gravity produce relativity large observed motions.

Stable-type meters provide gravity readings which are linear over relatively wide ranges of scale reading. Unstable meters are usually nonlinear but generally have higher sensitivities.

The sensitivity of an unstable gravimeter can be increased by lengthening its natural period of oscillation. The elongation l of a spring from a zero-load length (Fig. 5.1) is:

$$l = \frac{mg}{k} \tag{5.1}$$

in which m is the mass on the spring and k is the spring constant (or stiffness constant, a strain-per-unit stress on the spring). Forces acting on a spring in simple oscillation or with simple harmonic motion can be written:

$$m\ddot{u} + ku = 0 \quad \text{or:} \quad \ddot{u} + \omega^2 u = 0 \tag{5.2}$$

in which u is the displacement and $\ddot{u}$ the acceleration of mass m oscillating with an angular frequency ω. Since $\omega = k^{1/2} m^{-1/2}$, the period of oscillation is:

$$\tau = 2\pi\sqrt{\frac{m}{k}} \tag{5.3}$$

We substitute eq. 5.3 into 5.1 to obtain for the spring elongation:

$$l = \frac{g\tau^2}{4\pi^2} \tag{5.4a}$$

A change in spring length δl corresponding to a small change δg can be written:

$$\delta l = \frac{\delta g \tau^2}{4\pi^2} \tag{5.4b}$$

and the meter sensitivity is given by:

$$\frac{\delta l}{\delta g} = \frac{\tau^2}{4\pi^2} \tag{5.5}$$

We find, on substituting, that to measure a change in gravity of 0.01 mgal by a change in spring length of 10^{-5} cm, the instrument must have an oscillation period of 6 sec. To measure a change in gravity of 0.001 mgal with the same 10^{-5} cm change in length, the required period is 20 sec.

A considerable variety of gravimeters have been constructed since about 1932. Books by Nettleton (1976, chapter 3), Dobrin (1975, chapter 11), Pick et al. (1973, pp. 148—170), as well as earlier ones on geophysical prospecting, describe common types. On pp. 60—63 of his book, Nettleton lists 29 gravimeters described in patents or in the literature, although the list is not complete. We consider here only representative types of land meters and the more common types of shipborne meters.

GRAVIMETER DRIFT

The spring and associated connectors in gravimeter sensors cannot be made perfectly stable, even where temperature and barometric pressure are kept constant within the meter case. All gravimeters are therefore affected by small mechanical changes in the course of measuring gravity differences. Resultant time variations in measurement are usually slow, but can be rapid. These variations are called meter drift. If the variation exhibits a sudden offset, it is called a tare. Either drift or tares produce measurement errors which can be large compared with gravity variations. Corrections for drift can be made, even as they are encountered during field surveys. Tares are identified as to the successive times of measurement between which they occurred.

Drift corrections, which can be applied to gravity measurements, either when not corrected or when corrected for tidal variations, are obtained from drift curves. These are plots of gravity at a base station (or base stations) as a function of time (e.g., time of day). Fig. 5.2 illustrates a drift curve for measurements which have not been adjusted for tidal variations, and also a calculated tidal curve (e.g., based on eqs. 3.30). During a gravity survey, base station readings are made periodically, usually at intervals not exceeding three hours. Drift curves obtained from laboratory measurements usually

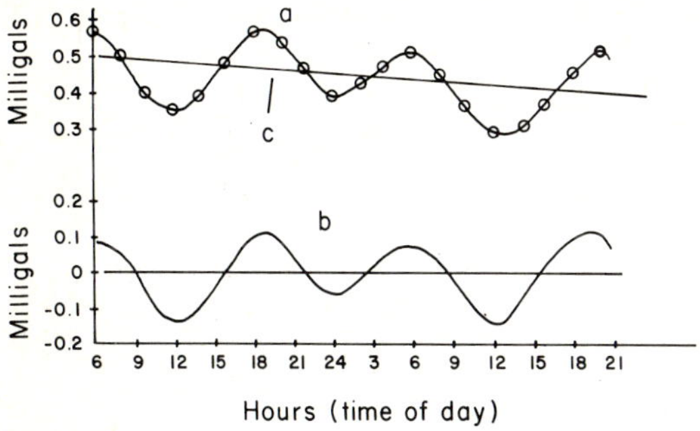

Fig. 5.2. Illustration of drift and tide curves. *a* = field-observed drift curve, including tidal attractions; *b* = corresponding theoretically computed tidal variations; *c* = instrument drift curve obtained by subtracting from *a* or by plotting tide-corrected base-station measurements.

include tide corrections; such measurements can be made at relatively long intervals. Drift rates are obtained from the slopes of drift curves (e.g., milligals per minute or per month). The rate which applies at the time of a field measurement is used to correct an observed value.

In marine investigations, average drift rates are based on measurement misties obtained at succesive port readings. Where the ship returns to the same port station, average drift is the difference in station values divided by the time between the readings. Where the ship stops at different ports, drift is the discrepancy between gravimeter readings in the ports, which have been tied to values at nearby base stations, divided by the respective time interval. Where drift appears to be significantly larger than the drift history of a given meter, a tare may have occurred (as during a cruise leg). Most modern gravimeters (land and sea) exhibit linear, low drift rates.

LAND GRAVIMETERS

Stable-type gravimeters

The Hartley (1932) meter is perhaps the earliest gravimeter constructed for geophysical exploration. It consists of a mass suspended on a spring (Fig. 5.3), where the position of the mass relative to the instrument case is determined by an optical lever. The mass is brought back to a null position by adjusting a weak spring. The meter can detect changes in gravity of about 1 mgal; the meter has not had wide application.

The Hoyt (1938) or Gulf gravimeter is one of the earliest high-sensitivity field instruments (Fig. 5.4). Gravity differences are measured by the unwinding or rotation of a flat-ribbon spring, rather than by spring

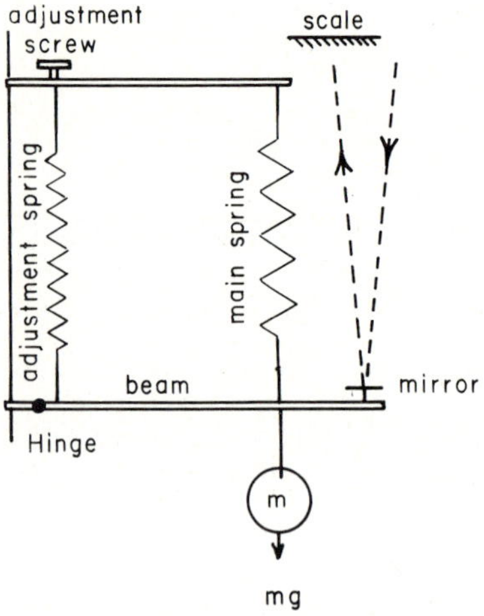

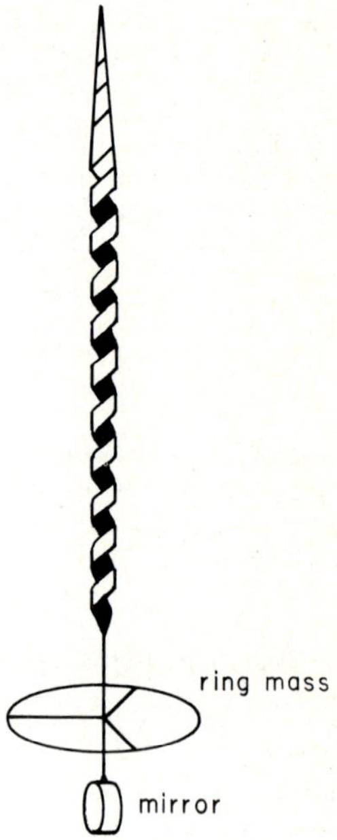

Fig. 5.3. Hartley gravimeter (stable-type meter).

Fig. 5.4. Hoyt or Gulf gravimeter (stable-type meter). An increase in weight of the ring mass unwinds a flat-ribbon spring; the angle of rotation is a measure of the change in gravity.

elongation. For the same spring elongation, a ribbon spring rotates through a considerably larger angle than a round wire. The field instrument can measure gravity differences to about 0.04 mgal. The range of gravity differences which can be read without resetting the reading dial is relatively small. Because of this limitation and the relatively large weight, about 80 lb, this meter has been replaced by lighter, smaller, and more easily read gravimeters.

The Graf-Askania gravimeter was designed circa 1937, then having an accuracy of approximately 0.2 mgal. A number of improved models have been developed since then. A current model (Gs-15) has a measuring accuracy of 0.01 mgal in field use, the drift being less than 0.05 mgal/h. Used at an undisturbed station at constant temperature, the meter can record variations of 1 μgal and has a drift of 0.1 mgal/month. The sensing element (Fig. 5.5) consists of a mass on a horizontal beam attached to two horizontal helical (main) springs. The torsion angle of the springs is proportional to the gravity change producing the beam deflection. Measurements are obtained by bringing the beam to a null position with a vertical compensating spring, the null position being determined by a capacitative-transducer measurement. Temperature compensation is obtained with two thermostats. The

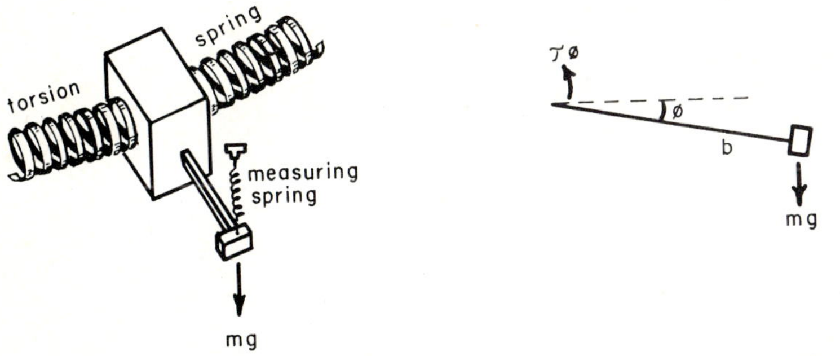

Fig. 5.5. Graf-Askania land gravimeter (stable-type meter); a horizontal torsion spring (left figure) counteracts the torque produced by weight mg at end of beam of length b (right figure). For stable balance, $mgb = \tau\phi$, where τ is the spring torsion and ϕ the angle of spring rotation.

outer one has a sensitivity of 0.01°C and can be adjusted for 25°, 35°, 40°, or 45°C, in accordance with the outside temperature. The inner thermostat provides continuous regulation. The meter has a dial range of about 800 mgal, which can be reset to provide a total adjustable range of 6,000 mgal, making measurements possible at any land station on the earth. The meter has wide application, particularly outside of North America.

The Norgaard gravimeter is a Swedish-constructed torsion-fiber instrument developed circa 1940. The fiber is made of quartz and the operating principle is similar to the Mott-Smith and Worden meters (see next section), but with novel spring calibration and temperature compensation (e.g., Nettleton, 1976, p. 39). Gravity is measured to about 0.2 mgal. The instrument has been used primarily by European and USSR investigators.

Unstable-type gravimeters

The Holweck and Lejay (1930) instrument consists of an inverted pendulum, the mass being at the upper end of a flat spring. The size and shape of the mass are adjusted and the spring constant selected so that the pendulum approaches instability, resulting in the period becoming very sensitive to small changes in gravity. Compared with ordinary pendulums, the ratio of changes in period to change in gravity in inverted pendulums is relatively large. The measurement accuracy is only 1 to 2 mgal (Nettleton, 1976, p. 30), however, and hence this instrument has had limited use.

The Truman or Humble gravimeter, used as early as 1932, appears to be the first adequate field gravimeter. The mass (Fig. 5.6) is attached on the horizontal arm of a triangular member hinged at the opposite end. A relatively strong spring acts on the bottom vertex of the triangular piece so that the sensing element is unstable. To measure gravity differences, instrument deflections are nulled by a weak spring. With proper adjustment,

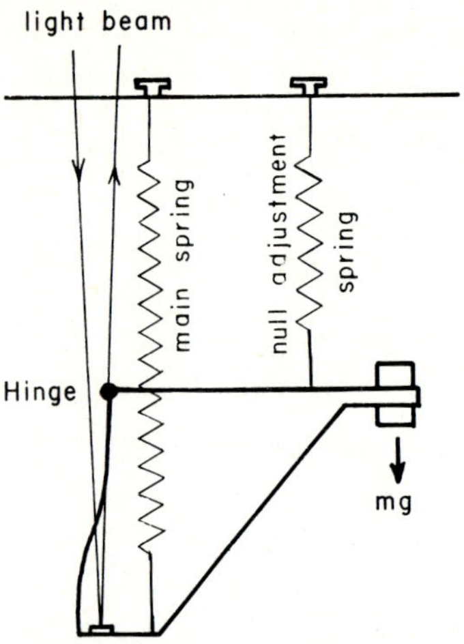

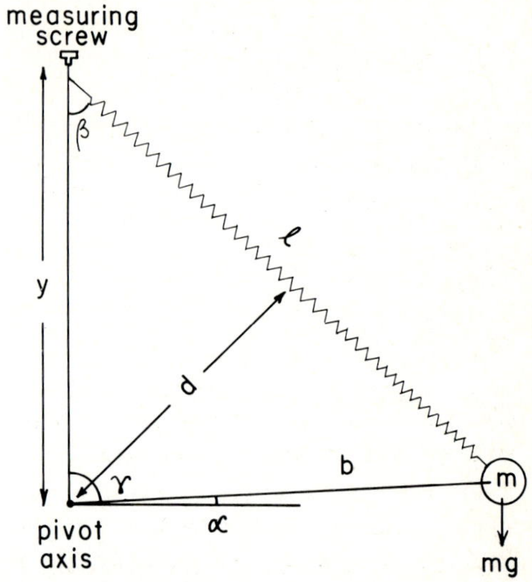

Fig. 5.6. Truman or Humble gravimeter (unstable-type meter).

Fig. 5.7. LaCoste and Romberg land gravimeter (unstable-type meter). A zero-length spring counters the torque which weight *mg* produces about a beam pivot axis. See text for meaning of symbols.

this instrument can measure gravity to about 0.2 mgal. The meter is no longer commonly used.

The LaCoste and Romberg (L & R) gravimeter is one of the most accurate and widely used field meters. It is in principle a long-period vertical seismograph which employs a zero-length tension spring. This type of spring has zero length under no load, or would be zero if the turns did not interfere with each other. The force acting on a coiled spring is $F = k(l - l_0)$, where k is the spring constant, l the spring length under load, and l_0 the unloaded length. In the zero-length spring, $l_0 = 0$ and $F = kl$. The spring is made by winding prestressed coils; an example is a typical screen-door spring in which the coils open only when a force is applied. Gravimeters with zero-length springs can be made to have long periods, and thus high sensitivities.

In the L & R meter, a weight mg on a beam of length b produces a gravitational torque L_g about a pivot axis (Fig. 5.7), where:

$$L_g = mgb \cos \alpha \tag{5.6}$$

and the spring produces an opposing torque L_s of:

$$L_s = -kld \tag{5.7}$$

We substitute, from the law of sines:

$$\frac{l}{\sin \gamma} = \frac{b}{\sin \beta} \quad \text{or:} \quad l = b\,\frac{\sin \gamma}{\sin \beta} \tag{5.8}$$

into eq. 5.7 to obtain (Fig. 5.7):

$$L_s = -kby \sin\gamma = -kby \cos\alpha \tag{5.9}$$

At equilibrium the beam is at rest; the torque acting on the pivotal point is zero. Thus:

$$L = (mbg - kby)\cos\alpha = 0 \tag{5.10}$$

and:

$$y = \left(\frac{m}{k}\right)g \tag{5.11}$$

The gravitational force on m is thus proportional to the distance y between the adjustment screw and the pivotal axis.

The sensing element is not sensitive to angles α, γ, and β (Fig. 5.7); it can be in equilibrium over small ranges of these angles. When the beam is displaced slightly from an equilibrium point, there is no restoring force. The beam can be adjusted to natural periods of 15 or 20 sec. The meter, thermostatically controlled, can measure difference of 0.01 mgal.

The LaCoste and Romberg geodetic meter is a recent modification of the above-described meter. It can measure gravity to 0.01 mgal over a 7,000 mgal range, that is for the entire earth, with a single adjustment screw. Thus, this meter, unlike the exploration unit, does not require dial resettings to match gravity values in different areas. To provide the required measuring accuracy, the geodetic meter has a low drift rate and is carefully calibrated for its entire operating range. The meter is also considerably smaller and lighter (weighing about 5lb) than the exploration model, and is therefore an invaluable regional, gravity-surveying tool.

The Mott-Smith gravimeter (Mott-Smith, 1938) appears to be the first all quartz-fiber instrument (Fig. 5.8). A horizontal weight arm, attached to a

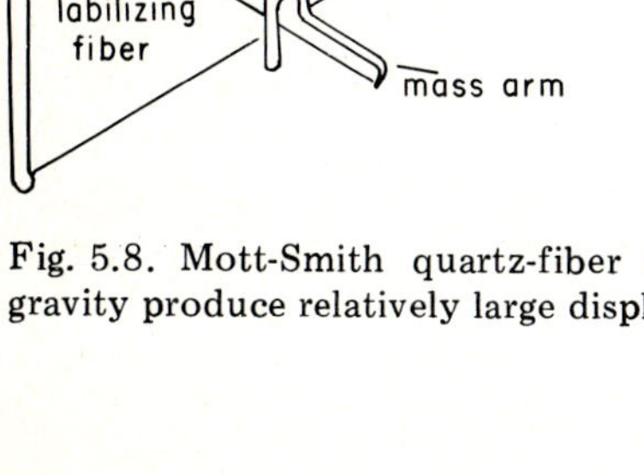

Fig. 5.8. Mott-Smith quartz-fiber gravimeter (unstable-type meter). Small changes in gravity produce relatively large displacements of the mass arm.

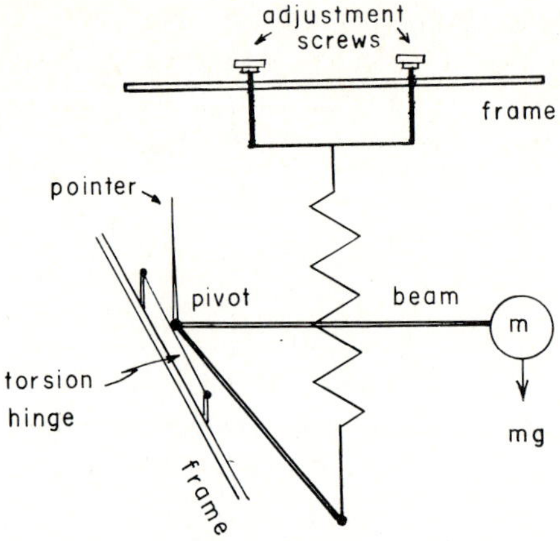

Fig. 5.9. Worden all-fused quartz gravimeter (unstable-type meter). Temperature variations are compensated by a tungsten fiber (not shown); the instrument need not be maintained at constant temperature.

torsion fiber, is connected to a labilizer fiber, which is attached to a spring. The torsion on the fiber and tension of the labilizer fiber are adjusted so that small gravity differences produce large displacements of the horizontal weight arm. The elastic constants of fused quartz are exceptionally uniform. The instrument is maintained at a constant temperature to further enhance a stable response. The sensitivity is better than 0.1 mgal.

The Worden gravimeter (Fig. 5.9) consists of all fused-quartz parts (frame, springs, and connecting arms), which are temperature compensated by a tungsten fiber. The instrument need not be kept at constant temperature. A zero-length spring activates a horizontal torsion-fiber bar, which is the primary moving part of the sensing element. The torque produced by the mass on a weight arm rotates the torsion fiber, which, in turn, rotates a vertical arm containing a pointer. A measuring screw operates a null-adjusting spring. The rotational period of the sensing element is about 6 to 8 sec, providing a meter sensitivity of 0.01 mgal. The sensing element is very small and light. Mounted in a partial vacuum, this system requires no clamping or instrument damping. The meter has been used as a geodetic meter, although its drift rates are generally not so low as those in the LaCoste and Romberg or the Askania geodetic meters.

SEA GRAVIMETERS

The beam-type sea gravimeters commonly used are adaptations of land meters. Other types of sea meters were specifically designed for use aboard ship, however. The earliest of these are the gas-pressure devices of Hecker

(1903) and Haalck (1931). Later, vibrating-string-type meters were developed by Gilbert (1949) and improved upon by others. Recently, axially symmetric meters were developed that are free of the cross-coupling of horizontal and vertical ship accelerations that affect beam-type meters.

LaCoste and Romberg (L & R) beam gravimeters

The sensing element in these sea meters is an overdamped version of that in the L & R land meter, with increased beam stiffness and with added fine-wire suspensions to prevent horizontal components of beam motion. The literature (e.g., LaCoste, 1967; and LaCoste and Romberg, Inc., brochures) contains descriptions of the L & R meter systems. We describe here its operating principles as being representative of beam-type meters.

The first successful tests of L & R sea meters were made in submarines in 1955. The meter was suspended from a gimbal joint. This system was modified to operate on a surface ship in 1958, and later was tested in airplanes (Nettleton et al., 1962) and then in helicopters. The system was designed for accelerations up to about 50 gal at normal ship periods (about 5—15 sec). Accelerations at sea are commonly larger than this, however, so that the gimbal-suspended system operated successfully in low sea states only. In aircraft, the system readily compensated short-period accelerations, but not the long-period vertical accelerations. Long-period motions can be minimized by use of a good autopilot.

L & R introduced a substantial system modification in 1965 when the meter was made to operate on a stabilized platform. This system measures accurately in moderate sea states, although the cross-coupling of vertical and horizontal accelerations acting on the meter case are difficult to correct for adequately at high sea states. By 1974 L & R constructed an inertial navigation system which measures the Eötvös effect (eq. 6.42), instrumentally correcting measured gravity to its equivalent stationary-position value.

To minimize beam oscillations aboard ship, the beam is highly overdamped (using air damping). The period is made very long to produce a near-zero net-restoring force. The high damping does not produce a slow response time (that is, the time in which the velocity of beam motion reaches approximately 0.6 its steady-state value, where the velocity depends upon the distance of the mass from its null position) as we might expect. LaCoste (1967) pointed out that increased damping can actually decrease the response time.

An oscillatory beam displacement u, with no forced vibrations, is given by:

$$u = b \sin(\bar{\alpha} + \alpha) \doteq b(\bar{\alpha} + \alpha) \tag{5.12}$$

for small average angle $\bar{\alpha}$ from the horizontal and small instantaneous angle α

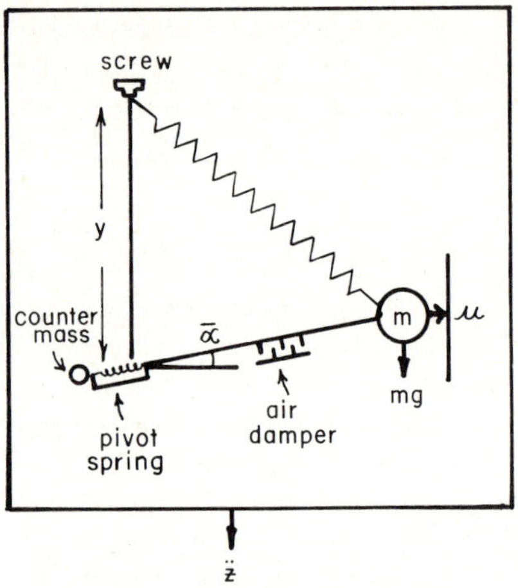

Fig. 5.10. LaCoste and Romberg beam-type sea meter, a highly overdamped modification of the L & R land meter. Vertical accelerations $\ddot{z}$ acting on the meter case cause the overdamped beam to oscillate slowly about a main beam position (at angle $\bar{\alpha}$ from the horizontal).

from this average; b is the beam length (Fig. 5.10). Acting on mass m are the force of acceleration, a damping force, and a spring-restoring force. The sum of these is described by the well-known differential equation:

$$m\ddot{u} + R\dot{u} + ku = 0 \qquad (5.13)$$

in which R is a damping term and k an equivalent spring constant. High overdamping makes the damping force $R\dot{u}$ considerably larger than the other forces.

Outside forces acting on the beam must be added to describe the total forces. The sum of all forces is given by (LaCoste, 1967):

$$m\ddot{u} + R\dot{u} + ku = m(g + \ddot{z}) - Cy \qquad (5.14)$$

in which the first term on the right is the sum of the gravitational and vertical acceleration forces acting on the meter case, and Cy the vertical force which the spring exerts on the center of the mass when the mass is nulled (C is a constant). We write eq. 5.14 in terms of accelerations:

$$\ddot{u} + 2h\omega_0\dot{u} + \omega_0^2 u = g + \ddot{z} - cy \qquad (5.15)$$

in which h is the damping coefficient, having a very large value, ω_0 is the natural angular frequency of oscillation, and $c = C/m$.

The damping force is designed to be very large and the restoring force very small (k small). The large damping and resultant slow beam motion make the beam acceleration small. Rapid changes in ship acceleration or in gravity thus principally affect a change in beam velocity. In contrast to the land meter, which measures beam displacement (eq. 5.11), the sea meter measures rate of beam displacement.

Eq. 5.15 can be solved for g if u and its derivatives and $\ddot{z}$ and y are known. To obtain g we integrate both sides over a time interval:

$$\int_{t_1}^{t_2} (\ddot{u} + 2h\omega_0\dot{u} + \omega_0^2 u)\,dt = \int_{t_1}^{t_2} (g + \ddot{z})\,dt - cy \int_{t_1}^{t_2} dt \tag{5.16}$$

The integral on the left side is approximated by its middle term. The average of the first term on the right side, for time T, is written:

$$\frac{1}{T}\int_0^T (g + \ddot{z})\,dt = \bar{g} + \frac{\dot{z}}{T}\Bigg|_0^T \tag{5.17}$$

The average gravity $\bar{g}$ must be corrected by the difference in vertical velocity between the beginning and the end of the averaging interval, divided by the time interval. At sea there is no net change in $\dot{z}$ over an interval, the vertical accelerations produced by tides being negligible. Eq. 5.16 therefore reduces to:

$$\int_{t_1}^{t_2} g\,dt = \int_{t_1}^{t_2} (\ddot{u} + 2h\omega_0\dot{u} + \omega_0^2 u)\,dt + cy \int_{t_1}^{t_2} dt \tag{5.18a}$$

and average gravity over the interval can be written:

$$\bar{g} = \frac{d}{dt}\int_{t_1}^{t_2} (\ddot{u} + 2h\omega_0\dot{u} + \omega_0^2 u)\,dt + cy \tag{5.18b}$$

This equation can be approximated by:

$$\bar{g} = 2h\omega_0\bar{\dot{u}} + cy \tag{5.18c}$$

in which $\bar{\dot{u}}$ is the average beam velocity over the interval.

The L & R meter system includes several-stage filtering to provide average beam-velocity values and circuits which provide output readings of $\bar{g}$. Instrument controls permit changing the averaging period, usually between 1 and 5 min. The averaging period is commonly made short at low sea accelerations and long at higher accelerations.

In the original form, suspended from a gimbal joint, the gravimeter measured total acceleration along the meter axis. This is the vector sum of gravity and the horizontal accelerations causing the meter case to tilt or swing through a small vertical angle (Fig. 6.10 and eq. 6.45a). This old L & R system used two perpendicular horizontal inertial bars, each rotating freely with a period of 2 min about a perpendicular horizontal axis, to measure the vertical deflection of the suspended meter case. Acting as long-period levels, the bars provided a reference for the deflection measurement. To obtain gravity the system included an automatic computer which squared the values of the two horizontal acceleration components and then subtracted their sum from the measured value along the meter axis.

In a gimbal suspension the meter-sensing element is essentially unaffected by the cross-coupling of vertical and horizontal accelerations acting on the meter case (Harrison, 1960). The old L & R system was designed for average swing angles of less than 2° (so that, in eq. 6.45a, $\overline{\cos\theta}$ can be approximated by $1-0.5\bar{\theta}^2$), corresponding to horizontal accelerations up to 50 gal. To correct for larger swing angles, that is larger accelerations, requires new instrument design.

The present system is mounted on a stabilised platform, such that the sensing element is essentially unaffected by horizontal accelerations. Deviations from horizontality of the platform (off-leveling) produces errors in measured gravity (eq. 6.46), although in modern platforms these errors are usually less than 1 mgal. The L & R platform is stabilized by two gyroscopes which drive torque motors to prevent precession relative to inertial space. The gyros are referenced to two horizontal accelerometers which act as long-period levels. Negative feedback from the accelerometers precesses the gyros about the vertical such that the platform is maintained horizontal as the ship moves over the earth's surface.

In a highly overdamped beam-type meter mounted on a horizontal platform, apparent gravity can be distorted over each cycle of ship motion when horizontal and vertical components of acceleration have equal periods and are out of phase. A cross-coupling of the acceleration components then occurs, reaching maximum effects for 90° phase differences (as indicated in eq. 6.56). This effect is generally not averaged out over a ship period. Over a cycle of circular wave motion, illustrated in Fig. 6.11, the beam moves down at a relatively greater rate when the ship is in the upward part of the cycle than when at the bottom or top of the cycle. The rate of beam motion is increased in the "3 o'clock" position of the cycle (Fig. 6.11) and in the opposite direction in the "9 o'clock" position. Since the rate of the beam displacement measures the change in gravity, the resultant apparent gravity is too high while the ship moves upward and too low while it moves downward.

Outputs of the two horizontal accelerometers are used to compute cross-coupling corrections. These corrections are small at low sea states and can usually be corrected for adequately at moderate sea states. At high sea states however, the corrections commonly include errors unless the meter is performing perfectly. LaCoste (1973a) developed a method for identifying inadequate meter performance and then for making appropriate corrections to measured gravity. Measured components of velocity and acceleration of ship motion are described as power series, and the terms are then cross-correlated with observed gravity. Good correlations with any of the motion terms indicate errors in measured gravity; gravity is then corrected so that it no longer correlates with the corresponding terms.

To make an inertial navigation system of its platform, L & R added a third gyro, fixed to the meter case. This gyro stabilizes the platform in a north—south direction. A servoamplifier drives a torque motor to maintain a north heading, thus making the platform a gyrocompass. 84 min Schuler oscillations (e.g., Bell, 1969), which are equivalent to those of a pendulum with length equal to the earth's radius and thus has a period of 84 min (which affect all inertial systems), are physically, but not mathematically avoided by using damped erection loops (period about 12 min). The gyros are also compensated for drift, although drift rates need to be checked periodically.

Computations of the Eötvös corrections are obtained from the outputs of the three gyros and the two horizontal accelerometers (the stabilized platform is not used as a reference in the computations, to avoid possible errors from platform off-leveling). Valliant and LaCoste (1976) developed differential equations for computing Eötvös corrections and errors in the corrections from the applicable instrumental measurements.

Tests with the three-axis inertial platform (Valliant and LaCoste, 1976) produced gravity observations accurate to 3 mgal, compared with values based on more accurate electronic navigation methods.

The inertial platform has the potential for providing ship-position information, as well as obtaining Eötvös corrections. Positioning requires determination of ship-velocity components, however, which means determination of error angle in the gyrocompass and wandering of the platform (Valliant and LaCoste, 1976). It is not yet known whether the L & R system can provide the desired accuracy without upgrading the gyros and accelerometers. Sophisticated systems do exist which can provide adequate accuracy.

Graf-Askania Gss-2 gravimeter

This meter is a sea-going modification (Graf, 1958) of the earlier Graf-Askania static land meter. The sensing element (Fig. 5.11) consists of a flat bar which two helical torsion springs maintain in a nearly horizontal position, the bar rotating about the axis of the springs. The torsion of the springs critically balances the moment resulting from the weight of the bar, thereby producing a long-period motion (about 6 sec). A set of ligaments connect the four corners of the bar to the ends of the torsion springs, which eliminate horizontal motions of the bar. The bar moves in the field of a strong permanent magnet; the resultant eddy currents produce strong magnetic damping. In later models damping was greatly increased (Graf and Schulze, 1961) to permit accurate gravity measurements at vertical accelerations up to 0.1 *g*.

Gss-2 meters operate on stabilized platforms. Like beam-type meters constrained in horizontal positions, measurements include cross-coupling accelerations. These accelerations must be removed, which is done by measuring horizontal accelerations with accelerometers attached to the

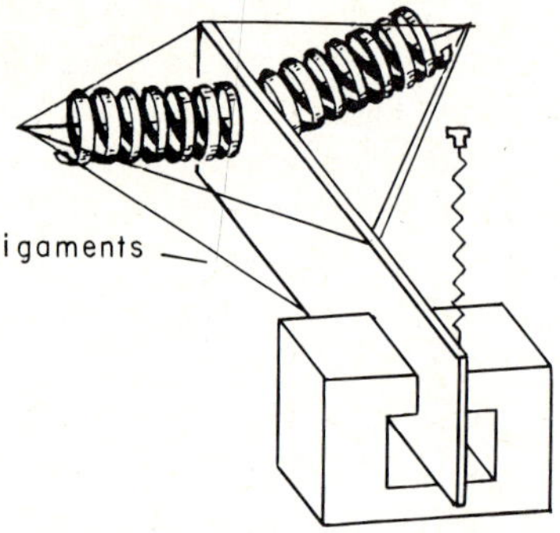

Fig. 5.11. Graf-Askania Gss-2 gravimeter, a highly overdamped modification of the Graf-Askania land meter. Vertical accelerations acting on the meter case cause the beam to oscillate slowly in the field of a strong permanent magnet. Ligaments prevent horizontal motions of the beam.

platform and then removing the effect with the use of automatic computers. Cross-coupling effects are negligible in calm seas; they can be removed satisfactorily at moderate sea states, but usually not adequately at high sea states.

Extensive experience has shown that these meters usually provide measurements accurate to within a few milligals, and better at low sea states.

The Askania Gss-3 gravimeter

This is an axially symmetric sea gravimeter that is essentially unaffected by cross-coupling accelerations; it is therefore capable of accurate measurements in relatively high sea states. The sensing element is a force-balanced type, that is, accelerations causing displacement of the mass-spring system are counteracted so as to maintain the mass in a near-null position.

The mass-spring system (Fig. 5.12) consists of a vertical tube which houses the main spring. Five ligaments permit frictionless motion of the tube constrained to the vertical direction, so that horizontal accelerations do not affect the motion. The mass weighs about 30 g and the system has an undamped period of 0.25 sec.

A capacitative-displacement transducer is mounted at the top of the tube, where it detects deflections of the mass system. The transducer voltage is fed to two windings of a moving coil situated at the base of the tube, the coil moving in the field of a permanent magnet. The output of one winding goes into a proportional feedback circuit, whereby the mass displacement is reduced by the feedback amplification factor; the output of the other goes into an integral feedback circuit which drives the mass displacement to zero, and which also suppresses ship accelerations. The current generated in the moving coil is thus a measure of gravity variations. The output from this coil

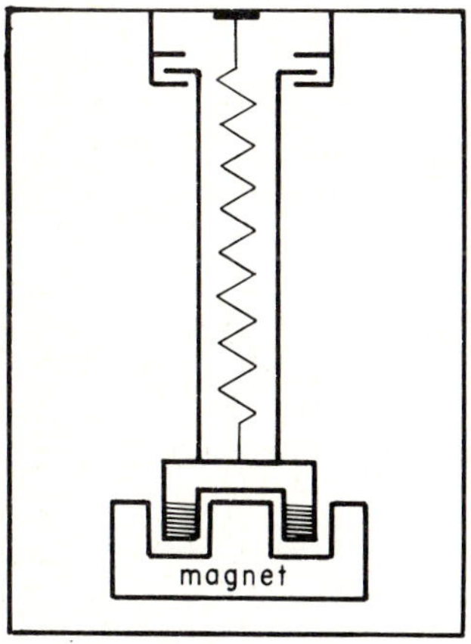

Fig. 5.12. Askania Gss-3 gravimeter, an axially symmetric meter. The mass is a tube constrained to move in the vertical direction; the motion of the mass is insensitive to cross-coupling accelerations. Voltage generated in a capacitative-displacement transducer (top of the tube) measures mass displacements and is fed through a winding on movable coils in the field of a strong permanent magnet (bottom of tube) which, by negative feedback circuits, greatly decreases the mass displacement. Currents generated in separate windings of the coils measure changes in gravity.

is heavily filtered to average the relatively short period (5—15 sec) vertical components of ship acceleration.

The sensor is claimed to be insensitive to horizontal accelerations up to 0.3 *g*. The meter is mounted on a stabilized platform. The manufacturer (Bodenseewerk Gerätetechnik GMBH, Überlingen, W. Germany) states that, mounted on their new platform Kt 10, the over-all effect of horizontal accelerations will not produce errors greater than 0.05 mgal. The claimed accuracy of the combined Gss-3 and Kt 10 system (the Kss 10) is ± 0.1 mgal for a vertical acceleration of 0.05 *g*, ± 0.75 mgal for an acceleration of 0.12 *g*, and ± 1 mgal for an acceleration of 0.18 *g*.

The Gss-3 gravimeter has been used by European investigators.

LaCoste and Romberg (L & R) axially symmetric gravimeter

L & R designed an axially-symmetric sea meter (LaCoste, 1973b), which operates on a stabilized platform, having a high sensitivity at large sea states. Being axially symmetric, the sensor is not subject to the cross-coupling affecting beam-type meters. The instrument has been tested successfully in the laboratory, but not yet at sea. The sensing element (Fig. 5.13) includes a mass constrained to move only along its sensitive axis, that is, in the vertical direction. The mass is highly damped, causing the mass to move slowly. This makes it possible to achieve high sensitivities without concern

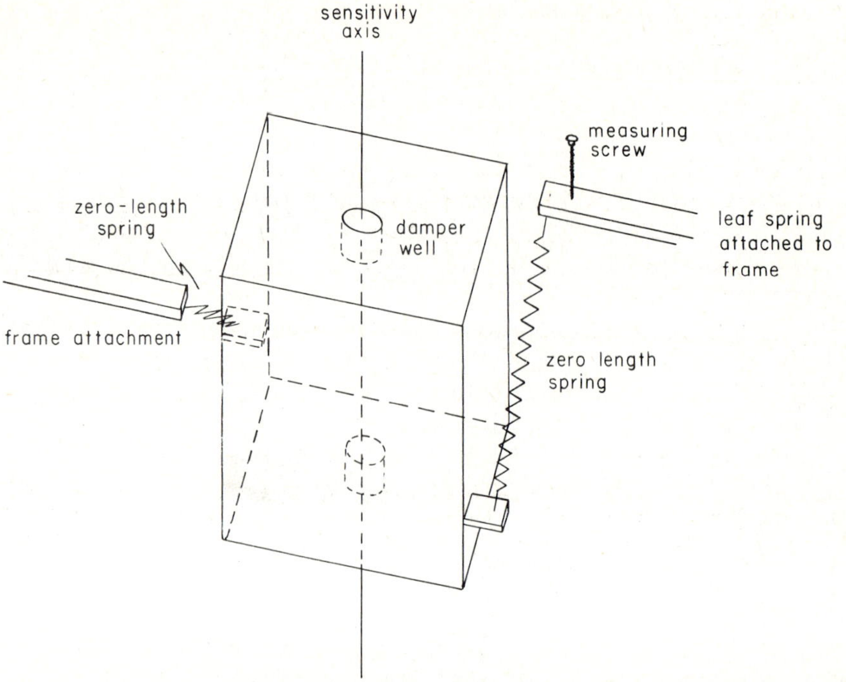

Fig. 5.13. Sensing element of a LaCoste and Romberg axially symmetric gravimeter. The mass is constrained to move in the direction of the sensitivity axis, making the sensor insensitive to cross-coupling accelerations. Changes in gravity acting on the mass are balanced by two diagonal zero-length springs located on opposite sides of the mass.

for possible motion instabilities during high sea states. The restoring force is made small to produce a high-displacement sensitivity.

The mass is suspended from the meter case by a number of links. These are so placed and adjusted that the mass is translated only along its sensitive axis, accompanied by a slight rotation. Forces orthogonal to the sensitive axis do not affect the motion of the mass. High air damping of the mass is achieved by cylindrical plugs supported on the meter case and recessed into cylinders in the top and bottom of the mass. The restoring force in the suspension can be controlled by adjusting the lengths of the individual links.

The force of gravity on the movable mass is balanced by two zero-length springs. The period of the mass is made long, the displacement sensitivity thereby being made very high, by using appropriate spring constants and carefully adjusting the lengths of the links. A zero-restoring force means an infinite sensitivity.

Meter tilts produced by platform off-leveling do not appreciably affect the sensor period or sensitivity.

The Gilbert vibrating-string gravimeter

Gilbert (1949) constructed the first vibrating-string gravimeter, which he operated in a submarine. The string consists of a thin, flat, conducting ribbon of beryllium copper about 5 cm long; a mass is suspended at the lower end. The ribbon is vibrated at a resonant frequency, about 1,000 Hz, the frequency depending on the mass-per-unit length and on the tension produced by the weight of the suspended mass. Variations in gravity change the weight of the mass, hence the tension on the ribbon and the resonant frequency.

The wavelength of the fundamental mode of a standing wave in a rigidly clamped string is twice the string length (nodes being at the clamped ends). Since frequency varies inversely with wavelength, the resonant frequency varies inversely with twice the string length.

Rayleigh (1945) showed that the frequency of the fundamental mode of a wire clamped at both ends is (see also Wing, 1969):

$$f = \frac{1}{2L}\sqrt{\frac{T}{\rho_1}}\left(1 + \frac{2X}{L}\sqrt{\frac{\mu A + T}{T}}\right) \tag{5.19}$$

in which L is the string length, T the tension, ρ_1 the mass per unit length, x the radius of gyration of the wire cross section about its mid plane, A the area of cross section, and μ the modulus of rigidity. A flat ribbon minimizes the contributions that rigidity makes to the resonant frequency, in this case to about 1%. The term in brackets in eq. 5.19 can be ignored with a ribbon, treating it as a constant bias in the gravity measurements. Accordingly, Gilbert (1949) used for the frequency:

$$f = \frac{1}{2L}\sqrt{\frac{mg}{\rho_1}} \tag{5.20}$$

in which mg is the weight of the suspended mass producing the tension in the ribbon. We differentiate eq. 5.20 with respect to g to obtain a relative frequency of:

$$\frac{df}{f} = \frac{1}{2}\left(\frac{dg}{g}\right) \tag{5.21}$$

An oscillator vibrates the string or ribbon at its resonant frequency in a magnetic field; voltages developed in the ribbon are reinforced through positive feed-back circuits. The mechanical Q (the reciprocal of energy loss per cycle in a vibrating system to the total stored energy) of the ribbon is made very high, about 20,000, to ensure stable oscillations. The suspended mass is a conductor which moves in a magnetic field; damping is electromagnetic. The resonant frequency is compared with a crystal frequency standard and frequency changes are measured electronically to a small fraction of a cycle.

The Gilbert meter was reported to measure to about 1.5 mgal (Gilbert, 1949) in a submarine, relative to pendulum measurements. A severe limitation in all vibrating-string meters, when subjected to varying vertical accelerations in a ship, is that the measured mean vertical acceleration may be in error. This error is called a nonlinear rectification (Wing, 1969), which can become very large in high-sea states.

Gilbert (1952) modified the string gravimeter for measurements in boreholes to determine rock densities. Subsequent borehole models were made by Goodell and Fay (1964) and by Howell et al. (1966), which included greatly increased mechanical *Q*'s of the vibrating string to increase sensitivities of measured gravity.

The MIT or Wing vibrating-string gravimeter

Wing (1969) constructed a gravimeter (at the Massachusetts Institute of Technology) using two oppositely acting vibrating strings (Fig. 5.14). With perfect string geometry, the double string essentially eliminates the nonlinear rectification produced in all single strings.

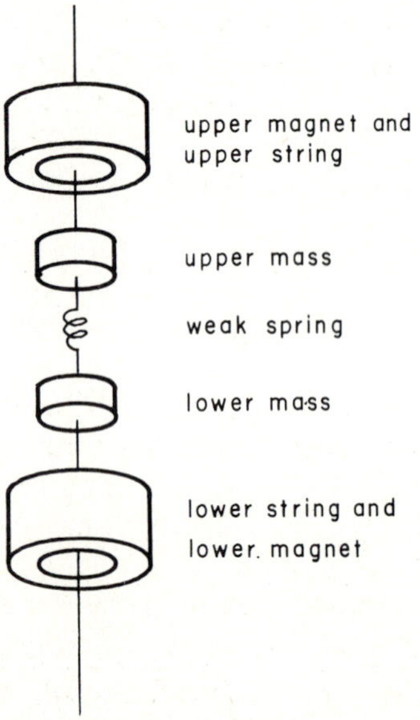

Fig. 5.14. Sensing element of the double vibrating-string accelerometer. The upper mass increases tension on the string vibrating in the field of the upper magnet; the lower mass decreases tension on the string vibrating in the field of the lower magnet. Ligaments (not shown) constrain the mass to move only along the sensor axis. (Adapted from Wing, 1969.)

The tension in a string is $T = T_0 + mg$, where T_0 is the pre-tension and mg the weight of mass m and g the apparent gravity (measured acceleration).

Ignoring the small rigidity term, eq. 5.19 reduces to:

$$f = \frac{1}{2L}\sqrt{\frac{T_0}{\rho_1}\left(1 + \frac{mg}{T_0}\right)} \tag{5.22}$$

We expand in binomial series to obtain (Wing, 1969):

$$f = \frac{1}{2L}\sqrt{\frac{T_0}{\rho_1}}\left[1 + \frac{1}{2}\left(\frac{mg}{T_0}\right) - \frac{1}{8}\left(\frac{mg}{T_0}\right)^2 + \frac{1}{16}\left(\frac{mg}{T_0}\right)^3 + \ldots\right] \tag{5.23}$$

which can be written:

$$f = K_0 + K_1 g + K_2 g^2 + K_3 g^3 + \ldots \tag{5.24}$$

in which the K's are instrumental constants.

In the two strings shown in Fig. 5.14, the upper mass increases the tension in the upper string, and the lower mass decreases the tension in the lower string. We therefore write the series expansion for the upper (1) and lower (2) frequencies:

$$\begin{aligned} f_1 &= K_{01} + K_{11}g + K_{21}g^2 + K_{31}g^3 + \ldots \\ f_2 &= K_{02} - K_{12}g + K_{22}g^2 - K_{32}g^3 + \ldots \end{aligned} \tag{5.25}$$

The output of the double-vibrating string is the difference frequency $\Delta f = f_1 - f_2$. Thus:

$$\Delta f = (K_{01} - K_{02}) + (K_{11} + K_{12})g + (K_{21} - K_{22})g^2 + \ldots \tag{5.26a}$$

which is obtained by heterodyning the two string frequencies. All even-order terms are the nonlinear terms; they approach zero for near-perfect string geometry. The difference frequency is then a measure of g, that is:

$$\Delta f = (K_{11} + K_{12})g \tag{5.26b}$$

The sensor used is an American Bosch Arma Vibrating String Accelerometer. The strings are thin beryllium ribbons, about 1 cm long, carrying electric currents in oscillating circuits so that the strings vibrate at their natural frequencies. The mechanical Q is about 2,000.

The sensor is mounted on a gyrostabilized platform, so that the sensor axis is kept vertical and measures the sum of gravity and vertical acceleration. Ligaments keep the masses accurately aligned to the sensor axis so as to eliminate horizontal accelerations. Cross-coupling of vertical and horizontal accelerations is eliminated by careful instrument design and adjustment of cross-support ligaments.

The meter system has reliabilities of better than 1 mgal in moderate sea states (Bowin et al., 1972).

Other vibrating-string gravimeters

Lozinskaya (1959) described a meter similar to Gilbert's, mounted in a gimbal frame. It was used in the Caspian Sea. Tsuboi et al. (1961) constructed a string-type meter in Japan. Tomoda and Kanamori (1962) mounted this type of meter on a gyrostabilized platform and obtained many measurements with it in the Western Pacific. They measured vertical accelerations to correct for meter nonlinearities, obtaining accuracies of 2 mgal. Seto (1968) described a three-vibrating-string gravimeter which measures total gravity at sea.

The Bell gravimeter

The sensitive element of this meter, Fig. 5.15, is a modification of force-balanced accelerometers developed for inertial navigation systems (Bell Aero-Space Company, 1971). The sensor consists of a coil moving vertically between two permanent magnets. Variations in both gravity and vertical acceleration change the weight of the coil and its supporting form. Coil displacements are sensed by capacitor rings in a bridge circuit. Unbalances in the bridge produce currents in the coil in a direction which restores the balance. Varying currents in the coil thus measure changes in apparent gravity, that is in gravity plus vertical acceleration.

The accelerometer is very small, housed in a cylindrical case less than 5 cm high and 2 cm thick. The sensor operates in a constant-temperature case to main uniform fields in the permanent magnets. Lateral springs constrain the

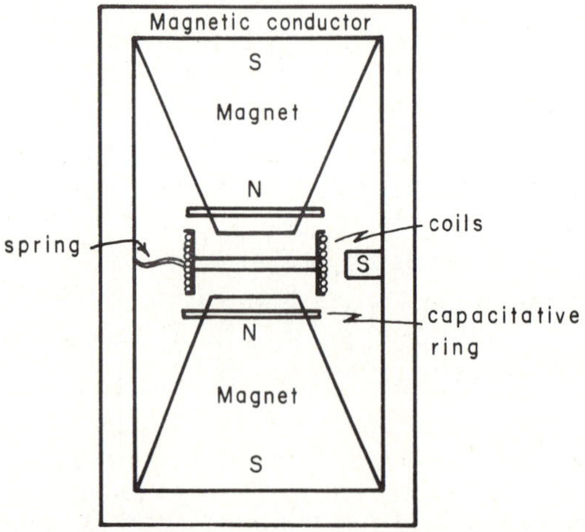

Fig. 5.15. Bell force-balanced gravimeter. A mass with coil windings moves vertically between two permanent magnets. Capacitor rings sense coil displacements; currents thus produced in a bridge circuit restore the mass to a balance position.

sensor to allow motion only along its sensitive axis. The unit is mounted on a stabilized platform and, hence, horizontal accelerations do not affect the coil motion.

Because the meter is axially symmetric, cross-coupling of horizontal and vertical accelerations is negligible in even high-sea states.

The balancing current, which reads out digitally, is proportional to apparent gravity (vertical acceleration and gravity). Gravity variations are separated from ship accelerations by electronic filtering of the output signal.

The Bell meter has been used primarily by the U.S. Navy and by oil companies. Measurements are reported to be reliable to about 1 mgal (Coons and Smalet, 1967).

EARTH-TIDE GRAVIMETERS

Earth-tide gravimeters were developed in the 1940's. These instruments measure gravity variations at a fixed station with the high accuracies needed to make analyses of the various tidal components. Tidal gravimeters most commonly used have beam-type sensors.

The LaCoste and Romberg (L & R) earth-tide gravimeter

The L & R tidal meter, a modification of the L & R land meter, was developed in 1947. It was modified by 1952 so that its accuracy is better than 1 μgal. These modifications include a null method; a lever system maintains the sensor weight in the same position so as to eliminate spring-hysteresis errors. Measuring screw errors were reduced to less than 0.1 μgal, an error reduction of over 300 from those in land meters. The spring calibration factor is stabilized by use of a lever system which acts on the main spring; the calibration factor of the lever system remains constant with time. A capacitance bridge readout determines the position of the mass with high sensitivity. The nulling device used a flat response for all periods of ground motion greater than 10 min. The meter case is sealed against barometric changes and is accurately thermostated by a thermistor-transistor heater control system at a uniform operating temperature (usually 50°C). Meter drift rates are normally less than 0.03 mgal per month. The meter is fully automated and requires almost no attention when set up.

The Askania (Gs-25) tidal gravimeter

This instrument is a modification of the Graf-Askania land meter; it can measure to about 1 μgal. The meter is (1) doubly thermostated, to assure constant temperature; (2) it has a linear drift rate of less than 0.1 mgal per month, which is achieved, in part, by using an electromagnetic calibration device; (3) it has a high reading sensitivity, which is obtained with a capacitative-displacement transducer and potentiometer recorder; and (4) it has a measuring range of 5,000 mgal (a direct measuring dial of 600 mgal).The meter includes devices such as a temperature-stable power supply, improved pick-up electronics, an amplitude-controlled supply of plate capacitor, a low-noise-level preamplifier, and a selective amplifier.

PENDULUMS

Pendulums were once the principal instruments for measuring gravity differences and absolute gravity.

The formula for pendulum swing is based on a fictitious or mathematical pendulum: a point mass hanging on a weightless string. The component of gravity acting on mass m is $g\sin\theta$ (Fig. 5.16); m is suspended by a string of length l and makes a swing angle θ, being decelerated an amount $\ddot{s} = l\ddot{\theta}$ (where dots indicate differentiation with respect to time) along the arc path. The accelerations acting on m can be written:

$$l\ddot{\theta} + g \sin \theta = 0 \qquad (5.27)$$

In practice the swing angle is so small that $\sin \theta \doteq \theta$; eq. 5.27 then becomes:

$$\ddot{\theta} + \left(\frac{g}{l}\right) \theta = 0 \qquad (5.28)$$

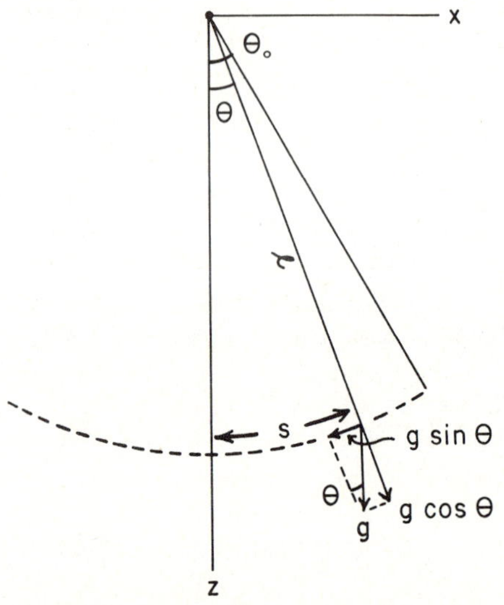

Fig. 5.16. A mathematical or fictitious pendulum (a point mass on a weightless string). The component of gravity along the swing arc, $g \sin \theta$, decelerates the mass by $l\ddot{\theta}$.

Solutions of this equation are of the form:

$$\theta = \theta_0 \sin \omega t = \theta_0 \sin \sqrt{\frac{g}{l}}\, t \tag{5.29}$$

in which θ_0 is the maximum swing angle and ω is the angular frequency. Since $\omega = g^{1/2} l^{-1/2}$, the swing period τ is:

$$\tau = 2\pi \sqrt{\frac{l}{g}} \tag{5.30}$$

similar to the period in eq. 5.3.

A physical pendulum with mass M has a moment of inertia I about its swing axis, which is a distance h above the center of mass. The reduced length l of the pendulum is:

$$l = \frac{I}{Mh} \tag{5.31}$$

and its period is:

$$\tau = 2\pi \sqrt{\frac{I}{Mgh}} \tag{5.32}$$

Assuming that the constants of the pendulum remain unchanged, observed periods or changes in period can be used to determine gravity differences. Thus:

$$\frac{\tau_1^2}{\tau_2^2} = \frac{g_2}{g_1} \tag{5.33}$$

or we differentiate eq. 5.32 to obtain:

$$\frac{\delta\tau}{\tau} = -\frac{1}{2}\frac{\delta g}{g} \tag{5.34}$$

Swing periods can be measured with high accuracy, as by reflecting light from the pendulum mass to a photocell and electrically counting the number of oscillations over a specified interval. A principal problem arising from moving a pendulum from station to station is that the forces between the knife edge and the plane surface on which it swings are not strictly constant. Microscopic irregularities in both surfaces restrict contacts to relatively few points. The result is that accuracies of field pendulums are about 0.5 mgal (corresponding to change in period of about 10^{-6} sec).

Sea pendulums

Pendulum swings are affected by the flexure and sway of the support mount. If the mount experiences horizontal accelerations $\ddot{x}$ eq. 5.28

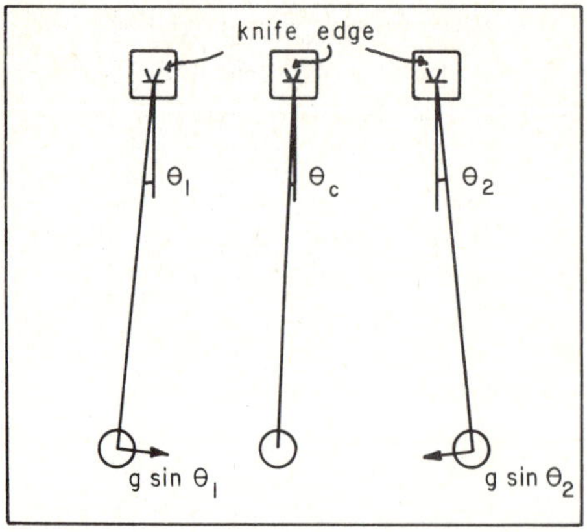

Fig. 5.17. Vening Meinesz three-pendulum sea apparatus. Averages of swing angles $(\theta_1 - \theta_c)$ and $(\theta_2 - \theta_c)$ of two fictitious pendulums remove first-order horizontal accelerations acting on the instrument case.

becomes:

$$\ddot{\theta} + \left(\frac{g}{l}\right)\theta + \left(\frac{1}{l}\right)\ddot{x} = 0 \tag{5.35}$$

To measure gravity differences at sea, Vening Meinesz constructed a three-pendulum apparatus in which the three pendulums are mounted in line on a common support (Fig. 5.17). The apparatus hangs in gimbals to reduce the effects of horizontal accelerations.

Eq. 5.35 describes swings of the individual pendulums. When identical outside pendulums swing with equal amplitude and opposite phase, the difference in swing angle, $\theta_1 - \theta_2$, is described by the equation:

$$(\ddot{\theta}_1 - \ddot{\theta}_2) + \frac{g}{l}(\theta_1 - \theta_2) = 0 \tag{5.36}$$

This represents a fictitious pendulum which is free of first-order horizontal accelerations acting on the pendulum mount.

In field measurements, the center pendulum is initially held stationary while the outside pendulums are put into equal and opposite oscillations. Motions of the outside pendulums cause the center pendulum to swing with angle θ_c. The three pendulum swings are combined in two pairs, producing two fictitious pendulums which have swing angles $\theta_1 - \theta_c$ and $\theta_2 - \theta_c$. A reference from which the swing angles are measured is provided by a short-period pendulum swinging perpendicular to the other pendulums, indicating deviations from the vertical of the main pendulum swing plane. First-order vertical accelerations are removed by averaging recordings over 30-min intervals.

Second-order horizontal and vertical accelerations are produced by waves with periods long compared with those of the pendulums. Browne (1937) analyzed these effects (eq. 6.45a), resulting in Browne corrections of $\overline{\ddot{x}}\,(2g)^{-1}$ and $\overline{\ddot{z}}\,(4g)^{-1}$ for average horizontal ($\overline{\ddot{x}}$) and average vertical ($\overline{\ddot{z}}$) accelerations. Second-order effects can range to tens of milligals in magnitude; they must be removed. Vening Meinesz used two low-damped, free-hanging pendulums having periods long compared with wave periods to measure horizontal components of ship acceleration, and thus determine tilt of the apparatus.

Sea pendulum apparatuses measure gravity differences to 1 or 2 mgal.

Absolute-gravity pendulums

Physical pendulums have two axes about which they can swing with the same period. These axes are a distance H apart (Fig. 5.18), and are at distances h_1 and h_2 from the pendulum center of gravity. In the reversible pendulum oscillation periods are measured about the two axes. It is difficult to locate the two swing axes precisely with respect to the center of gravity; hence, the observed periods τ_1 and τ_2, corresponding to these two axes, usually differ by a small amount. Reversible pendulums provide good measurements of absolute gravity because they are based on two separate determinations of period and swing length.

It can be shown (e.g., Cook, 1965) that a reversible pendulum has the same swing periods and value of absolute gravity as a fictitious pendulum with length equal to the distance between the two axes. The period of this pendulum length is thus (similar to eq. 5.30):

$$\tau = 2\pi\sqrt{\frac{H}{g}} \qquad (5.37)$$

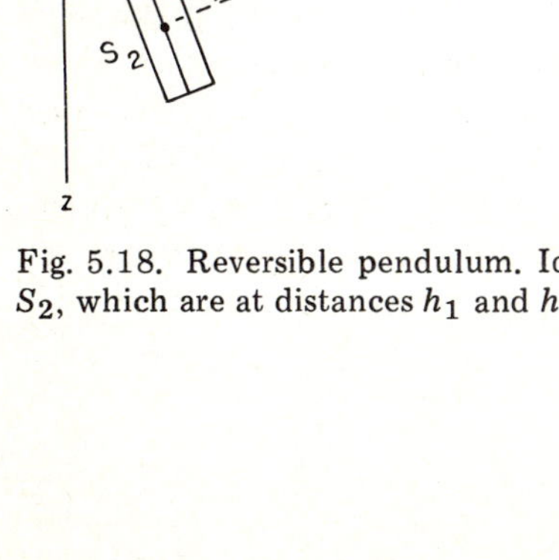

Fig. 5.18. Reversible pendulum. Identical swing periods are obtained about axes S_1 and S_2, which are at distances h_1 and h_2 from the center of gravity C.

where g is the value of absolute gravity at the station and $H = h_1 + h_2$. The period τ is related to physical periods τ_1 and τ_2 (e.g., Heiskanen and Vening Meinesz, 1958, p. 93) by:

$$\tau = \tfrac{1}{2}(\tau_1 + \tau_2) - \tfrac{1}{2}(\tau_1 - \tau_2)\left(\frac{H}{h_1 - h_2}\right) \tag{5.38}$$

Kater (1818) introduced the reversible pendulum for measuring absolute gravity, although his measurements did not allow for flexure of the platform. Helmert (1898) analyzed the effect of flexure, a factor which Kühnen and Furtwängler (1906) incorporated into their determination of absolute gravity (981,274 mgal) at Potsdam. Other absolute gravity determinations were made with reversible pendulums in 1936 and thereafter at various sites, particularly in North America and Europe. The three most significant determinations were made in Washington, D. C. (Heyl and Cook, 1936: $g = 980{,}081.6$ mgal, s.d. ~ 2 mgal), in Teddington, England (Clark, 1939: revised value of g is 981,183.1 mgal, s.d. $\sim$ 0.5 mgal), and Leningrad, USSR (Yanovsky, 1958: $g = 981{,}918.7$ mgal, s.d. ~ 1.0 mgal). These and other measurements are discussed by Cook (1965), who suggests that in the earliest Potsdam measurements, length and time could not be determined as precisely as in the later studies and that flexure may not have been fully calculated.

FREE-MOTION, ABSOLUTE GRAVITY APPARATUSES

Absolute gravity can be determined by precisely measuring the time during which a body falls in free motion through a known distance, or, conversely, precisely measuring the distance for a known time. The well-known relationship between time t and distance s is:

$$s = \tfrac{1}{2}gt^2 \quad \text{or:} \quad g = \frac{2s}{t^2} \tag{5.39}$$

Volet (1946) utilized the free-fall principle in the 1940's when it became possible to measure times with sufficient accuracy to determine g to about 1 part in 10^7. It is now possible to measure time and distances with accuracies of about 1 part in 10^9, such that g can be measured to a few parts in 10^9. Absolute determinations of gravity are about two orders of magnitude better than those obtained before 1966, indeed a major achievement.

The high accuracy of absolute-g determinations has resulted in more precise determinations of physical measurements, as Sakuma (1971) points out, such as force (weight), pressure, absolute volt, absolute ampere, and temperature at the boiling point of water. Sakuma concludes that "the aim of the dominant use of gravity in metrology has now been attained". The high accuracy has equally important geophysical applications. It is now possible to measure secular variations in gravity associated with crustal movements because free-motion apparatuses are not subject to drift. Similarly, long-period variations in the vertical produced by the 18.6 year nutation in the earth's axis (caused by the lunar orbit being 5°.1 out of the ecliptic) can be investigated by a free-motion apparatus operating continuously in the same location (the International Latitude Observatory in Mizusawa is studying latitude variations exhibiting an 18.6-year period). Even more far-reaching studies concern possible variations in the gravitational constant G; the results will have important implications in astronomy.

Two different free-motion apparatuses have been developed. One uses simple free fall, the other a symmetrical free motion produced by launching a body upward. The symmetrical method, developed more recently, provides nearly an order-of-magnitude improved determination of g (Sakuma, 1971).

Six values of absolute gravity, employing either free-fall or symmetrical apparatuses, have been determined since 1967 (Sakuma, 1971). The values were obtained at the National Bureau of Standards, Gaithersburg, Maryland; the National Physical Laboratory, Teddington, England; Bureau International de Poids et Mesures, Sèvres (Paris), France; Wesleyan University, Middletown, Connecticut; National Research Laboratory of Metrology, Kakioka, Japan, and Physikalisch-Technische Bundesanstalt, Braunschweig, Germany. The reported standard deviations of measurements at these sites vary considerably, ranging from 0.005 to 2 mgal, although values at all of the six stations are within about a milligal of the revised Potsdam system.

Free-fall experiments

It is difficult to determine accurately the values of initial velocity and position of a simple free-falling body. Accordingly, time intervals in the apparatus are measured which correspond to precisely known height intervals. Three horizontal planes (Fig. 5.19) are separated by distances h_1, the height between the upper two planes, and h_2, the height between the upper and lower planes. Corresponding times t_1 and t_2 are measured. The height-time relationships are:

$$h_1 = \tfrac{1}{2}gt_1^2 + v_0 t_1 \quad \text{and:} \quad h_2 = \tfrac{1}{2}gt_2^2 + v_0 t_2 \tag{5.40}$$

where v_0 is the velocity of the body as it passes the upper plane. We solve for g to obtain:

$$g = \left(\frac{2}{t_2 - t_1}\right)\left(\frac{h_2}{t_2} - \frac{h_1}{t_1}\right) \tag{5.41}$$

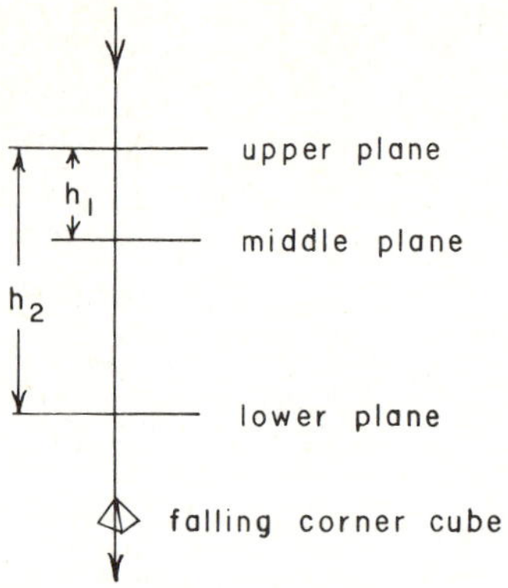

Fig. 5.19. Free-fall path. A body (e.g., a corner cube) falls distance h_1 in time t_1 and distance h_2 in time t_2.

Free-fall apparatuses were developed by Faller (1965), Hammond (1970), and other investigators. Hammond and Faller (1971) developed a portable free-fall apparatus, comprised of an optical interferometer illuminated by a He-Ne laser in which one mirror is fixed and the other, a corner cube, is the free-falling body. This apparatus is used to measure not only absolute g but also to check values of g previously measured at principal gravity stations with other apparatuses. Comparisons among the measurements obtained since about 1965 provided good agreements; the comparisons also showed that the 1906 Potsdam value is 13.8 mgal high, a high value having been suspected for some time. The Potsdam base has now been reduced by 14 mgal (to a value of 981,260 mgal). The principal sites in the revised Potsdam system are now estimated to be accurate to within 0.5 mgal (Sakuma, 1971).

Free-fall apparatuses experience only small mechanical shock when a body is released; larger shocks are encountered during launching in symmetrical experiments. However, resistance of residual air is a limiting factor in the fall. To evacuate too much air creates undesirable electrostatic effects. The precision of free-fall apparatuses is a few parts in $10^{-8}g$.

Symmetrical free-motion experiments

The symmetrical method, proposed initially by Volet (1947) and first used to measure absolute g by Cook (1967), is illustrated in Fig. 5.20. A body is catapulted vertically upward, decelerating as it crosses two horizontal planes separated by a known interval h, and accelerating as it crosses the planes again on the way down. Two independent time intervals, t_1 and t_2, are measured, corresponding to the upward and downward travel across each of the two planes. The height above the upper plane, h_0, at

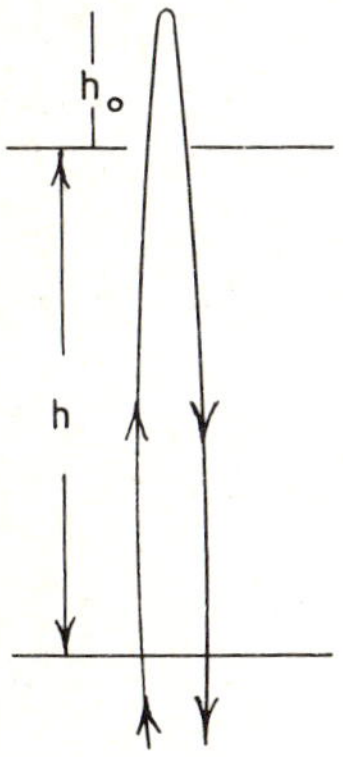

Fig. 5.20 Symmetric free-motion path. A body (e.g., a glass ball or a corner cube) catapulted upward travels time t_1 between crossings of the upper plane and time t_2 between crossings of the lower plane.

which the body reaches zero velocity is:

$$h_0 = \tfrac{1}{2}g\left(\frac{t_1}{2}\right)^2 = \tfrac{1}{8}gt_1^2 \tag{5.42a}$$

and the corresponding height above the lower plane is:

$$h_0 + h = \tfrac{1}{8}gt_2^2 \tag{5.42b}$$

Gravity is thus given by:

$$g = \frac{8h}{(t_2^2 - t_1^2)} \tag{5.43}$$

The symmetrical method has two important advantages over the free-fall method (Sakuma, 1971). Resistance of residual air, which is proportional to the velocity of the moving body, has essentially no effect on the time intervals t_1 and t_2. During the upward motion the resistance acts downward and during the downward motion it acts upward. Similarly, resistances produced by eddy currents due to variations in the earth's magnetic field also cancel out. The second advantage is that timing errors, which might result from time constants of chronographs, cancel out because the two signals at each level are produced by equal upward and downward velocities. A practical difficulty, however, is that upward launching produces mechanical disturbances which could affect the measurements.

In Cook's (1967) apparatus, the moving body is a glass ball and its passage across the two horizontal planes is timed by flashes of light which the ball produces when it passes between horizontal slits serving to define each plane optically. The height between the planes is measured interferometrically. The value of gravity which Cook obtained at the National Physical

Laboratory (Teddington, England) is 981,181.75 mgal, s.d. 0.13 mgal (1.4 mgal less than the pendulum-determined value of Clark, 1939). Microseismic ground motions were the principal limiting factor in the measurements.

Sakuma (1971) constructed a symmetrical apparatus at the Bureau International de Poids et Mesures (Sèvres, Paris, France) with which he obtained a value of gravity of 980,925.949 mgal ±0.0054 mgal. The projectile in the apparatus is a corner cube which acts as one mirror in the vertical beam of a Michelson interferometer. To achieve such a high accuracy in measured gravity requires the elimination of microseismic ground disturbances. This was done by operating the sensing element on a servocontrolled stabilized table.

GRAVITY GRADIOMETERS

Gravity gradiometers measure the first derivative of the gravitational field. Evjen (1936) was perhaps the first to indicate that vertical gradients can be used to interpret gravity anomalies. The gradient can be integrated, as described by Paterson (1961), to obtain gravity.

Gravity gradients are commonly given in mgal m^{-1} or in Eötvös units (E°), where:

$$1\ E^{\circ} = 10^{-9}\ \text{gal cm}^{-1} = 10^{-4}\ \text{mgal m}^{-1}$$

The normal vertical gravity gradient at the earth's surface (eq. 6.6) is approximately 0.3086 mgal m^{-1} (3086 E°), varying slightly with latitude and altitude (see eq. 6.8). To determine gravity with about a 1-mgal accuracy requires measuring gradients to about 1 part in 10^6 (Thyssen-Bornemisza, 1970), which has not yet been achieved. Gradient measurements are complicated by the fact that topographic variations have nearly as large an effect as do common buried structures. The topographic effects must be corrected for, as indicated by Chinnery (1961) and described by Hammer (1976).

Airborne gradiometers consist of two accelerometers. Because both are subject to the same craft accelerations, these gradiometers are not sensitive to vertical craft motions. The instruments, therefore, have considerable appeal, although they do not yet have the sensitivity needed. Several gradiometers have been constructed; possibly an instrument with sufficient accuracy can be developed within the next few years.

Various types of airborne gradiometers have been tested (Hood and Ward, 1969). One consists of balances (as designed by Boitnott, 1961, 1962, and tested by Lundberg, 1957). Another consists of immersible liquids (mercury and thallium salts) enclosed in a glass container (the liquids rise to different heights in two columns of the container because of their density differences;

the liquids are oscillated at their natural frequency and, in the presence of a vertical gradient, one liquid weighs slightly more than the other relative to their zero-gradient weights). Sensitivities of both instruments need to be increased for significant airborne gravity surveys.

Ames (1974) described a rotational gradiometer designed for operation on a moving craft. The instrument uses rotational properties of tensors to separate the effects of gravity from the gravity gradients. The sensor consists of four masses at the ends of four transversely vibrating arms which are spun in a vacuum. A gravity gradient produces differential torques on the arms, which are detached by strain transducers. The resultant torque alternates at twice the rotational frequency (sharply resonant) of the sensor. It has been claimed that this system can detect a gradient of approximately 1 E°. Sensitivities of about 0.1 E° (10^{-5} mgal m^{-1}) are needed for airborne surveying.

Gravity gradiometers can have application in orbiting satellites. (A gravimeter in the satellite center of gravity gives zero gravity.) Thompson et al. (1965) have described a gradiometer consisting of a double-string, double-mass vibrating string gravimeter developed by American Bosch Arma for use in satellites. The difference between the two string frequencies is a measure of the gravity gradient.

TORSION BALANCES

Eötvös (1896) invented and constructed the torsion balance during his investigations on the deviations of the earth's shape from a sphere. The balance measures distortions or curves in the earth's gravity field, in contrast to the conventionally measured intensity of the field. The torsion balance was used for geophysical prospecting from 1915 to the late 1930's, when it was replaced by gravimeters and thus made obsolete. The instrument cannot be used aboard ships.

Because the torsion balance is now primarily of historical interest, we consider only its general principles. Detailed descriptions of the theory and field operation appear in Heiland (1940, pp. 170—292), Jakosky (1940, pp. 196—246), and Nettleton (1940, pp. 63—99).

The torsion-balance sensor (Fig. 5.21) consists of two masses, commonly on a z-shaped beam supported on a torsion fiber. The masses are offset both vertically and horizontally from one another, so that each is acted upon by slightly different values of the earth's gravity field. In a curved field (Fig. 5.22), in which the direction of the vertical has a slight curvature, horizontal components of force act in different directions on the two masses. The differential force component exerts a torque on the beam, causing it to rotate by a slight amount. The resultant angle of twist depends on the

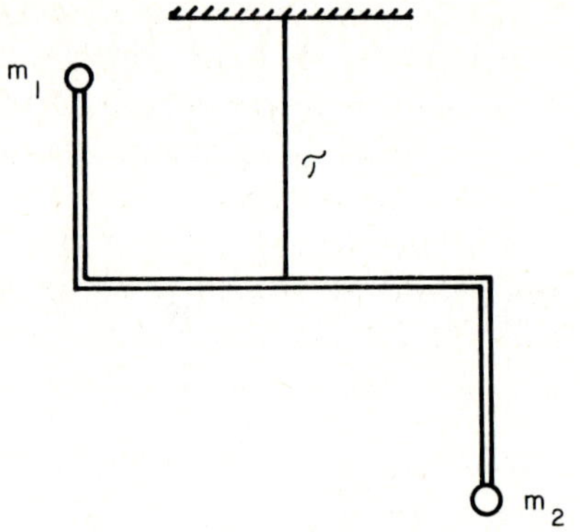

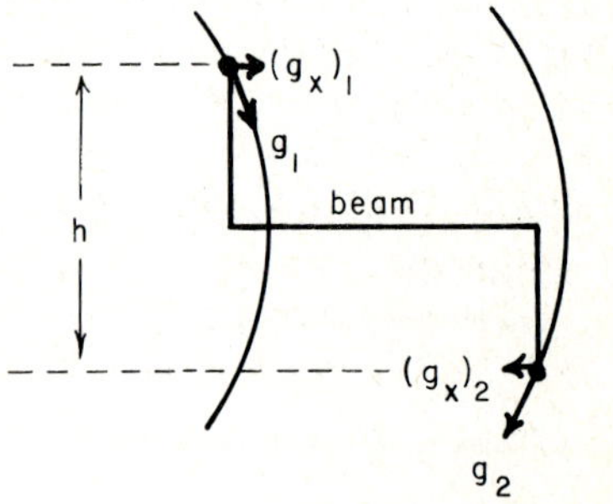

Fig. 5.21. Z-beam torsion balance. Masses m_1 and m_2 are at different geopotential levels. The balance rotates a small angle about the torsion fiber τ.

Fig. 5.22. Illustration of a curved gravity field (solid lines), in which the horizontal components $(g_x)_1$ and $(g_x)_2$ of the total gravity field (g_1 and g_2) are unequal, causing the beam to rotate.

magnitude of the horizontal forces, and hence on the gravity gradient extending across the z balance.

The torsion balance measures the angle of twist in the suspended fiber. The torque on the suspension fiber is expressed by (e.g. Nettleton, 1976, p. 68):

$$
\begin{aligned}
L = mhl \left(\frac{\partial^2 W}{\partial y\, \partial z} \cos\alpha - \frac{\partial^2 W}{\partial x\, \partial z} \sin\alpha \right) \\
+ ml^2 \left[2 \frac{\partial^2 W}{\partial x\, \partial y} \cos 2\alpha + \left(\frac{\partial^2 W}{\partial y^2} - \frac{\partial^2 W}{\partial x^2} \right) \sin 2\alpha \right]
\end{aligned}
\qquad (5.44)
$$

where α is the angle between the beam and a fixed direction, as north; m the mass of the weights on the beam; h the height difference between the two masses; $2l$ the horizontal offset of the masses, and W the earth's gravitational potential. The unknowns in the equation are the three second-order spatial derivatives W_{yz}, W_{xz}, and W_{xy}, and the quantity $(W_{yy} - W_{xx})$. A fifth unknown is the direction of the undisturbed beam position. The procedure for solving the equation is to set up five simultaneous equations for L by measuring the beam rotation (angle α) for five different azimuths of the instrument. The field instruments usually consisted of two parallel beams, with lower weights on opposite ends of the beams, thus providing two measurements of α differing by 180° for each instrument position. An additional unknown is introduced, the undisturbed position of the second beam. The double balance provides for three azimuthal measurements (120° apart), instead of five.

Two of the measured quantities are horizontal gravity gradients, namely $W_{xz} = \partial g/\partial x$ and $W_{yz} = \partial g/\partial y$; their vector sum is [illegible] s, the total horizontal gradient of the earth's gravity field. Gradients can be integrated to obtain gravity differences between field stations; increasing gradients indicate increasing gravity, and hence the direction toward a mass excess.

The two quantities W_{xy} and $(W_{yy} - W_{xx})$ are measures of the curvature of the equipotential field. The relationship between these quantities and the minimum and maximum curvatures of the field, R_1 and R_2, is given by (e.g., Nettleton, 1976, p. 69):

$$g\left(\frac{1}{R_1} - \frac{1}{R_2}\right) = [(W_{yy} - W_{xx})^2 + 4W_{xy}^2]^{1/2} \tag{5.45}$$

The quantity in this equation is called the differential curvature of the field.

Both the gradient and curvature quantities are given in Eötvös (E°) units. Sensitivities of field torsion balances were about 1 E°, that is 0.1 mgal km^{-1}.

Chapter 6

GRAVITY REDUCTIONS, CORRECTIONS, AND ANOMALIES

A gravity anomaly, the difference between gravity on the geoid and reference spheroid, is produced by mass distributions which cause the geoid to deviate from the spheroid. The anomaly includes the effect of the height difference N_0 between the geoid and spheroid (second term in eq. 2.76). The height effect is relatively small; therefore, it is conventionally omitted in computing mass distributions corresponding to observed anomalies. Where the geoid lies outside the spheroid, the height effect is approximately equal to the free-air correction (eqs. 2.75 and 6.6), about 3 mgal for a geoidal height of 10 m. The largest geoidal undulations are about 100 m (Fig. 2.19).

Published values of gravity anomalies are nearly all based on a sea-level datum. In geophysical prospecting, relative anomalies (usually privileged information) rather than absolute values are obtained. Such variations refer to a conveniently placed datum, usually other than sea level, which may be a plane or smooth curved surface near ground elevations. Use of this type of datum minimizes the size of gravity reductions, thus reducing effects of erroneously assumed rock densities between the station and datum surface.

Measurements made at sea level give gravity on the geoid, except if obtained on a moving platform, where corrections are applied. Land measurements are commonly made above sea level; measured gravity must then be reduced to the sea-level equivalent before an anomaly can be obtained by subtracting a value for normal gravity on the spheroid. This procedure is the universally accepted one for computing anomalies; otherwise normal gravity would have to be corrected to the elevation of each gravity station.

Reductions made to observed gravity to obtain sea-level equivalents are based on various assumptions, although none truly represent actual conditions. In one method the measurements are reduced to sea level by correcting only for station elevation. This procedure is equivalent to replacing masses above sea level with a thin coating of variable surface density just below sea level. The sea-level surface, in effect, is then a level of isostatic compensation. In a second method the effect of masses above sea level is subtracted from measured gravity, reducing gravity to sea level on the assumption that these masses do not exist. Because they do exist the reduction is artificial. The corresponding level of compensation is at great depth in the earth. In a third method the effects of masses above sea level are subtracted and then reinserted at depth directly underneath, thereby eliminating "roots" of mountains (or "antiroots" of ocean basins) which compensate for topography. The corresponding depth of isostatic compensation lies between that of the two mentioned above; specifically, it lies in the upper mantle, near the Mohorovicic discontinuity or down to a depth of about 100 km, depending on the compensation mechanism assumed. Comparable reductions apply for measurements made at sea.

The three basic reduction methods are the free-air reduction, from which free-air anomalies are obtained; the Bouguer reduction which, combined with the free-air reduction and terrain corrections, leads to Bouguer anomalies; and the isostatic reduction, which leads to isostatic anomalies.

Corrections to observed gravity are also made, where needed. These include corrections for moving platforms (principally ships), for terrain, for earth curvature, for earth tides, and for assumed geological structures.

Standard forms of reductions and corrections are applied to observed gravity, g_{obs}, to obtain conventional types of gravity anomalies. Any anomaly can be described as:

$$\Delta g = \left(g_{obs} + \sum_l \delta_l + \sum_n \epsilon_n \right) - \gamma_\Phi \tag{6.1}$$

where $\Sigma_l \delta_l$ is the sum of l reductions and $\Sigma_n \epsilon_n$ the sum of n corrections, and γ_Φ normal gravity at latitude Φ.

GRAVITY REDUCTIONS

Free-air, Bouguer, and isostatic reductions are described below; others are of historic interest.

Free-air reduction

In the free-air, also called the Faye reduction, observed gravity at elevation h is reduced to its sea-level equivalent on the assumption that topographic masses are displaced vertically down to sea level, that is to a variable-density sheet, but are not removed. The reduction is therefore the difference in gravity on the geoid and at elevation h in free air (e.g., as would be obtained in a balloon).

To sufficient accuracy, gravity at sea level on a spherical earth, g_0, is:

$$g_0 = G \frac{M}{r^2} \tag{6.2}$$

in which M is the mass and r the radius of the earth; the corresponding gravity at elevation h, g_h, is:

$$g_h = G \frac{M}{(r + h)^2} \tag{6.3}$$

We express g_h in terms of g_0 by:

$$g_h = g_0 \left[1 - 2\left(\frac{h}{r}\right) + 3\left(\frac{h}{r}\right)^2 + \ldots \right] \tag{6.4}$$

in which the effect of the earth's curvature is usually neglected. The free-air reduction δ_F is:

$$\delta_F = (g_0 - g_h) = 2g_m \left(\frac{h}{r}\right) \left[1 - \frac{3}{2}\left(\frac{h}{r}\right) + \ldots \right] \tag{6.5}$$

in which g_m is a mean value of gravity over the earth.

The second term in eq. 6.5 has a value of approximately 0.3 mgal for an elevation of 2 km; the term can be omitted except for stations at elevations above about 700 m. The commonly applied free-air reduction is:

$$\delta_F = 0.3086\, h \quad \text{mgal} \tag{6.6}$$

where h is in meters. The reduction is added for stations above sea level, where, owing to the greater distance from the earth's center, gravity is smaller.

At high elevations, where the second term in eq. 6.5 becomes appreciable, the free-air reduction can be written:

$$\delta_F = 0.3086\, h - 0.073\, H^2 \quad \text{mgal} \tag{6.7}$$

in which h is in meters and H in kilometers.

Where still greater accuracy is desired, corrections are made for latitude as well as altitude; a formula for this case is (Heiskanen and Vening Meinesz, 1958, p. 54):

$$\delta_F = (0.30877 - 0.00044 \sin^2\Phi)\, h - 0.073\, H^2 \quad \text{mgal} \tag{6.8}$$

in which Φ is the geographic latitude.

Bouguer reduction

Gravity is increased by the attraction of the mass between station elevation and datum (usually sea level). The vertical component of attraction exerted by this mass is the Bouguer reduction, which is subtracted from observed gravity to reduce the value to sea level. The magnitude of the Bouguer reduction is about one-third that of the free-air reduction.

Attractions produced by masses above sea level can be approximated on three different assumptions: the simple Bouguer reduction, in which the mass above sea level is represented by an infinite horizontal slab whose thickness is the height of station above datum; the Bouguer reduction with terrain correlations, in which the top surface of the horizontal slab has the shape of the surrounding terrain; and the expanded Bouguer reduction, in which a slab, whose upper surface represents the actual terrain, has the curvature of the earth.

Simple Bouguer reductions, δ_{BS}, are obtained by subtracting the effect of an infinite horizontal slab from observed gravity. The vertical attraction produced by a horizontal slab of thickness equal to the height of station is obtained by integrating eq. 2.4 (also eq. 7.9a). In cylindrical coordinates an element of volume $d\tau$ is $r\,dr\,dz\,d\theta$; we replace $\cos \zeta$ in eq. 2.4 by (z/q) and the simple Bouguer reduction δ_{BS} for a constant density slab can be written (with origin of coordinates at the station):

$$\delta_{BS} = G\rho \int_0^{2\pi} \int_0^{h} \int_0^{\infty} \frac{zr\,dr\,dz\,d\theta}{(r^2 + z^2)^{3/2}} \tag{6.9}$$

We integrate over the limits to obtain the slab effect:

$$\delta_{BS} = 2\pi G\rho h \tag{6.10}$$

in which h, the slab thickness, is the station elevation or height above datum.

Bouguer reductions (complete with terrain corrections), δ_{Bt}, are obtained by subtracting the effect of an infinite horizontal slab whose upper surface follows the terrain. Corrections are applied to eliminate the effects of terrain, usually computed out to distances where the earth's curvature becomes significant. The reduction δ_{Bt} consists of a simple slab plus a terrain correction ϵ_t to account for the irregularity of the upper slab surface; thus:

$$\delta_{Bt} = \delta_{BS} - \epsilon_t \tag{6.11}$$

Terrain corrections are always added to observed gravity (subtracted from the slab effect). This is seen from the fact that terrain at elevations higher than the station exert an upward attraction on the gravimeter, making observed gravity too low, and terrain lower than the station has already been subtracted out in the applied slab effect (eq. 6.10), and hence must be reinserted.

Terrain corrections can be determined in various ways. The classical procedure is to compute attractions produced by segments of hollow right-circular cylinders, with the station on the cylinder axis (Fig. 6.1). Cylinders are divided into segments of such size that the sum of corrections for each segment, $\Delta\epsilon_t$, are within prescribed limits. The height Δh of each segment is its average elevation above or below that of the station. The terrain correction corresponding to a segment of a hollow cylinder of inner and outer radii r_1 and r_2, subtended angles θ_1 and θ_2 and terrain height Δh can be written (similar to eq. 6.9):

$$\Delta\epsilon_t = G\rho \int_{\theta_1}^{\theta_2} \int_0^{\Delta h} \int_{r_1}^{r_2} \frac{zr\,dr\,dz\,d\theta}{(r^2 + z^2)^{3/2}} \tag{6.12}$$

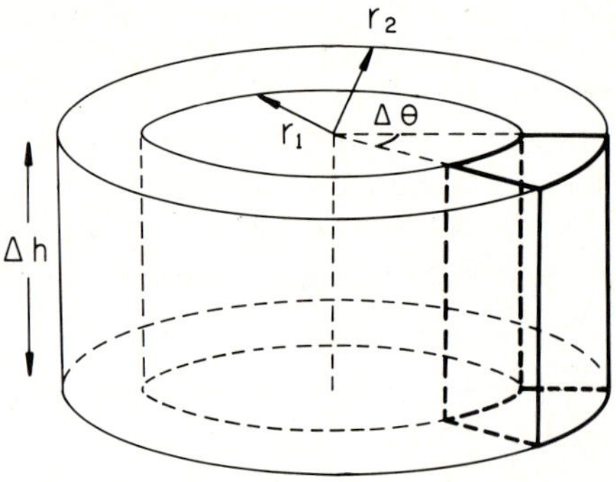

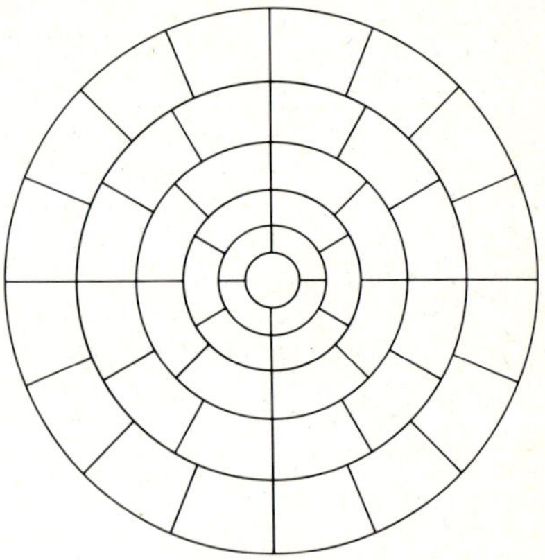

Fig. 6.1. Hollow right-circular cylinder with segment of inner and outer radii r_1 and r_2, included angle $\Delta\theta$, and height Δh above (below) station.

Fig. 6.2. Schematic representation of zones and compartments as may be used to compute terrain corrections at a station positioned in the center of the grid.

which, by integration, becomes:

$$\Delta\epsilon_t = G\rho(\theta_2 - \theta_1)[(r_1^2 + \Delta h^2)^{1/2} - (r_2^2 + \Delta h^2)^{1/2} + (r_2 - r_1)] \qquad (6.13)$$

The total terrain correction, ϵ_t (eq. 6.11), is the sum of individual segment corrections.

Terrain corrections are commonly obtained by dividing a region into segmented circular zones (Fig. 6.2), with the station at the center of the grid. Average terrain heights Δh are assigned to each segment and corresponding values of $\Delta\epsilon_t$ are obtained from eq. 6.13.

A similar procedure can be developed if we use right rectangular prisms of various sizes and densities. Nagy (1966a) developed a digital computer program using a subroutine that calculates the vertical attraction at a station produced by a unit prism (Fig. 6.3). The prism is described by its vertical and horizontal coordinates, if we use equations of the form eq. 7.39. The sum of the attractions of all compartments comprising the terrain provides the terrain correction. The prism technique can be applied to buried bodies as well as to terrain; the method is described in Chapter 7.

Terrain correction can also be calculated by the line-integral methods developed by Hubbert (1948b), if we apply equations of the form eq. 7.63. Fig. 6.4 illustrates several Hubbert curves, which provide estimates of attractions at a station produced by approximately two-dimensional terrain. The curves permit the selection of station locations at which terrain effects are kept below specified values, or are at least minimized. Line-integral methods can also be applied to three-dimensional terrain, if we use eq. 7.68. Digital computers can be programed to correct for two- and three-dimensional terrain. The computations are similar to those used for buried

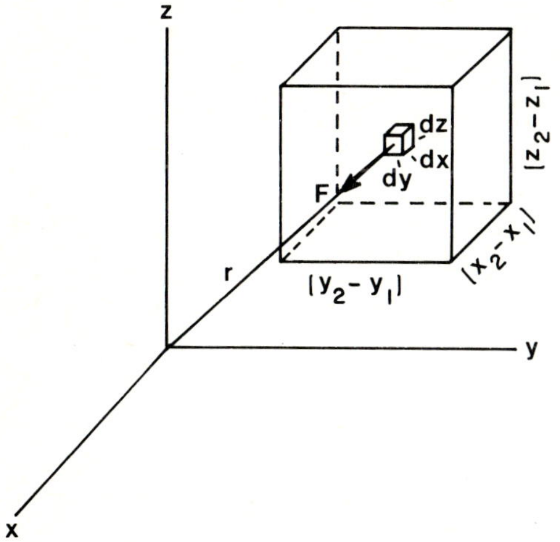

Fig. 6.3. Right-rectangular prism, with volume element, as used by Nagy to compute the attraction force *F* produced by a "unit" building element in calculating terrain corrections and attractions of buried bodies. (Modified after Nagy, 1966a.)

bodies, described in Chapter 7. Whether segments of cylinders, or prisms, or other methods are used to compute ϵ_t, terrain configurations are nearly always obtained from topographic maps.

Expanded Bouguer reductions, δ_{BE}, incorporate the effect of the earth's curvature, ϵ_s, in addition to those of the slab and terrain. The total reduction is:

$$\delta_{BE} = \delta_{BS} - (\epsilon_t + \epsilon_s) \tag{6.14}$$

Expanded Bouguer reductions are not commonly used; hence, procedures for computing curvature effects are not considered. They are similar to those used in the determination of isostatic anomalies, described in the next section.

Isostatic reduction

Isostatic reductions require the determination of attractions produced by topographic masses above sea level, allowing for earth curvature, and then reinserting equivalent masses directly beneath, according to specified procedures. Reductions corresponding to the Pratt concept of isostasy were developed by Hayford (Hayford, 1909; Hayford and Bowie, 1912; also see Heiskanen and Vening Meinesz, 1958, p. 155 ff); those corresponding to the Airy concept were developed by Heiskanen (1938b; also see above reference to Heiskanen and Vening Meinesz). A third reduction was developed by Vening Meinesz, but has not been applied much.

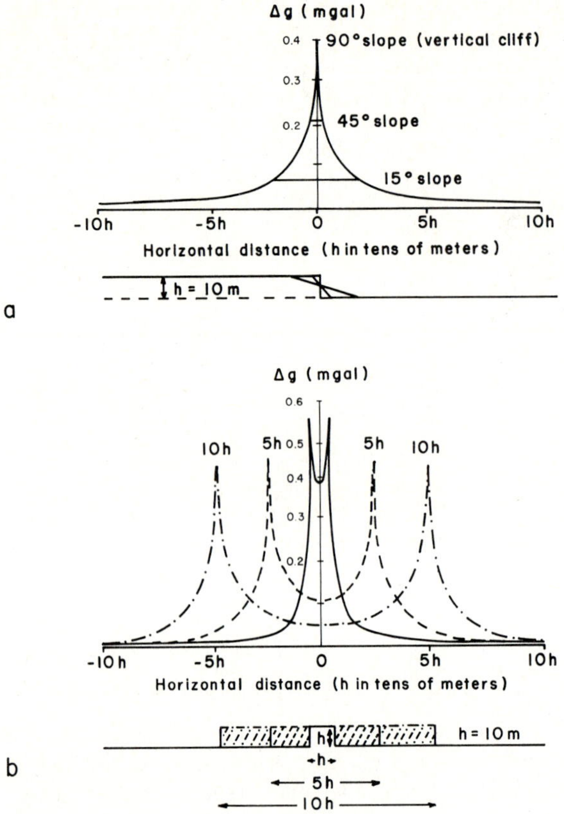

Fig. 6.4. Examples of Hubbert's terrain-correction curves for: a. semi-infinite layer with edge slopes of 90°, 45°, and 15°; b. vertical ridge having a width to height ratios of 1, 5 and 10. Gravity values are based on $\rho = 2.0\ \mathrm{g\ cm^{-3}}$ and a height of 10 m. The gravity value is proportional to the height per 10 m and density per 2 g cm^{-3}. (Modified after Hubbert, 1948b.)

Computations of the attractions produced by topographic masses in both the Pratt-Hayford and Airy-Heiskanen methods use plane formulas out to distances of 166.7 km from the station and spherical formulas for larger distances. Spherical formulas are also used to compute compensation effects at distances greater than 166.7 km.

The topographic-compensation reductions of both the Pratt-Hayford and Airy-Heiskanen methods are of the form:

$$\delta_i = -\delta_{i\mathrm{T}} + \delta_{i\mathrm{C}} \tag{6.15}$$

in which δ_i is the isostatic reduction, $\delta_{i\mathrm{T}}$ the topographic reduction, subtracted from observed gravity, and $\delta_{i\mathrm{C}}$ the compensation reduction, added to gravity.

TABLE 6.I

Hayford zones and compartments (From Heiskanen and Vening Meinesz, 1958, p. 161)

Zone	Outer radius (m)	Number of compartments	Zone	Outer radius (degrees)	Number of compartments
A	2	1	18	1°41′13″	1
B	68	4	17	1°54′52″	1
C	230	4	16	2°11′53″	1
D	590	6	15	2°33′46″	1
E	1280	8	14	3°03′05″	1
F	2290	10	13	4°19′13″	16
G	3520	12	12	5°46′34″	10
H	5240	16	11	7°51′30″	8
I	8440	20	10	10°44′	6
J	12400	16	9	14°09′	4
K	18800	20	8	20°41′	4
L	28800	24	7	26°41′	2
M	58800	14	6	35°58′	18
N	99000	16	5	51°04′	16
O	166700	28	4	72°13′	12
			3	105°48′	10
			2	150°56′	6
			1	180°00′	1

Pratt-Hayford reductions concern methods for determining the topographic mass above sea level in any segmented area and in reducing the density of the underlying material between sea level and depth of compensation (usually taken to be a depth of 113.7 km).

Two columns of the same cross-section have approximately equal weights (Fig. 4.12) at a depth of compensation D below sea level where (eq. 4.25) $\rho_h(D+h)=\rho_n D$, that is, when the density reduction of a column (eq. 4.22) is $\Delta\rho = -(h/D)\rho_h$.

Hayford developed formulas for computing the attraction at a station due to the topography and corresponding isostatic compensation for zones throughout the earth. Table 6.I shows the Hayford zones and compartments. Topographic attractions produced by compartments out to zone O (outer radius 166.7 km) ignore earth curvature (except for minor correction terms in the formulas). The formulas can be obtained similarly to eq. 6.12, where $\bar{h}$ is the mean elevation of a compartment and h that of the station. Thus:

$$\Delta\delta_{iT} = G\rho_h \int_{\theta_1}^{\theta_2}\int_h^{\bar{h}}\int_{r_1}^{r_2} \frac{zr\,\mathrm{d}r\,\mathrm{d}z\,\mathrm{d}\theta}{(r^2+z^2)^{3/2}} \tag{6.16}$$

which becomes:

$$\Delta\delta_{iT} = G\rho_h(\Delta\theta)[(r_1^2+\bar{h}^2)^{1/2}-(r_1^2+h^2)^{1/2}-(r_2^2+\bar{h}^2)^{1/2}+(r_2^2+h^2)^{1/2}] \tag{6.17}$$

The station height is small compared to outer radius r_2, so that h can be taken as zero without significant error. The equation then becomes:

$$\Delta\delta_{iT} = G\rho_h(\Delta\theta)[(r_1^2 + \bar{h}^2)^{1/2} - (r_2^2 + \bar{h}^2)^{1/2} + (r_2 - r_1)] \qquad (6.18)$$

The mass deficiency in the same compartment, according to the Pratt-Hayford procedure, is $(\Delta\delta_{iC})_{PH}$, obtained in a manner similar to eq. 6.16:

$$(\Delta\delta_{iC})_{PH} = -G\rho_h\left(\frac{\bar{h}}{D}\right)(\Delta\theta)[(r_1^2 + D^2)^{1/2} - (r_1^2 + h^2)^{1/2} - (r_2^2 + D^2)^{1/2} + (r_2^2 + h^2)^{1/2}] \qquad (6.19)$$

The station height is again small compared with r_2, giving:

$$(\Delta\delta_{iC})_{PH} = -G\rho_h\left(\frac{\bar{h}}{D}\right)(\Delta\theta)[(r_1^2 + D^2)^{1/2} - (r_2^2 + D^2)^{1/2} + (r_2 - r_1)] \qquad (6.20)$$

Similar reductions are made in oceanic provinces, where the ocean water is assumed to be compensated by heavier masses at depth. The density of the mass deficiency ρ' resulting from the presence of ocean water is $\rho' = (\rho_h - 1.027)$ (eq. 4.27), where 1.027 g cm^{-3} is the density of sea water and ρ_h the density of rock replacing the water of depth h'. The density of the mass excess (antiroot), $\Delta\rho'$, to depth D, compensating for water depth h', is:

$$\Delta\rho' = \frac{h'}{D}\rho' \qquad (6.21)$$

A comparison of eqs. 4.22 and 6.21 shows that a mass deficiency $\Delta\rho$ beneath mountains (roots of mountains) and a mass excess $\Delta\rho'$ beneath oceans (antiroots) are related by:

$$\Delta\rho' = \left(\frac{\rho_{h'} - 1.027}{\rho_{h'}}\right)\Delta\rho \qquad (6.22a)$$

For $\rho_{h'} = 2.67$ g cm^{-3}:

$$\Delta\rho' = 0.615\ \Delta\rho \qquad (6.22b)$$

which shows that the magnitude of the mass excess in an ocean compartment is about 61% of the mass deficiency in a continental compartment.

A comparison of the magnitudes of the topographic and compensation effects is obtained by an examination of eqs. 6.18 and 6.20. Topographic effects are relatively larger near the station and compensation effects are relatively larger at greater distances; Fig. 6.5 shows that the two are about the same at a distance of 15 km (zone K) when $D = 113.7$ km.

Plane formulas (eqs. 6.18 and 6.20) are used to determine topographic and compensation effects to zone L (29 km) and, with small corrections for

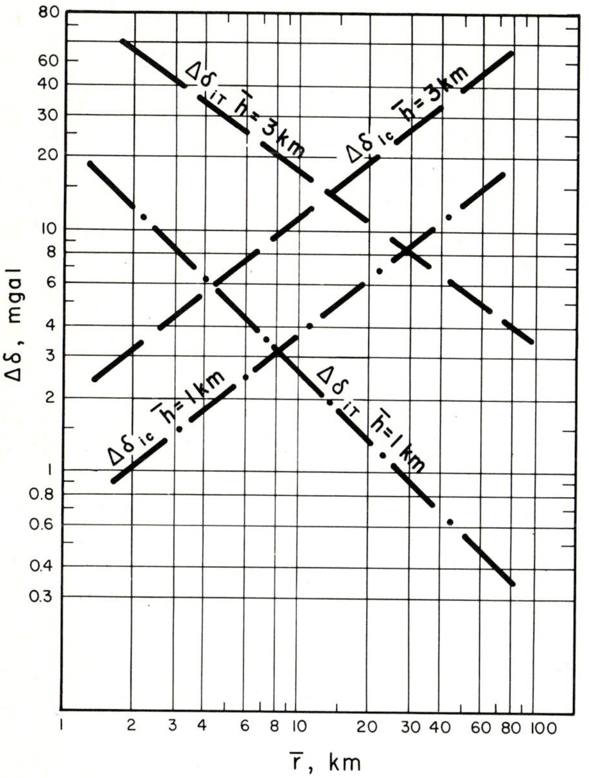

Fig. 6.5. Curves showing variations in vertical attractions of $\Delta\delta_{iT}$, produced by topographic masses in the shape of a ring with heights of 1 km and 3 km and density 2.67 g cm^{-3}, and in $\Delta\delta_{ic}$, produced by a corresponding compensating mass, according to the Pratt-Hayford concept.

earth curvature, to zone O (166.7 km). At greater distances (zones 18 to 1), spherical formulas are used to compute both topographic and compensation effects.

Following the methods of Heiskanen and Vening Meinesz (1958, p. 162), spherical formulas are obtained as follows. In Fig. 6.6 the vertical attraction dg_c due to compensation at P produced by an element of mass dM, is:

$$dg_c = G \frac{\sin \zeta}{q^2} dM \tag{6.23}$$

in which $q^2 = l^2 + f^2 - 2lf \sin(\psi/2)$ and ψ is the angular distance between P and dM, l the cord corresponding to angle ψ, f the depth below station of the compensating mass dM, and r the radius of the earth. Fig. 6.6 shows

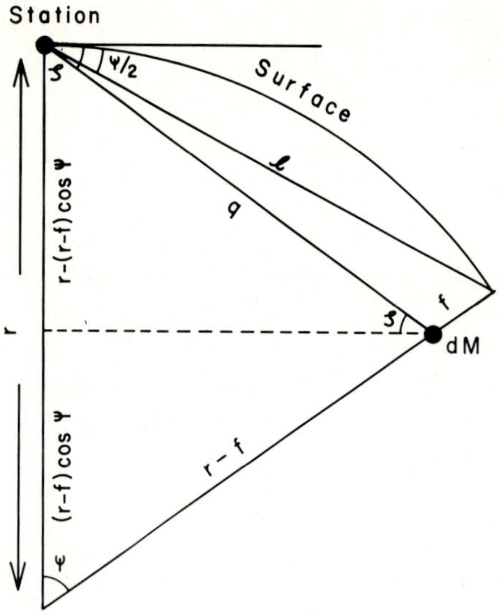

Fig. 6.6. Diagram showing quantities used in eqs. 6.25 to 6.31 to compute effects of earth curvature on topographic and on compensating masses acting at a gravity station.

that:

$$\sin \zeta = \frac{r-(r-f)\cos\psi}{q} \tag{6.24}$$

and also that:

$$r-(r-f)\cos\psi = r(1-\cos\psi) + f\cos\psi \tag{6.25}$$

As:

$$r(1-\cos\psi) = l\sin(\psi/2) \doteq \frac{l^2}{2r} \tag{6.26}$$

eq. 6.24 can be written:

$$\sin\zeta = \frac{\left(\dfrac{l^2}{2r}\right) + f\cos\psi}{q} \tag{6.27}$$

and eq. 6.23 becomes:

$$\mathrm{d}g_c = \frac{G\left[\left(\dfrac{l^2}{2r}\right) + f\cos\psi\right]\mathrm{d}M}{\left[l^2 + f^2 - 2lf\sin\left(\dfrac{\psi}{2}\right)\right]^{3/2}} = GE_c\mathrm{d}M \tag{6.28}$$

in which E_c is the compensating effect, which is:

$$E_c = \frac{\left(\frac{l^2}{2r}\right) + f \cos \psi}{\left[l^2 + f^2 - 2lf \sin \left(\frac{\psi}{2}\right)\right]^{3/2}} \tag{6.29}$$

E_c provides usable approximate values of compensation in each of zones 18 to 1 (Table 6.I), so that integration of eq. 6.28 is unnecessary.

Comparable spherical formulas can be developed for topographic corrections. The vertical component of attraction dg_T at P produced by an element of topographic mass dM at height h can be developed, similar to the case of a compensating element of mass; the formula is:

$$dg_T = G \frac{\left[\left(\frac{l^2}{2r}\right) - h \cos \psi\right] dM}{\left[l^2 + h^2 + 2lh \sin \left(\frac{\psi}{2}\right)\right]^{3/2}} = GE_T\, dM \tag{6.30}$$

in which:

$$E_T = \frac{\left(\frac{l^2}{2r}\right) - h \cos \psi}{\left[l^2 + h^2 + 2lh \sin \left(\frac{\psi}{2}\right)\right]^{3/2}} \tag{6.31}$$

Topographic and compensating masses are equal in any compartment or ring. The sum of topographic and isostatic reduction, E_R, in a cylindrical ring is obtained from:

$$2\pi G\rho E_R = 2\pi G\rho(E_T - E_c) \tag{6.32}$$

Hayford and Bowie (1912) published tables for the spherical terms E_T, E_c, and E_R for values of $1.5° < \psi < 180°$. Values of E_R decrease rapidly as ψ increases, becoming quite small for $\psi > 45°$.

Heiskanen pointed out (Heiskanen and Vening Meinesz, 1958, p. 164) that because Hayford computed effects produced by compartments within zones rather than the attraction due to an entire zone, errors up to several milligals can be introduced. The Hayford tables should be used with awareness of this potential error.

Another modification to the Hayford tables now commonly made corrects for the fact that Hayford's computed topographic effects in zones A to O are based on elevations above sea level rather than heights above station. Computations are simplified by computing effects of compartmental heights above station, that is, computing terrain corrections and adding them to the simple Bouguer reductions. Bullard (1936) developed a formula based on the simple Bouguer slab (δ_{BS}), a correction ϵ_s for earth curvature outside zone O, and the terrain correction, which has the form:

$$\delta_{BLD} = \delta_{BS} - (\epsilon_s + \epsilon_t) \tag{6.33}$$

In the presently applied Hayford reduction, effects of topographic masses and corresponding compensations are determined for each zone (or compartment), so that the reduction δ_{BLD} includes the complete Bouguer reduction.

Airy-Heiskanen reductions are obtained by subtracting the effect of attractions produced by masses above sea level and replacing them for assumed roots beneath mountains and antiroots beneath oceans. A root of thickness t extends from a depth of compensation T (usually the Mohorovicic discontinuity in continental areas) to depth $T + t$, and under oceanic zones the antiroot extends from depths $T - t$ to T. Values for t can be obtained from eqs. 4.31 and 4.32.

The effect of Airy-Heiskanen isostatic compensation $(\delta_{iC})_{AH}$ of continental zones is obtained by integrating an equation of the form in 6.16 over cylindrical rings. This gives:

$$(\Delta\delta_{iC})_{AH} = -2\pi G\Delta\rho \, \{[r_1{}^2 + (T+t)^2]^{1/2} - (r_1{}^2 + T^2)^{1/2} + (r_2{}^2 + T^2)^{1/2} - [r_2^2 + (T+t)^2]^{1/2}\} \tag{6.34a}$$

The corresponding effect over oceanic zones is:

$$(\Delta\delta_{iC})_{AH} = 2\pi G\Delta\rho \, \{(r_1^2 + T^2)^{1/2} - [r_1^2 + (T-t)^2]^{1/2} + [r_2^2 + (T-t)^2]^{1/2} - (r_2^2 + T^2)^{1/2}\} \tag{6.34b}$$

Compensations in zones 18 to 1 are obtained by the use of spherical formulas, that is, eq. 6.29 for E_c, eq. 6.31 for E_T, and eq. 6.32 for E_R. Effects of topographic masses are based on those previously computed by Hayford for zones in Table 6.I.

Heiskanen (1938b) published tables of computed topographic-isostatic reductions, based on the Hayford zones, for assumed crustal thicknesses of $T = 20$, 30, 40, and 60 km. The tables are for stations at sea-level elevations; additional tables give corrections for stations at elevations of 1,000, 2,000, 3,000, and 4,000 m.

Vening Meinesz reductions are modifications of the Airy-Heiskanen method, in which Vening Meinesz's system of regional isostatic compensation is used. The procedure, developed in 1931, includes three sets of tables. One gives regional isostatic reductions based on specific assumptions regarding vertical compensating processes. The second gives isostatic reductions for a homogeneous crust of density 2.67 g cm^{-3} on a subcrust of density 3.27. The third provides means for investigating isostatic compensation as indicated by two-dimensional topographic features, such as across a continental margin or a mountain range. The Vening Meinesz method is not commonly used, so reduction procedures will not be described here; they are given by Vening Meinesz (1931, 1939, and 1941a).

World-wide isostatic reduction computational procedures have been developed which substantially decrease the time required to compute the above-described reductions applied to individual stations. These procedures are described by Heiskanen and Vening Meinesz (1958, p. 175 ff). One procedure is to construct isostatic-reduction maps, which give reductions for the zone groups according to the Pratt-Hayford, the Airy-Heiskanen, or the Vening Meinesz systems. The maps can be used for topographic-isostatic reductions which are added to the combined free-air and Bouguer reductions out to distances of 166.7 km (zone O). The maps are considered correct for zones 10 to 1, but may contain errors in the nearer zones.

Heiskanen (1953) introduced a method using digital computers in an attempt to reduce the time and work required in making isostatic reductions. He replaced Hayford zones with squares (actually spherical trapezoids) of varying sizes, such as $5' \times 5'$, $10' \times 10'$, $0.5° \times 0.5°$, $1° \times 1°$, $2° \times 2°$, $5° \times 5°$, etc. Mean elevations or depths of the squares are estimated and used for zones A through 11 (Table 6.I). The method can also be applied for zones 10 to 1, although isostatic-reduction maps provide satisfactory values for these zones. Heiskanen and Vening Meinesz (1958, p. 181) describe procedures for computing reductions for the squares.

Other reduction methods

Two other reduction methods developed at the turn of the century have been replaced by isostatic reductions; they are now of historic interest only.

Helmert condensation reductions attempt to provide a more realistic geoid than is obtained from Bouguer reductions; a geoid obtained from Bouguer anomalies can be as much as 500 m in error. Stokes, and later Helmert (1909), introduced the condensation reduction, whereby topographic masses are transferred to a very thin coating inside the geoid. The objective was to obtain the shape of the true geoid from gravity measurements. Helmert's "ideal geoid" differs by less than 3 m from the actual geoid. The method is described by Helmert (1909) and summarized by Heiskanen and Vening Meinesz (1958, p. 156).

Rudzki inversion reductions are based on moving topographic masses vertically so as not to change the shape of the geoid. The concept is similar to the condensation-reduction method. Procedures are described by Rudzki (1905) and summarized by Heiskanen and Vening Meinesz (1958, p. 159).

GRAVITY CORRECTIONS

Corrections include effects of moving platforms, terrain, earth curvature, known geologic structures, and earth tides. Terrain and curvature corrections are applied as part of the Bouguer and isostatic reductions, and are described in the sections on Bouguer reductions and isostatic reductions.

Moving-platform corrections

Moving platforms, such as ships and aircraft, encounter extraneous accelerations. Some accelerations can be corrected and others can be eliminated by appropriate instruments.

The extraneous accelerations include: (1) the Eötvös effect, which gives a spurious gravity measurement because a moving meter experiences modified centripetal accelerations about the earth's axis; (2) horizontal and vertical accelerations produced by three-dimensional platform motions; and (3) cross-coupling accelerations experienced by beam-type gravimeters, usually when rigidly attached to a stabilized platform.

The Eötvös correction, ϵ_{EC}, when added algebraically to gravity measured on a moving platform, gives the corresponding stationary value. The centripetal acceleration a_E in an easterly direction, counteracting the effect of the earth's rotation about its axis (Fig. 6.7), is:

$$a_E = \frac{V_\phi^2}{r \cos \phi} \tag{6.35}$$

in which ϕ is the (geocentric) latitude, V_ϕ the linear velocity due to rotation on the earth's surface at latitude ϕ, r the earth's radius, and $r \cos \phi$ the perpendicular distance from the earth's axis. The outward (vertical)

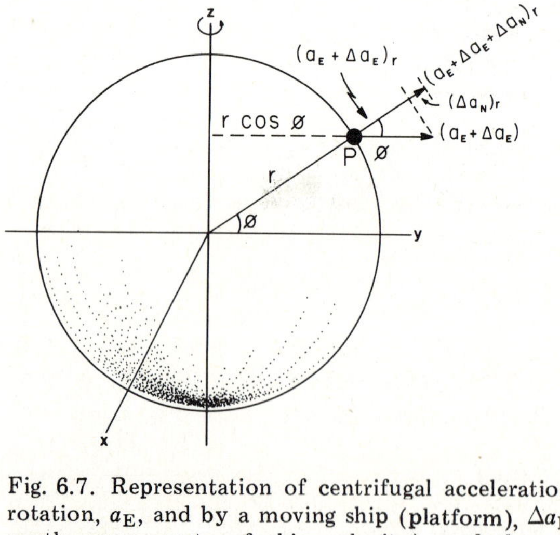

Fig. 6.7. Representation of centrifugal acceleration at point P produced by the earth's rotation, a_E, and by a moving ship (platform), Δa_E and Δa_N (corresponding to east and north components of ship velocity), and the respective components in the vertical direction.

component of acceleration is:

$$(a_E)_r = \frac{V_\phi{}^2}{r} \tag{6.36}$$

The centripetal acceleration experienced on a ship (platform) heading east with speed v_E is:

$$(a_E + \Delta a_E) = \frac{(V_\phi + v_E)^2}{r \cos \phi} \tag{6.37}$$

and the outward component of acceleration is:

$$(a_E + \Delta a_E)_r = \frac{(V_\phi + v_E)^2}{r} \tag{6.38}$$

Similarly, the centripetal acceleration about an equatorial axis experienced by a ship heading north with speed v_N is:

$$(\Delta a_N)_r = \frac{v_N^2}{r} \tag{6.39}$$

The total outward acceleration experienced at ship speed v is the sum of the last two formulas; thus:

$$(a_E + \Delta a_E + \Delta a_N)_r = \frac{V_\phi^2 + 2V_\phi v_E + v^2}{r} \tag{6.40}$$

in which $v^2 = v_E^2 + v_N^2$. Gravity on the platform, which is reduced by the effect of the ship velocity, must be added to observed gravity. The Eötvös correction is, accordingly:

$$\epsilon_{EC} = \frac{2V_\phi v_E + v^2}{r} \tag{6.41}$$

We substitute $V_\phi = r\omega \cos \phi$, where ω is the earth's angular velocity, and $v_E = v \sin \alpha$ to obtain:

$$\epsilon_{EC} = 2\omega v \cos \phi \sin \alpha + \frac{v^2}{r} \tag{6.42}$$

where α is the ship course (measured clockwise from north). When ship speed is given in knots, the equation is usually written (illustrated in Fig. 6.8):

$$\epsilon_{EC} = 7.5\, v \cos \phi \sin \alpha + 0.004\, v^2 \text{ mgal} \tag{6.43}$$

At ship speeds less than about 10 knots, which is commonly the case, the

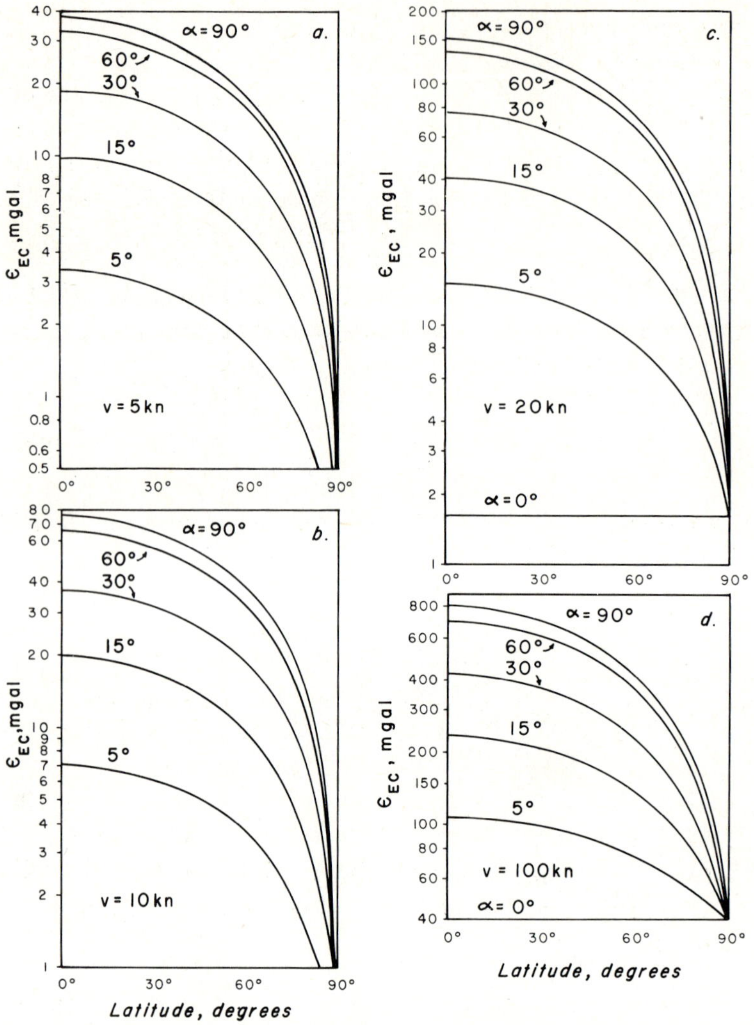

Fig. 6.8. Variations in Eötvös correction ϵ_{EC} as functions of latitude for ship speeds v of 5, 10, 20 and 100 knots and ship's course α of 0°, 5°, 15°, 30°, 60°, and 90°.

second term in eq. 6.43 is smaller than 1/2 mgal and is omitted in computations where accuracies are less than a milligal.

Easterly ship speeds increase the centripetal acceleration acting on the meter, and westerly speeds decrease it; hence, the Eötvös correction is added for easterly and subtracted for westerly components of ship speed.

Errors in ship speed and course produce Eötvös errors. We differentiate eq. 6.43 to obtain:

$$d\epsilon_{EC} = \frac{\partial \epsilon_{EC}}{\partial v} dv + \frac{\partial \epsilon_{EC}}{\partial \alpha} d\alpha + \frac{\partial \epsilon_{EC}}{\partial \phi} d\phi \qquad (6.44a)$$

which becomes:

$$d\epsilon_{EC} = 7.5 (\cos\phi \sin\alpha\, dv + v \cos\phi \cos\alpha\, d\alpha - v \sin\phi \sin\alpha\, d\phi) + 0.008\, v dv \text{ mgal} \qquad (6.44b)$$

Fig. 6.9 illustrates the magnitude of this error. At middle latitudes an error of 0.2 knot in speed at 10 knots in an east—west direction produces an

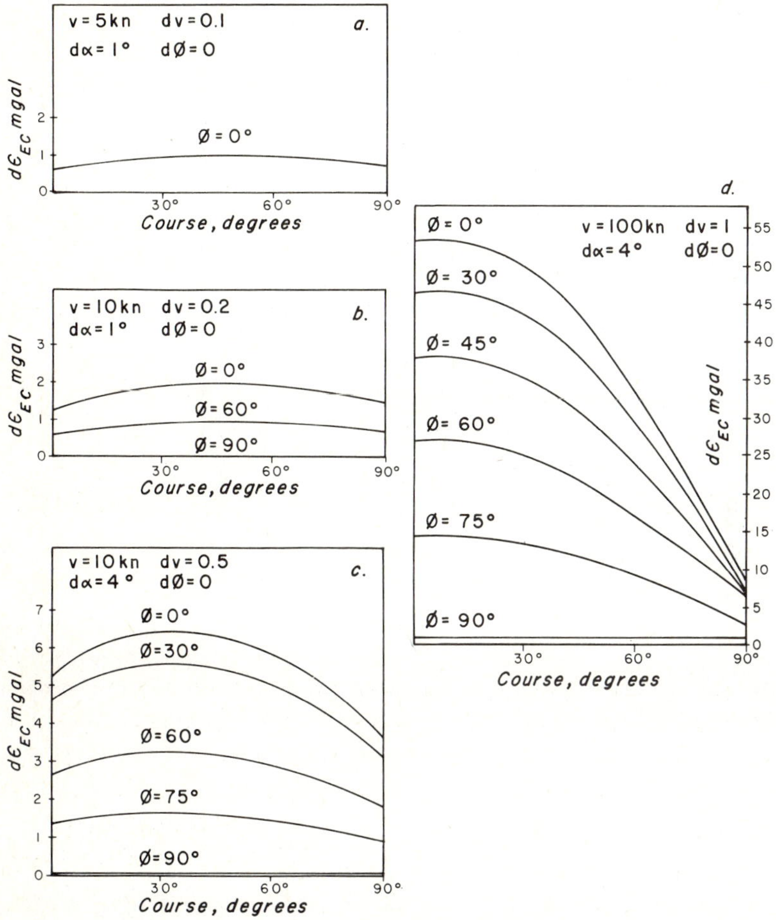

Fig. 6.9. Variations in the error in Eötvös correction for indicated errors in ship speed (dv) or in course ($d\alpha$) for ship speeds of 5, 10 or 100 knots (error in latitude position is assumed to be negligible).

Eötvös error of approximately 1 mgal. An error of 1° in course for a north—south heading at 10 knots also produces an Eötvös error of about 1 mgal. The last two terms in eq. 6.44b usually produce negligible errors at normal ship speeds.

Attempts have been made to measure gravity in helicopters and in fast-flying aircraft, but they have not been satisfactory because of excessive uncertainties in measurements. Eötvös corrections are large at aircraft speeds. At 200 knots corrections range up to 1,600 mgal. A 1% (2 knots) error in plane speed produces Eötvös errors up to 18 mgal, and a 1° course error produces errors up to 25 mgal. Also important, however, is the effect of vertical craft accelerations with periods longer than the averaging period in which continuous gravity measurements are made. These long-period accelerations cannot be readily corrected and they produce erroneous gravity data.

Horizontal accelerations essentially do not affect gravity measurements because instruments are designed to be insensitive to these motions, but they are not always removed completely. The accelerations are produced by ocean waves, fishtailing of the ship (platform), or because the ship does not maintain constant speed and course (as may result when crossing ocean currents). Ocean waves produce ship periods ranging from 4 to 15 sec and sometimes longer. Ship fishtailing commonly has periods of 1/2 to several minutes; it results in small horizontal accelerations which act on platforms. This has been shown theoretically by LaCoste (1967) for sinusoidal ship motions. The small effect that fishtailing has on gravity measurements has also been repeatedly observed at sea (e.g., Dehlinger et al., 1966). It has been noted that the largest fishtailing occurs in following seas, when observed errors in gravity are usually quite small, and fishtailing is usually small when the ship heads into the sea, conditions which often produce the largest measurement errors.

Effects of horizontal accelerations acting on the meter-sensing element are essentially eliminated in stabilized-platform operations; they are measured and then corrected for in gimbal-suspended meters. Horizontal-acceleration corrections applied to a gimbal system are the classical Browne (1937) correction (eq. 6.45b). Both the stabilized platform and gimbal-suspended systems have inherent limitations, however, so that measurements have uncertainties of about 1 mgal or more. Platform systems are used almost universally because they are relatively simple to maintain, provide a high accuracy in moderate seas, and permit measurements in rather heavy seas. Gimbal-suspended systems have not been used aboard surface ships for a decade or more.

Gimbal-suspended systems were designed and constructed by LaCoste and Romberg, initially for use in submarines and later aboard surface ships. In theory the meters measure total gravity along the instantaneous axis and therefore experience negligible cross-coupling accelerations. By contrast, beam-type meters are rigidly mounted on stabilized platforms and can experience substantial cross-coupling accelerations.

Horizontal accelerations acting on a gimbal-suspended meter are measured by two perpendicular horizontal accelerometers. The LaCoste and Romberg system used Vening Meinesz-type horizontal inertial bars with a period of one or two minutes. Horizontal accelerations measured by the accelerometers are subtracted from meter measurements by analog computers.

Gravity in a gimbal-suspension system can be obtained, as illustrated in Fig. 6.10, from acceleration g_a measured along the meter axis and average swing angle $\bar{\theta}$, which is a measure of horizontal acceleration. Averaged gravity is obtained from:

$$\bar{g} = \bar{g}_a \ \overline{\cos\theta} = \bar{g}_a \left(1 - \frac{\bar{\theta}^2}{2} + \ldots\right) \tag{6.45a}$$

for small $\bar{\theta}$. This formula, except for fourth-order terms, is the same as the Browne correction (Browne, 1937):

$$g_a \doteq g \left(1 + \frac{\bar{\theta}^2}{2}\right) \tag{6.45b}$$

The LaCoste and Romberg gimbal-suspension system used analog computers to obtain gravity (eq. 6.45a) for measured $\bar{\theta}$ less than about 2°, which

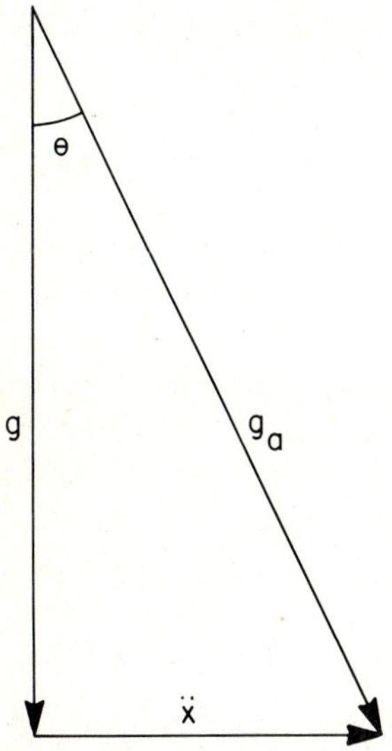

Fig. 6.10. The acceleration measured by a gravimeter free to measure along its sensing axis is g_a, the vector sum of gravity g and the horizontal acceleration $\ddot{x}$; θ is the angle of the meter axis from the vertical.

corresponds to horizontal accelerations of about 50,000 mgal. These corrections proved to be adequate when accelerations are within design limitations. Gimbal systems were not built to correct for larger accelerations; the simpler-operating stabilized platforms provide adequate measurements at large accelerations.

In platform systems slight off-leveling of the platform results in horizontal accelerations which can affect the sensing element. LaCoste (1967) showed that the error in gravity, ϵ_{OL}, produced by off-leveling is identical (to second-order terms) to that produced by an error in the measurement of swing angle $\bar{\theta}$ in a gimbal system. In both cases:

$$\epsilon_{OL} = -[(\overline{\Delta\theta})\ddot{x} + \tfrac{1}{2}g(\overline{\Delta\theta})^2] \quad (6.46)$$

where $\overline{\Delta\theta}$ is the averaged error in off-leveling or in swing angle (eq. 6.45a).

Platforms are constructed in such a way that off-leveling errors are less than a few minutes of arc for ships traveling in a straight line (LaCoste, 1967). Off-leveling errors are thus small except at large ship accelerations. Corrections for these errors can be applied by attaching horizontal accelerometers on a platform to measure the off-leveling accelerations; computers then calculate and remove the off-leveling error from measured gravity.

Vertical accelerations of the ship (platform), which the meter cannot differentiate from gravity, can exceed 0.1 g. As the ship oscillates about "sea level", the effect of vertical accelerations $\ddot{z}$ can be eliminated by averaging the continuously measured gravity for a time T, which is long compared to ship period. The meter is subjected to vertical accelerations, so that gravity is expressed by:

$$g = \frac{1}{T}\int_0^{T}(g + \ddot{z})\,dt \quad (6.47)$$

For the condition that there is no net change in vertical ship accelerations over T, that is, no vertical accelerations act which have periods nearly equal to T or longer:

$$\frac{1}{T}\int_0^{T}\ddot{z}\,dt \doteq 0$$

Vertical accelerations acting on the platform are removed by averaging or filtering for times T, so that g can be measured reliably. Filtering periods of 1—5 min (and even longer) are commonly used.

In practice, two factors complicate the averaging of vertical accelerations. One results from nonlinearity in the meter or recording system, which means that the beam system does not swing about its null position, so that the average of swings is not the true null value. Nonlinearities are eliminated through careful instrument design and adjustment. Corrections for nonlinearities can be determined by laboratory-determined empirical relationships between gravity and applied accelerations (vertical or horizontal). Such corrections have been applied to sea measurements by Dehlinger et al. (1966), using laboratory-determined empirical formulas provided by the meter manufacturer.

The second error results from inadequately filtering out vertical accelerations when gravity varies over the filtering period T, an effect analyzed by Talwani (1966). Usually the distortion of a gravity-anomaly wavelength caused by filtering can be computed when we record digitally. Talwani (1970) developed a correction based on the difference in vertical ship velocity at the beginning and end of a filter interval. With sharp cut-off filters, no correction is required.

Vertical accelerations with periods longer than time T are not averaged out by the meter system, making observed gravity unreliable. For example, measurements obtained during the passage of a tsunami wave (period of about 1 hour) will be distorted. Similarly, helicopter or aircraft gravity measurements have not been of ship-measurement quality primarily because

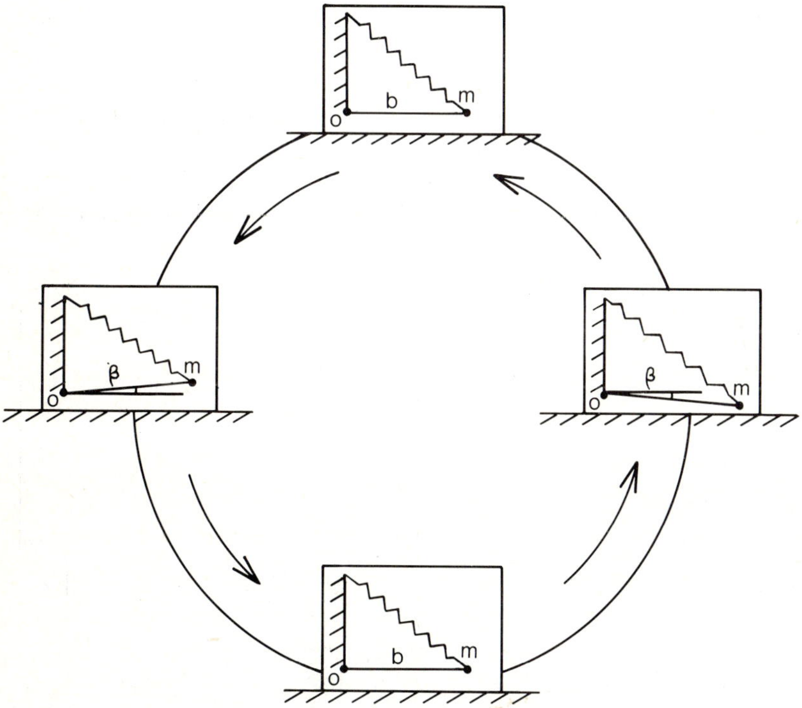

Fig. 6.11. Illustration of upward motion of a beam-type sensing element with lever arm b when the ship descends, and its downward motion while the ship comes back up, which results in cross-coupling of horizontal and vertical accelerations acting on the beam. The cross-coupling effect is greatest when ship motion is circular.

accelerations produced by long-period vertical motions are difficult to correct.

Cross-coupling of horizontal or vertical components of ship acceleration occurs in beam-type meters mounted on stabilized platforms when horizontal and vertical components of ship acceleration have equal periods and have a phase difference; the effect is to produce errors in measured gravity, called cross-coupling errors. Fig. 6.11 illustrates beam positions of a highly damped sensing element over a wave period. In the "12-o'clock" and "6-o'clock" positions, the rates of beam displacement are a measure of true gravity (see eq. 5.18c). During the up part of the cycle (3-o'clock position in the figure), the rate of beam displacement is greater than in the 12- and 6-o'clock positions; apparent gravity is then greater than true gravity. During the down part of the cycle (9-o'clock position) apparent gravity is too low. The resultant errors in gravity do not usually average out over a wave period.

Harrison (1960) and LaCoste (1967) analyzed cross-coupling in over-damped beam-type meters. The acceleration acting perpendicular to the meter beam (Fig. 6.11) is $(g + \ddot{z})$ when the beam is horizontal. When the beam makes an angle β with the horizontal, however, the acceleration acting perpendicular to the beam consists of a component of the vertical acceleration, which is $(g + \ddot{z})\cos\beta$, and a component of the horizontal acceleration, which is $-\ddot{x}\sin\beta$ (the negative sign indicating that the component due to the horizontal acceleration acts in the opposite direction to that due to the vertical component).

The torque L acting on the mass m (Fig. 6.11) about a pivotal point O, where the lever arm of length b makes an angle β with the horizontal, is given by:

$$L = mb[(g + \ddot{z})\cos\beta - \ddot{x}\sin\beta] \tag{6.48}$$

If we do not include the horizontal term in eq. 6.48 and instead replace gravity by g_e, where $g - g_e = \Delta g_{cc}$, the cross-coupling error, we write in place of eq. 6.48,

$$L = mb(g_e + \ddot{z})\cos\beta \tag{6.49a}$$

which can be written:

$$L = mb(g - \Delta g_{cc} + \ddot{z})\cos\beta \tag{6.49b}$$

We divide eqs. 6.48 and 6.49b by $mb\cos\beta$ and obtain for Δg_{cc}:

$$\Delta g_{cc} = \ddot{x}\tan\beta \doteq \ddot{x}\beta \tag{6.50}$$

since β is made to be small in beam meters.

The beam oscillation, for any one period, can be written:

$$\beta = \beta_0 + \beta_1 \sin(\omega_v t + \chi) \tag{6.51a}$$

in which $\beta_0 \sim 0$, χ is a phase angle and ω_v is the angular velocity of beam motion, which approximately equals the angular frequency of the vertical accelerations that drive the beam.

Horizontal accelerations, for any one period acting on the mass m, can similarly be written:

$$\ddot{x} = \ddot{x}_0 \sin(\omega_H t + \xi) \quad (6.51b)$$

in which ξ is a phase angle and ω_H is the angular frequency of the horizontal accelerations acting on the beam.

The cross-coupling error is, upon expanding the functions of sums of angles and substituting into eq. 6.50:

$$\Delta g_{cc} = \ddot{x}_0 \beta_0 (\sin \omega_H t \cos \xi + \sin \xi \cos \omega_H t) + \ddot{x}_0 \beta_1 (\sin \omega_H t \cos \xi + \sin \xi \cos \omega_H t)(\sin \omega_v t \cos \chi + \sin \chi \cos \omega_v t) \quad (6.52)$$

And the average cross coupling $\overline{\Delta g}_{cc}$ over a wave period T can be written:

$$\overline{\Delta g}_{cc} = \frac{1}{T}\int_0^T \ddot{x}_0 \beta_0 \cos \xi \sin \omega_H t \, dt + \frac{1}{T}\int_0^T \ddot{x}_0 \beta_0 \sin \xi \cos \omega_H t \, dt$$
$$+ \frac{1}{T}\int_0^T \ddot{x}_0 \beta_1 (\cos \chi \cos \xi \sin \omega_V t \sin \omega_H t + \sin \chi \sin \xi \cos \omega_v t \cos \omega_H t) \, dt$$
$$+ \frac{1}{T}\int_0^T \ddot{x}_0 \beta_1 (\sin \chi \cos \xi \cos \omega_v t \sin \omega_H t + \cos \chi \sin \xi \sin \omega_V t \cos \omega_H t) \, dt \quad (6.53)$$

The first two integrals in eq. 6.53 are zero (over a wave period). The other integrals also vanish, except when $\omega_H = \omega_v$, in which case:

$$\Delta g_{cc} = \frac{\ddot{x}_0 \beta_1}{T} \int_0^T \cos \chi \cos \xi \sin^2 \omega_v t dt + \frac{\ddot{x}_0 \beta_1}{T} \int_0^T \sin \chi \sin \xi \cos^2 \omega_v t dt$$
$$+ \frac{\ddot{x}_0 \beta_1}{T} \int_0^T \sin (\chi + \xi) \sin \omega_v t \cos \omega_v t dt \quad (6.54)$$

The last integral in eq. 6.54 is seen to vanish even when $\omega_H = \omega_v$. We obtain, upon integrating the first two integrals:

$$\overline{\Delta g}_{cc} = \tfrac{1}{2} \ddot{x}_0 \beta_1 (\cos \chi \cos \xi + \sin \chi \sin \xi) \quad (6.55a)$$

$$\overline{\Delta g}_{cc} = \tfrac{1}{2} \ddot{x}_0 \beta_1 \cos(\chi - \xi) \quad (6.55b)$$

Average cross-coupling errors in beam-type meters are seen to occur when horizontal and vertical accelerations acting on the beam have identical periods. The magnitude of the error is proportional to the average maximum horizontal acceleration and the average maximum beam oscillation. The error is largest when the phase difference $\chi - \xi$, between the beam motion and horizontal acceleration, is 0 or π.

We can write:

$$(\chi - \xi) = (\chi - \epsilon) + (\epsilon - \xi) \tag{6.56}$$

in which ϵ is the phase angle in vertical accelerations, similar in meaning to ξ in eq. 6.51b. In highly overdamped meters the damping coefficient (defined in eq. 5.15) is very large, such that the beam oscillation β_1 is approximately proportional to the vertical acceleration $\ddot{z}$. The phase difference $(\chi - \epsilon)$ is then approximately $\pi/2$. When $\ddot{z}$ and $\ddot{x}$ have a phase difference of $(\epsilon - \xi) = \pm\,\pi/2$ (which is the case for circular wave motion), $(\chi - \xi) = 0$ or π. Then $\Delta g_{cc} = \pm\frac{1}{2}\ddot{x}_0\beta_1$, its maximum value (which can be either positive or negative). As the phase difference between $\ddot{z}$ and $\ddot{x}$ becomes smaller or larger than $\pi/2$, Δg_{cc} decreases in value, becoming zero when $\ddot{z}$ and $\ddot{x}$ are in phase or 180° out of phase.

LaCoste (1967) calculated that cross-coupling errors can reach values up to about 50 mgal with beam-type meters mounted rigidly on stabilized platforms. Errors up to 20 mgal have been reported at sea with such platform-mounted meters (e.g., Bower, 1966; Talwani et al., 1966). Cross-coupling errors are small to negligible with beam-type meters mounted in a gimbal suspension, although such meter systems encounter errors in measured gravity when ship accelerations are in excess of 50 gal.

Cross-coupling effects can be viewed as errors produced by deflections of the sensing element from the direction along which accelerations are greatest. Effects of such deflections do not vanish over a wave period. LaCoste (1967) showed that cross-coupling errors are analogous to off-leveling errors.

Sea-meter systems usually include instruments required to make direct corrections for cross-coupling accelerations; hence, cross-coupling accelerations need not be computed separately. Horizontal accelerometers are attached to the stabilized platform (or the meter case); these measure $\ddot{x}$ (eq. 6.52). Computers calculate resultant cross-coupling accelerations in real time so that the meter reading is true gravity. Talwani (1966) and LaCoste (1967) developed analog computers for this purpose. Others have also developed instruments which correct for cross-coupling effects in real time.

Geologic corrections

The Pratt-Hayford and Airy-Heiskanen isostatic reductions make simplistic assumptions regarding compensating structures for topographic features. In areas where known geologic structures deviate markedly from the assumed compensation, such as over deep sedimentary basins or across continental margins, corrections can be applied to the isostatic reductions. Corrections can similarly be applied to Bouguer reductions.

Gravity corrections owing to large geological deviations from normally assumed structures are called geologic corrections. These corrections should be applied in zones or compartments where assumed densities of the crust or mantle in the Airy-Heiskanen system or assumed column densities or column depths of compensation in the Pratt-Hayford system differ significantly from those which actually exist. An example is a thick sedimentary section in which the desired anomaly is to exclude the effect of the section. The attraction of such a body can be removed in a manner similar to the removal of topography, on the assumption that the masses are compensated isostatically.

A geologic correction can be computed for each zone in a grid, as in Table 6.I, by replacing the geologic mass with an equivalent overlying topographic mass, wherein its effect is added to that of the actual topography. The equivalent elevation increase Δh_g due to a geologic feature of height h_g and density ρ_g is related to a body of assumed "normal" density ρ_n (e.g., 2.67 g cm^{-3}) of the same height by (similar to eq. 4.25):

$$\rho_n(h_g + \Delta h_g) = \rho_g h_g \tag{6.57a}$$

We obtain for the equivalent topographic increase:

$$\Delta h_g = \left(\frac{\Delta\rho}{\rho_n}\right) h_g \tag{6.57b}$$

in which $\Delta\rho = \rho_g - \rho_n$.

Woollard (1938) computed geologic corrections and used them to interpret gravity anomalies in North America. These geologic features may consist of density variations in near-surface structures, configurations in the top of the crystalline basement, configurations of the sub-basement, and other bodies producing deviations from isostatic equilibrium. Evans and Crompton (1946) computed the thicknesses of sedimentary layers in India and Burma, obtaining geologic corrections of —50 to —75 mgal.

During the past several decades it has become evident that the upper mantle as well as the crust are laterally inhomogeneous. Crustal and subcrustal structures have, in many instances, been identified as characteristic of specific types of geologic features, such as ocean ridges and trenches. Even for known features geologic corrections can be rather complex. In other areas, as Cenozoic-age mountains, actual roots of mountains can be quite different from those assumed in the Airy-Heiskanen reductions (see Chapter 4 for a discussion of larger geologic features). Hence, an alternative to applying geologic corrections to standard reductions is to analyze anomalies with inverse methods or to construct hypothetical geologic models (including the crust and upper mantle) and test the model against the observed anomalies by using digital computers (described in Chapter 7). A model is adjusted on the basis of known geological and geophysical data until a good fit is obtained.

Earth-tide corrections

Formulas which can be computer programed to give earth-tide corrections as they apply to gravity measurements are given in Appendix C. The formulas apply to any point and for any time on the earth's surface, as functions of orbital parameters.

GRAVITY ANOMALIES

Gravity anomalies are obtained from eq. 6.1. Specific anomalies are defined by the gravity reductions applied. The commonly used anomalies are described below.

Free-air (Faye) anomaly

The definition of the free-air anomaly, Δg_F, is:

$$\Delta g_F = (g_{obs} + \delta_F) - \gamma_\Phi = \left(g_{obs} + \frac{2gh}{r}\right) - \gamma_\Phi \tag{6.58}$$

where δ_F is the free-air reduction (eq. 6.5) and γ_Φ is normal gravity. The anomaly is computed from the formula:

$$\Delta g_F = (g_{obs} + 0.3086h) - \gamma_\Phi \quad \text{mgal} \tag{6.59}$$

in which height h is in meters.

Observed gravity has been reduced to sea level by correcting for station elevation without considering attractions resulting from topographic masses and terrain variations. The anomaly is what would be obtained if the measurement were made in a balloon at height h above a flat sea-level area.

The free-air is the simplest type of anomaly because no assumptions are made about rock densities, either above or below sea level. The anomalies are also significant because an average free-air anomaly value over an area (size depending upon terrain variations) approximately equals the isostatic value. This is to be expected in areas that are nearly in isostatic equilibrium. The major disadvantage in free-air anomalies is their dependence upon topographic variations.

Bouguer anomaly (simple)

A simple-Bouguer anomaly, Δg_{BS}, is defined by:

$$\Delta g_{BS} = (g_{obs} + \delta_F - \delta_{BS}) - \gamma_\Phi \tag{6.60}$$

which becomes, if we use eq. 6.10:

$$\Delta g_{BS} = \left(g_{obs} + \frac{2gh}{r} - 2\pi G\rho h\right) - \gamma_\Phi \tag{6.61a}$$

and which, in turn, is commonly written:

$$\Delta g_{BS} = [g_{obs} + (0.3086 - 0.0419\rho)h] - \gamma_\Phi \tag{6.61b}$$

Observed gravity is reduced to sea level by making a correction for station elevation and removing the mass above sea level, as if the mass consisted of an infinite horizontal slab of density ρ with thickness equal to the station elevation. No correction is applied for terrain variations.

Simple Bouguer anomalies provide rapid but crude determinations of sea-level anomalies on the assumption that no mass exists above sea level. Clearly, the anomaly is not a good indication of the deviation of the geoid from the spheroid. However, it has application in determining local buried masses because the major effect of the topographic attractions has been eliminated.

Bouguer anomalies usually have large negative values at higher land elevations and large positive values over oceans. This is: (1) because the anomalies exclude attractions actually produced by topographic masses, or because the anomalies include the artificial effect of filling the ocean with rock; and (2) because topographic masses and oceanic regions are compensated at depth.

Bouguer anomaly (complete)

A Bouguer anomaly is terrain-corrected unless otherwise indicated. The definition of a Bouguer anomaly, Δg_{Bt}, is (eq. 6.11):

$$\Delta g_{Bt} = (g_{obs} + \delta_F - \delta_{BS} + \epsilon_t) - \gamma_\Phi \tag{6.62}$$

which is commonly written:

$$\Delta g_{Bt} = [g_{obs} + (0.3086 - 0.0419\rho)h + \epsilon_t] - \gamma_\Phi \tag{6.63}$$

in which $\epsilon_t = \Sigma\Delta\epsilon_t$ in eq. 6.13.

Observed gravity is reduced to sea level by making a correction for station elevation and removing the mass above sea level, which is assumed to consist of an infinite horizontal slab with upper surface the shape of the surrounding terrain. The Bouguer anomaly incorporates attractions which result from terrain variations, especially near the station, where effects are greatest. The anomaly is useful for determining local-anomaly variations in irregular terrain, particularly for investigating structures commonly encountered in prospecting for petroleum or for ore-deposit.

The Bouguer (including the simple-Bouguer) anomaly has been used to relate station elevation roughly with isostatic compensation. Woollard has correlated gravity anomalies with surface elevation, crustal structure, and geology (Woollard, 1936, 1943, 1962, 1969), and with isostatic compensation (Woollard and Strange, 1966; Woollard, 1968). Woollard's (1969) analyses include relationships between station elevation and Bouguer anomaly, both being averaged over $1^\circ \times 1^\circ$, $2^\circ \times 2^\circ$, and also $3^\circ \times 3^\circ$ squares. Fig. 4.18 shows such curves for $3^\circ \times 3^\circ$ averages in various areas of the world.

Bouguer anomaly (expanded)

The expanded Bouguer anomaly, Δg_{BE}, is defined as:

$$\Delta g_{BE} = (g_{obs} + \delta_F - \delta_{BS} + \epsilon_t + \epsilon_s) - \gamma_\Phi \tag{6.64}$$

which is the Bouguer anomaly (eq. 6.62) corrected for earth curvature, ϵ_s, the term in eq. 6.14, which is similar to the correction in eq. 6.32. Expanded Bouguer anomalies are seldom calculated, because local anomaly variations rather than absolute values are of primary interest in analyzing Bouguer anomalies. Curvatures have a relatively small effect on local anomaly variations.

Pratt-Hayford isostatic anomaly

To obtain Pratt-Hayford isostatic anomalies, the sum of topographic corrections for all zones is subtracted from observed gravity and the sum of the compensation reductions for all zones is added. In the original Hayford topographic computations, topographic effects for each compartment or zone refer to a sea-level datum. The current practice is to refer topographic corrections for compartments in zones A to O (out to 166.7 km, Table 6.I) to the station elevation, using the complete Bouguer anomaly out to zone O (with allowance for earth curvature), and to apply spherical formulas to compute topographic effects, referred to a sea-level datum, and isostatic compensations for zones 18 to 1.

The Pratt-Hayford isostatic anomaly $(\Delta g_i)_{PH}$ is then given by:

$$(\Delta g_i)_{PH} = [g_{obs} + \delta_F - \delta_{BS} + (\epsilon_t + \epsilon_s)_{A-O} - (\delta_{iT})_{18-1} + (\delta_{ic})_{PH}] - \gamma_\Phi \tag{6.65}$$

in which subscripts A—O refer to zones A to O, and subscript 18—1 to zones 18 to 1: and δ_F is the free-air reduction; δ_{BS} the slab reduction; ϵ_t the total terrain correction for zones A to O; ϵ_s the total curvature correction out to zone O; $(\delta_{iT})_{18-1}$ the total topographic correction for zones 18 to 1; and $(\delta_{ic})_{PH}$ the total Pratt-Hayford compensation reduction for all zones (A to 1).

Airy-Heiskanen isostatic anomaly

To obtain Airy-Heiskanen isostatic anomalies, the topographic corrections for zones A to O are computed just as for the Pratt-Hayford anomalies.

The Airy-Heiskanen isostatic anomaly, $(\Delta g_i)_{AH}$, is, similar to eq. 6.65:

$$(\Delta g_i)_{AH} = [g_{obs} + \delta_F - \delta_{BS} + (\epsilon_t + \epsilon_s)_{A-O} - (\delta_{iT})_{18-1} + (\delta_{ic})_{AH}] - \gamma_\Phi \tag{6.66}$$

in which symbols identical to those in eq. 6.65 have the same meaning, and $(\delta_{ic})_{AH}$ is the total Airy-Heiskanen compensation reduction for all zones (A to 1).

Vening Meinesz isostatic anomaly

This isostatic anomaly is more complex to compute than the other two types of isostatic anomalies. Because of this, and the fact that the anomaly was computed only by the late Vening Meinesz, the computational procedure is not reproduced here (see Vening Meinesz, 1931, 1939).

Two-dimensional, pseudo-isostatic anomaly

A two-dimensional, pseudo-isostatic anomaly has been used by investigators at the Lamont-Doherty Geological Observatory of Columbia University to analyze structures across continental margins which are overshadowed by an edge effect that may be of secondary interest. Two-dimensional line-integral procedures are used to compute the attraction produced by an ocean filled with rock and to compute the attraction resulting from the corresponding antiroot (Fig. 4.17).

The two-dimensional anomaly $(\Delta g_i)_{2D}$ can be written:

$$(\Delta g_i)_{2D} = [g_{obs} + \delta_F + (\delta_B)_{2D} - (\delta_{ic})_{2D}] - \gamma_\Phi \tag{6.67}$$

in which δ_F, the free-air reduction, vanishes at sea; $(\delta_B)_{2D}$ is the two-dimensional line-integral value of the ocean filled with rock; and $(\delta_{ic})_{2D}$ is the two-dimensional line-integral value of the antiroot compensating the 2D filled-in ocean.

Chapter 7

ANALYTICAL METHODS FOR INTERPRETING GRAVITY ANOMALIES

A principal purpose for obtaining gravity anomalies is to locate and determine distributions of buried masses. The total mass producing an anomaly can be calculated; however, because gravity is a potential function, the mass distribution cannot be determined uniquely. Hence, to calculate distributions assumptions must be made. If either the density or shape of the body is assumed (or known), the other can be obtained directly under somewhat simplifying conditions. For relatively complex structures, both the density and shape can be assumed, especially when control points are available, and the resultant model readily improved with the use of digital computers.

This chapter describes both indirect and direct methods for analyzing structures producing anomalies which have the wavelengths and amplitudes of the observed. Indirect methods use hypothetical models for which corresponding attractions are computed and compared with the observed. Direct methods attempt to solve equations which represent the anomaly.

Hypothetical models can range from simple mathematically described bodies, for which the attraction can be computed by direct integrations (e.g., a sphere, a horizontal slab, a rectangular prism), to complex structures having irregular shapes or consisting of numerous layers with different density contrasts (e.g., continental margins, archipelagoes) for which attractions are computed by either two- or three-dimensional numerical methods. Clearly, models need to incorporate known seismic, borehole, and other available information. After each computed anomaly, the model is adjusted until it adequately matches the observed anomaly.

Direct methods are used to determine total mass, applying Gauss's theorem (eq. 2.23b), without specifying its size or shape. Total mass is of interest when concentrated, as in a dense ore body, but seldom otherwise. For irregular unknown mass distributions, direct methods usually refer to inverse methods, where integral equations are solved that satisfy the observed anomaly. In inverse methods, either the shape of the body is assumed and the density distribution solved for, or the density contrast(s) assumed and the shape solved for (e.g., the shape of a sedimentary basin of assumed density contrast).

Indirect methods allow the use of wide ranges of density and layer assumptions, whereas inverse methods usually provide more rapid solutions for simplified models. Hypothetical models are used more commonly when structures are analyzed because they allow for greater complexity of layering and density contrasts, and also because the additional required computations usually mean only added computer time.

REGIONAL AND RESIDUAL ANOMALIES

Because all anomalous masses within a given volume produce a composite gravity anomaly, it is often desirable to separate attractions caused by regional features from those caused by pertinent structures. This can be done by dividing the observed anomaly (Fig. 7.1) into a regional anomaly, usually representing effects of masses of little or no concern in the analysis, and a residual. The residual anomaly is supposedly produced by structures of interest, but it does not follow that all residuals represent such structures. Conversely, the regional is sometimes of interest, such as averaged anomalies over selected grids (e.g., $1^\circ \times 1^\circ$ square), and residuals are then considered as noise.

The residual Δg_{res} is defined as:

$$\Delta g_{res} = \Delta g_{obs} - \overline{\Delta g} \tag{7.1}$$

in which Δg_{obs} is the total observed anomaly and $\overline{\Delta g}$ the regional value. The regional is commonly constructed to include effects of all masses below some arbitrary depth, with the residual then representing shallower bodies. Where masses at all depths are of interest, as in analyses of crustal and subcrustal structures, the entire anomaly is used and the regional then vanishes. Commonly, however, pertinent structures are situated in the upper crust. This is particularly true in geophysical prospecting for economic deposits and in investigations of certain larger features, such as sedimentary basins or major faults. The objective then is to apply regional values which most nearly eliminate effects of undesired deeper masses.

Residuals can also be defined to represent deep structures where a specifically defined regional effect has been removed from the observed anomaly. An example is a residual obtained by Cochran and Talwani (1977) in which the effect of a mid-ocean ridge system is removed from observed anomalies so as to determine anomaly patterns that may be related to thermal models of the upper mantle and crust.

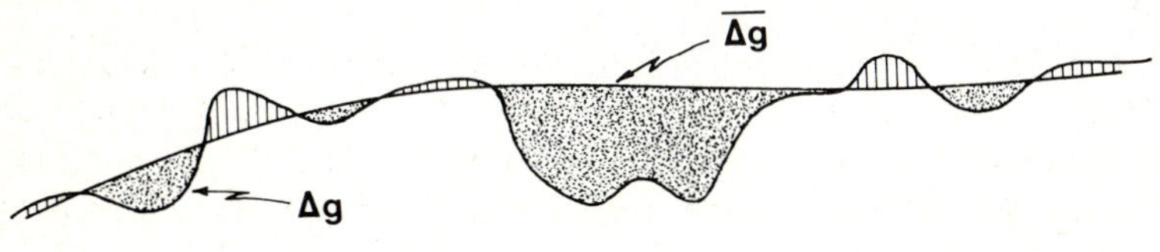

Fig. 7.1. Illustration of an observed gravity anomaly Δg and an assumed regional anomaly $\overline{\Delta g}$; the residual anomaly is $\Delta g - \overline{\Delta g}$.

Quantitative analysis of a structure from residuals clearly requires that the residuals are not distorted by effects of other masses. In practice it is difficult adequately to isolate residuals produced by specific structures, except in simple cases, such as salt domes or seamouts. More commonly, residuals provide rough estimates of the effects of a specific structure.

Residuals are usually obtained from Bouguer anomalies, which are approximately independent of topographic effects and are often obtained at sufficiently short station intervals to indicate variations produced by shallow structures. Free-air anomalies are not used to obtain residuals except where regional calculations include effects of topography (e.g., the above-cited regional across an ocean-ridge system). Isostatic anomalies are commonly obtained at station intervals too wide to provide meaningful residuals; otherwise they could be so used.

Regional-field anomalies

Regionals are commonly smoothed averages of the observed anomaly field, and hence representative of deep-seated bodies. In effect, regionals can be viewed as the attraction produced by a model comprised of structures below some depth. However, the depth is not specifically defined, being delineated rather by the curvature of the regional field.

Perhaps the simplest type of regional is obtained by drawing an arbitrary smooth surface through the total Bouguer field (Fig. 7.1). Such regionals can be constructed to permit making educated guesses regarding the attractions produced by specific types of anticipated structures (illustrated, again, by salt domes or seamounts). Where such readily identifiable structures are not present, however, the regional construction is often quite arbitrary. To provide some objectivity, Griffin (1949) proposed determining $\overline{\Delta g}$ as the arithmetic mean of surrounding observed anomaly values (e.g., 6 to 12) on a specified circle or grid. More elaborate averaging methods can also be used (e.g., Pick et al., 1973, p. 375). For some problems the averaged values are of interest, as previously mentioned, in which case they are averaged for a specified grid size, such as a $1^\circ \times 1^\circ$ square.

Regionals are sometimes obtained by least-squares methods, the regional being the least-squares value of surrounding observed anomalies. Thus

$$\overline{\Delta g} = \sum_{i=1}^{n} \sum_{j=1}^{m} C_{ij} x_i y_j \qquad (7.2)$$

in which coefficients C_{ij} are computed by least-squares methods in the coordinates x and y. Residuals so obtained are, in effect, random errors. In practice, terms to the second and sometimes third order in x and y are included in the computations. Orthogonal polynomials (Linsser, 1965) have also been used to replace the coordinates x and y in eq. 7.2.

Regionals based on mathematical determinations provide objectivity, although this can be apparent as well as real. Mathematically constructed regionals generally lack the flexibility often needed to differentiate between large and deep structures (e.g., a salt dome) and between small and shallow structures (e.g., fault trends).

Specially defined regionals are sometimes applied in particular analyses. An example is the Cochran and Talwani (1977) method for removing the effect of an ocean ridge from anomalies across ridge systems so as to bring out resultant anomaly patterns. Their regional is defined on a gravity—age relationship. Geologic ages are assigned to each $1^\circ \times 1^\circ$ square across a ridge, based on magnetic isochron maps. Average free-air anomalies for each $1^\circ \times 1^\circ$ isochron are then averaged along the isochrons, from which an empirical gravity—age curve is constructed relating averaged anomalies with isochron age. Residuals are obtained as the difference between the averaged $1^\circ \times 1^\circ$ free-air anomaly (of observed values) and the empirical gravity—age regional value.

Another example of a regional is a procedure devised by McKenzie and Bowin (1976). Filters are applied to observed bathymetry to compute a value of gravity which can be viewed as a regional. The difference between observed and the regional gravity is a residual that represents structures exclusive of those producing the bottom topography.

Second- and higher-derivative methods

Computing second or higher vertical derivatives of the gravity field amounts to removing a mathematically determined regional characterized by low curvature in the observed anomaly field. Second-derivative values, when plotted on a map, provide contours of equal curvature, which is a form of residual map. Second-derivative maps indicate trends of structures and isolate small structures, but they have limited application for quantitative structural analysis and are seldom used in analyzing sea gravity data.

The second-vertical derivative of the gravity anomaly is obtained from Laplace's equation for $\Delta g(x,y,z)$, which in cylindrical coordinates is:

$$\frac{\partial^2 \Delta g}{\partial z^2} + \frac{\partial^2 \Delta g}{\partial r^2} + \frac{1}{r}\frac{\partial \Delta g}{\partial r} - \frac{1}{r^2}\frac{\partial^2 \Delta g}{\partial \theta^2} = 0 \tag{7.3}$$

in which z is vertical downward, r the horizontal distance, and θ the polar angle. We let $\overline{\Delta g}$ be the mean value of anomalies on a circle of radius r, which is independent of θ, to obtain:

$$\frac{\partial^2 \overline{\Delta g}}{\partial z^2} = -\left(\frac{1}{r}\frac{\partial \overline{\Delta g}}{\partial r} + \frac{\partial^2 \overline{\Delta g}}{\partial r^2}\right) \tag{7.4}$$

In the neighborhood of the center of the circle, $\overline{\Delta g}$ takes the form:

$$\overline{\Delta g} = a + br^2 + \ldots \tag{7.5}$$

We substitute this equation into 7.4 to obtain for the second derivative:

$$\frac{\partial^2 \overline{\Delta g}}{\partial z^2} = -4b \tag{7.6}$$

The coefficient b is obtained from eq. 7.5 as the difference between Δg at the center of the circle ($r = 0$) and an average value on the circle. Thus:

$$b = \frac{1}{r^2} \left[\overline{\Delta g}(r) - \Delta g(0) \right] \tag{7.7}$$

and the second-vertical derivative becomes:

$$\frac{\partial^2 \Delta g}{\partial z^2} = -\frac{4}{r^2} [\overline{\Delta g}(r) - \Delta g(0)] \tag{7.8}$$

More elaborate methods in which second-derivatives are used can be obtained if we include several circles of different radii. Such methods are described by Elkins (1951), Rosenbach (1953), Pick et al. (1973, p. 379), and Agarwal and Lal (1969, 1971) and will not be described here.

Derivatives higher than the second, which can be obtained in similar manners, indicate still greater curvatures in the total-anomaly field and are used to locate relatively smaller and shallower structures.

INTEGRATION METHODS FOR MATHEMATICALLY DESCRIBED BODIES

The vertical component of attraction and the corresponding gravitational potential produced by an arbitrary mass can be written (Fig. 7.2a):

$$\Delta g = G \int_\tau \frac{\Delta\rho \cos \zeta \, \mathrm{d}\tau}{q^2} \tag{7.9a}$$

$$\Delta V = G \int_\tau \frac{\Delta\rho \, \mathrm{d}\tau}{q} \tag{7.9b}$$

in which $\Delta\rho$ is the density contrast between the body and the surrounding material, q the vector from P (station) to a point Q in the body, ζ the angle q makes with the vertical, and integration is over the volume of the body.

Eqs. 7.9a and b are commonly integrated in the coordinate system most appropriate to the shape of the body. Table 7.I lists formulas obtained from eqs. 7.9a and b for Δg and ΔV in rectangular, cylindrical, and spherical coordinates for cases both where P is and is not at the origin of coordinates. In rectangular coordinates z is positive downward and $\mathrm{d}x\,\mathrm{d}y\,\mathrm{d}z$ is the volume element; in cylindrical coordinates r is the horizontal vector, θ the polar angle, and $r\,\mathrm{d}r\,\mathrm{d}\theta\,\mathrm{d}z$ the volume element; in spherical coordinates, r is the

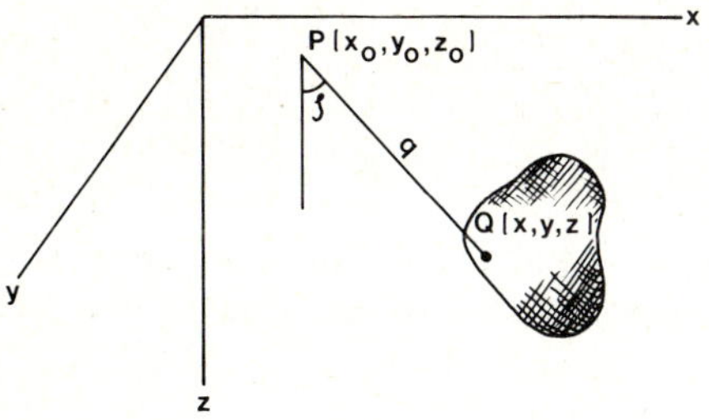

a.

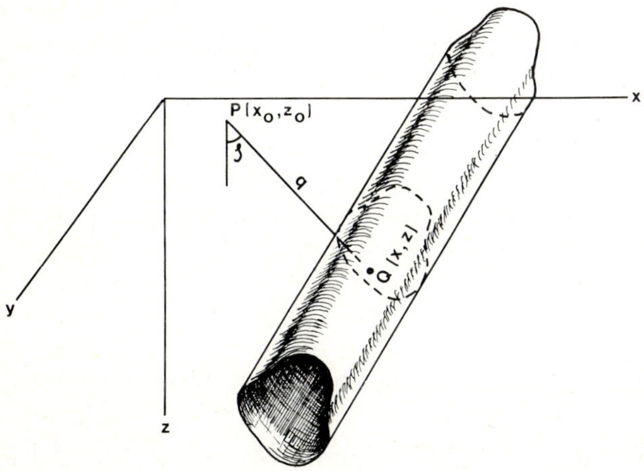

b.

Fig. 7.2. a. Three-dimensional body and, b. two-dimensional body producing attractions at point P; ζ is the angle which the vector q (from P to Q) makes with the vertical.

radius vector, ϕ the angle it makes with the vertical, θ the rotational angle in the horizontal plane, and $r^2 \sin \phi \, dr \, d\phi \, d\theta$ the volume element. The distance from P to Q is $[(x-x_0)^2 + (y-y_0)^2 + (z-z_0)^2]^{1/2}$ when P is not at the origin, and $(x^2 + y^2 + z^2)^{1/2}$ when at the origin.

Formulas for the attraction and potential produced by two-dimensional bodies (one direction extending to infinity, Fig. 7.2b) can be obtained by integrating eqs. 7.10a,b and 7.11a,b with limits of $\pm\infty$ in the y direction. The resultant two-dimensional formulas in rectangular and polar coordinates are listed in Table 7.II, where r is the polar radius and θ the angle from the horizontal.

The formulas in Tables 7.I and 7.II can be readily applied to compute the attraction or potential produced by mathematically described bodies, and

TABLE 7.I

Three-dimensional formulas for computing gravitational acceleration and potential produced by anomalous masses

Coordinate system	P not at origin	eq.	P at origin	eq.
Rectangular	$\Delta g = G \int_z \int_y \int_x \frac{\Delta\rho(z - z_0)\, dx\, dy\, dz}{[(x - x_0)^2 + (y - y_0)^2 + (z - z_0)^2]^{3/2}}$	(7.10a)	$\Delta g = G \int_z \int_y \int_x \frac{\Delta\rho z\, dx\, dy\, dz}{(x^2 + y^2 + z^2)^{3/2}}$	(7.10b)
	$\Delta V = G \int_z \int_y \int_x \frac{\Delta\rho\, dx\, dy\, dz}{[(x - x_0)^2 + (y - y_0)^2 + (z - z_0)^2]^{1/2}}$	(7.11a)	$\Delta V = G \int_z \int_y \int_x \frac{\Delta\rho\, dx\, dy\, dz}{(x^2 + y^2 + z^2)^{1/2}}$	(7.11b)
Cylindrical	$\Delta g = G \int_z \int_\theta \int_r \frac{\Delta\rho(z - z_0)\, r dr\, d\theta\, dz}{[(r - r_0)^2 + (z - z_0)^2]^{3/2}}$	(7.12a)	$\Delta g = G \int_z \int_\theta \int_r \frac{\Delta\rho z\, r dr\, d\theta\, dz}{(r^2 + z^2)^{3/2}}$	(7.12b)
	$\Delta V = G \int_z \int_\theta \int_r \frac{\Delta\rho\, r dr\, d\theta\, dz}{[(r - r_0)^2 + (z - z_0)^2]^{1/2}}$	(7.13a)	$\Delta V = G \int_z \int_\theta \int_r \frac{\Delta\rho\, r dr\, d\theta\, dz}{(r^2 + z^2)^{1/2}}$	(7.13b)
Spherical	$\Delta g = G \int_\theta \int_\phi \int_r \frac{\Delta\rho(r \cos\phi - r_0 \cos\phi_0) r^2 \sin\phi\, dr\, d\phi\, d\theta}{q^3}$	(7.14a)	$\Delta g = G \int_\theta \int_\phi \int_r \Delta\rho \sin\phi \cos\phi\, dr\, d\phi\, d\theta$	(7.14b)
	$\Delta V = G \int_\theta \int_\phi \int_r \frac{\Delta\rho\, r^2 \sin\phi\, dr\, d\phi\, d\theta}{q}$	(7.15a)	$\Delta V = G \int_\theta \int_\phi \int_r \Delta\rho r \sin\phi\, dr\, d\phi\, d\theta$	(7.15b)

TABLE 7.II

Two-dimensional formulas for computing gravitational acceleration and potential produced by anomalous masses

Coordinate system	P not at origin	eq.	P at origin	eq.
Rectangular	$\Delta g = 2G \int_z \int_x \frac{\Delta\rho(z - z_0)\, dx\, dz}{(x - x_0)^2 + (z - z_0)^2}$	(7.16a)	$\Delta g = 2G \int_z \int_x \frac{\Delta\rho\, z dx\, dz}{x^2 + z^2}$	(7.16b)
	$\Delta V = 2G \int_z \int_x \Delta\rho \ln[(x - x_0)^2 + (z - z_0)^2]^{-1/2} dx\, dz$	(7.17a)	$\Delta V = 2G \int_z \int_x \Delta\rho \ln(x^2 + z^2)^{-1/2} dx\, dz$	(7.17b)
Polar	$\Delta g = 2G \int_\theta \int_r \frac{\Delta\rho(r \sin\theta - r_0 \sin\theta_0) rd\, rd\theta}{q^2}$	(7.18a)	$\Delta g = 2G \int_\theta \int_r \Delta\rho \sin\theta drd\theta$	(7.18b)
	$\Delta V = 2G \int_\theta \int_r \Delta\rho \ln(q)^{-1} rdrd\theta$	(7.19a)	$\Delta V = 2G \int_\theta \int_r \Delta\rho \ln(r)^{-1} rdrd\theta$	(7.19b)

thus gravity profiles across a body. Applications of these formulas are illustrated below.

Sphere

The attraction at a point on or outside a uniform sphere (Fig. 7.3a) of radius a ($R > a$) is readily obtained by eq. 7.14a or 7.15a. For a point at distance R from the center of the sphere (origin of coordinates) with mass M, the attraction is :

$$\Delta g = G\Delta\rho \int_0^{2\pi} \int_0^{\pi} \int_0^{a} \frac{(R - r\cos\phi)r^2 \sin\phi \, dr \, d\phi \, d\theta}{(R^2 + r^2 - 2Rr\cos\phi)^{3/2}} = G\frac{M}{R^2} \tag{7.20a}$$

and the potential:

$$\Delta V = G\Delta\rho \int_0^{2\pi} \int_0^{\pi} \int_0^{a} \frac{r^2 \sin\phi \, dr \, d\phi \, d\theta}{(R^2 + r^2 - 2Rr\cos\phi)^{1/2}} = G\frac{M}{R} \tag{7.20b}$$

These formulas show that the attraction and potential at a point outside a uniform sphere are the same as if the entire mass were concentrated at its center.

It follows that gravity along a profile across a buried sphere, with the origin of coordinates at P (the station), is:

$$\Delta g = GM\left[\frac{z}{(x^2 + z^2)^{3/2}}\right] \tag{7.21}$$

in which z is the depth to center of the sphere and x the horizontal distance from P to a point above the center.

At an interior point in the sphere ($R < a$), we obtain the attraction in two steps. Within radius R, we obtain from eq. 7.20a:

$$\Delta g_1 = G\frac{M_1}{R^2} \tag{7.22}$$

in which M_1 is the mass within radius R. For the spherical layer outside R:

$$\Delta g_2 = G\Delta\rho \int_0^{2\pi} \int_0^{\pi} \int_R^{a} \frac{(r\cos\phi - R)r^2 \sin\phi \, dr \, d\phi \, d\theta}{(R^2 + r^2 - 2Rr\cos\phi)^{3/2}} = 0 \tag{7.23}$$

It follows that uniform spherical layers exert no gravitational attraction at interior points.

Gravity on or within a uniformly layered earth can be calculated within 1% by assuming spherical layers. Gravity within the earth can be estimated from eq. 7.20a, integrating from the inner (R_1) to outer (R_2) radii of each layer. We then get:

$$\Delta g = G\frac{\Sigma M_i}{R^2} \tag{7.24}$$

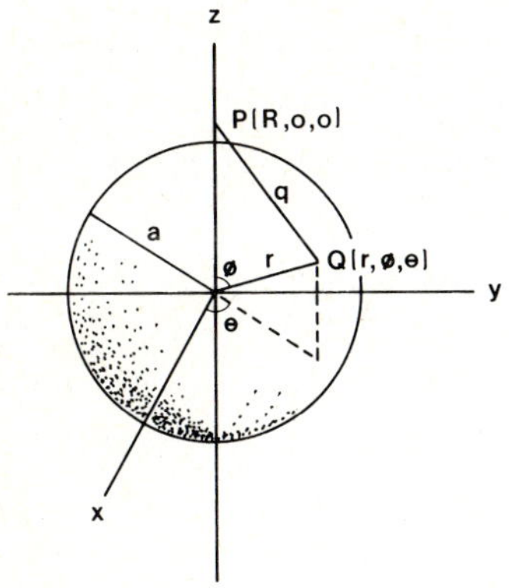
z
P(R,o,o)
q
a
ø
r
Q(r,ø,e)
y
e
x

a.

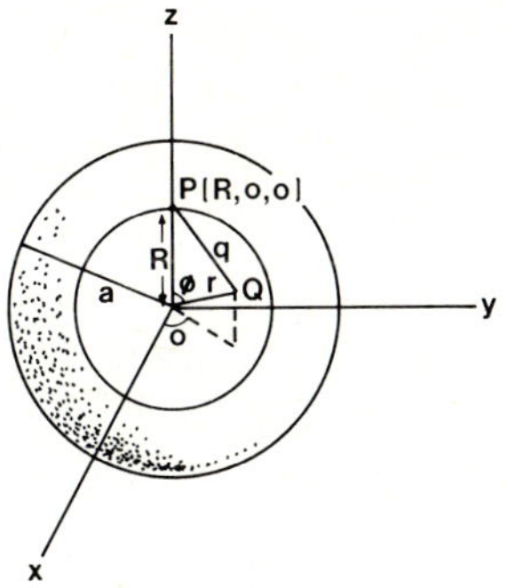
z
P(R,o,o)
q
R
a
ø
r
Q
o
y
x

b.

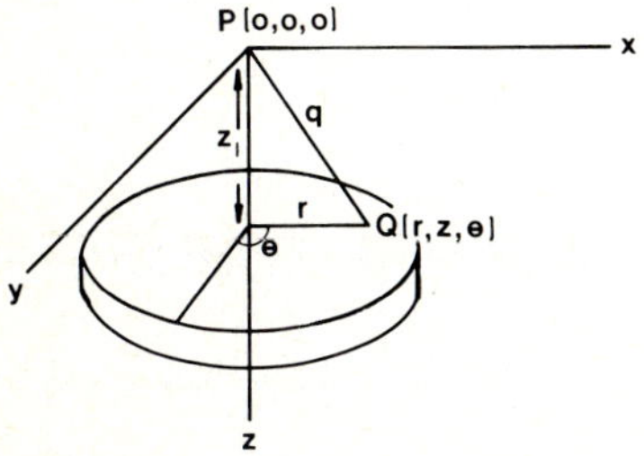
P(o,o,o)
x
q
z1
r
Q(r,z,e)
e
y
z

c.

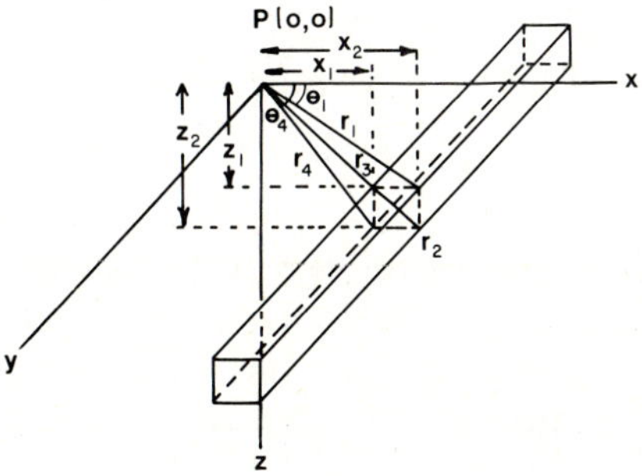
P(o,o)
x2
x1
x
e1
e4
z2
z1
r1
r4
r3
r2
y
z

d.

e.

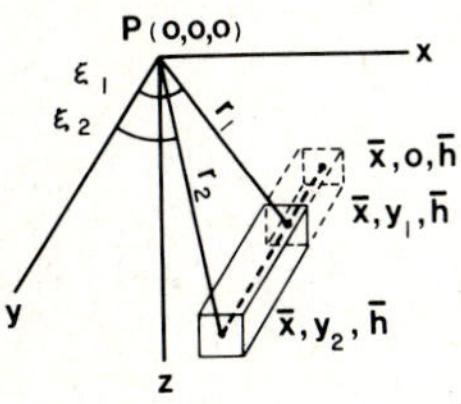
P(o,o,o)
x
ε1
ε2
r1
r2
x̄,o,h̄
x̄,y1,h̄
x̄,y2,h̄
y
z

f.

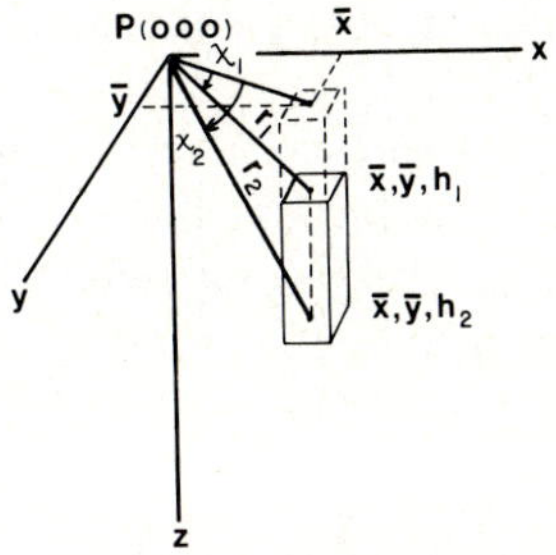
P(ooo)
x̄
x
χ1
ȳ
χ2
r1
r2
x̄,ȳ,h1
x̄,ȳ,h2
y
z

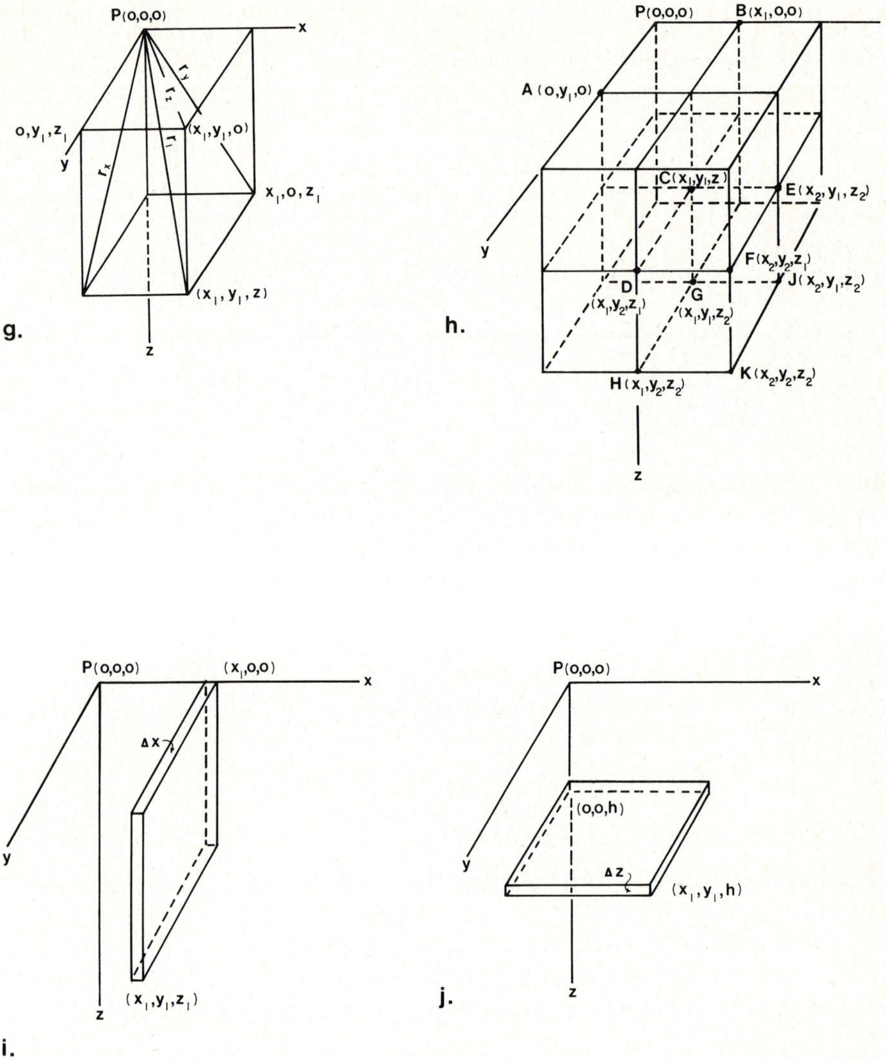

Fig. 7.3. Mathematically describable bodies for computing attractions at point P: a. sphere of radius a, P is at an exterior point; b. sphere, P at interior point; c. horizontal disk, P at exterior point; d. horizontal, two-dimensional prism; e. horizontal rectangular prism; f. vertical rectangular prism; g. rectangular parallelepiped with one corner at origin of coordinates; h. rectangular parallelepipeds, one of which does not have a corner at the origin; i. vertical lamina (x_1 is the distance to the median plane of the lamina); j. horizontal lamina (h is the depth to the median plane).

TABLE 7.III

Gravity at the outer surface of successive layers in the earth

	Radius		Average density	Mass per layer	M and g at surface of layer	
	inner (km)	outer (km)	(g cm^{-3})	(g)	(g)	(gals)
Center of earth	0	0	12.2	0	0	0
Inner core	0	1,290	12.1	$0.1088 \cdot 10^{27}$	$0.109 \cdot 10^{27}$	437
Outer core	1,290	3,470	11.0	$1.8263 \cdot 10^{27}$	$1.935 \cdot 10^{27}$	1,072
Lower mantle	3,470	5,370	5.1	$2.4155 \cdot 10^{27}$	$4.351 \cdot 10^{27}$	1,007
Transition zone	5,370	5,970	4.1	$0.9947 \cdot 10^{27}$	$5.345 \cdot 10^{27}$	1,001
Upper mantle	5,970	6,350	3.3	$0.5981 \cdot 10^{27}$	$5.943 \cdot 10^{27}$	983
Crust	6,350	6,370	2.9	$0.0295 \cdot 10^{27}$	$5.973 \cdot 10^{27}$	982

in which i is the number of interior layers to radius R. For example, gravity on the surface of the outer core is that due to the entire mass in the core. Table 7.III shows values of gravity so computed at various discontinuities in the earth, which agree well with Bullen's Model A″ (Fig. 4.3).

Spherical shell

At a point on or outside a spherical shell with radius a and uniform surface density σ, the attraction is:

$$\Delta g = G\sigma \int_0^{2\pi} \int_0^{\pi} \frac{(R - a \cos \phi) a^2 \sin \phi \, d\phi \, d\theta}{(R^2 + a^2 - 2Ra \cos \phi)^{3/2}} = G \frac{M}{R^2} \tag{7.25}$$

The attraction is the same as if all mass in the shell were concentrated at its center.

At a point inside the shell, the sign of the term $(R - a \cos \phi)$ is reversed, to obtain a positive attraction. Integration shows that $\Delta g = 0$, that is, there is no attraction anywhere at an interior point in a uniform hollow shell (a result similar to eq. 7.23).

Infinite horizontal cylinder

Outside an indefinite horizontal cylinder of uniform density, we use eq. 7.16b to compute the vertical attraction by integrating along a plane containing point P and the normal to the cylinder (see Fig. 7.2b), and obtain:

$$\Delta g = 2GM_l \left(\frac{z}{x^2 + z^2} \right) \tag{7.26}$$

in which M_l is the cylinder mass per unit length.

Infinite horizontal slab

The vertical attraction on or outside an infinite horizontal slab can be readily obtained from eqs. 7.10b, 7.12b, or 7.16b. We thus write for the attraction (similar to eq. 6.9 and illustrated in Fig. 7.3c):

$$\Delta g = 2G\Delta\rho \int_{z_1}^{z_1+\Delta z} \int_{-\infty}^{\infty} \frac{z\,\mathrm{d}x\,\mathrm{d}z}{x^2+z^2}$$

$$= G\Delta\rho \int_0^{2\pi} \int_{z_1}^{z_1+\Delta z} \int_0^{\infty} \frac{zr\,\mathrm{d}r\,\mathrm{d}z\,\mathrm{d}\theta}{(r^2+z^2)^{3/2}} \tag{7.27}$$

Integration produces the Bouguer slab attraction (eq. 6.10):

$$\Delta g = 2\pi G\,\Delta\rho\,\Delta z \tag{7.28}$$

Δg is seen to be independent of the height above the slab.

Infinite horizontal sheet

We can obtain the vertical attraction resulting from a horizontal sheet of surface density $\Delta\sigma$ from the slab attraction. If we substitute $\Delta\sigma = \Delta\rho\Delta z$ into eq. 7.28, we get:

$$\Delta g = 2\pi G\Delta\sigma \tag{7.29}$$

which is also independent of height above the sheet.

Two-dimensional rectangular prism (horizontal)

The vertical attraction produced by a horizontal two-dimensional prism (Fig. 7.3d) can be derived from eq. 7.16b. Integrating over z and then x results in the formula (e.g., Heiland, 1940, p. 151):

$$\Delta g = 2G\Delta\rho \left[x_2 \ln\left(\frac{r_2}{r_1}\right) - x_1 \ln\left(\frac{r_4}{r_3}\right) + z_2(\theta_2 - \theta_4) - z_1(\theta_1 - \theta_3) \right] \tag{7.30}$$

where the symbols are those indicated in the figure and:

$$\begin{aligned} x_1 &= z_1 \tan\theta_3 = z_2 \tan\theta_4 \\ x_2 &= z_1 \tan\theta_1 = z_2 \tan\theta_2 \end{aligned} \tag{7.31}$$

For prisms with depth to center larger than twice the width:

$$\frac{r_2}{r_1} \sim \frac{r_4}{r_3} \quad \text{and:} \quad \Delta\theta \sim (\theta_1 - \theta_3) \sim (\theta_2 - \theta_4)$$

Eq. 7.30 then reduces to a readily programmable formula (Tanner, 1967):

$$\Delta g = 2G\,\Delta\rho\,\Delta\theta\,\Delta z \tag{7.32}$$

where $\Delta z = z_2 - z_1$.

For prisms with cross section ΔA, the sides small compared to distance from P to the prism axis, eq. 7.30 can be written:

$$\Delta g = 2G \frac{\Delta\rho\ \Delta A}{(\bar{x}^2 + \bar{h}^2)} = 2G \frac{\Delta\rho\ \Delta A}{\bar{r}^2} \tag{7.33}$$

where $\bar{x}$, $\bar{h}$, and $\bar{r}$ are the horizontal, vertical, and total distances to the prism axis.

Horizontal rectangular prism of cross section ΔA (three dimensions)

Horizontal rectangular prisms can be used as building blocks in approximating the size of a body. For a prism of cross section ΔA, with sides small compared to distance from P to the prism axis, the vertical attraction at P (Fig. 7.3e) produced by a prism extending from y_1 and y_2 can be written, if we use eq. 7.11a:

$$\Delta g = G\ \Delta\rho\ \Delta A \int_{y_1}^{y_2} \frac{\bar{h}\ dy}{(\bar{x}^2 + \bar{h}^2 + y^2)^{3/2}} \tag{7.34}$$

in which, as before, $\bar{x}$ and $\bar{h}$ are distances to the prism axis. By integrating this equation we obtain (e.g., Talwani, 1973):

$$\Delta g = G \frac{\Delta\rho\ \Delta A}{(\bar{x}^2 + \bar{h}^2)} \left(\frac{y_2}{r_2} - \frac{y_1}{r_1} \right) = G \frac{\Delta\rho\ \Delta A}{(\bar{x}^2 + \bar{h}^2)} (\cos \xi_2 - \cos \xi_1) \tag{7.35}$$

in which ξ_2 and ξ_1 are the angles which the r_2 and r_1 vectors (to the ends of the prism) make with the y axis (see Fig. 7.3e).

Vertical rectangular prism of cross section ΔA

Vertical rectangular prisms can be used as building blocks in approximating the size of a body. The vertical attraction at P (Fig. 7.3f) produced by a prism of cross section ΔA (sides small compared to distance to prism axis) extending from depth h_1 to h_2 is, if we use eq. 7.11a:

$$\Delta g = G\ \Delta\rho\ \Delta A \int_{h_1}^{h_2} \frac{z\ dz}{(\bar{x}^2 + \bar{y}^2 + z^2)^{3/2}} \tag{7.36}$$

in which $\bar{x}$ and $\bar{y}$ are component distances to the prism axis. We integrate over z to obtain:

$$\Delta g = G \frac{\Delta\rho\ \Delta A}{(\bar{x}^2 + \bar{y}^2)^{1/2}} (\cos \chi_1 - \cos \chi_2) \tag{7.37}$$

where χ_1 and χ_2 are angles from the horizontal which vectors make from P to the centers of the top and bottom of the prism.

Rectangular parallelepiped (including a cube)

Rectangular parallelepipeds, especially cubes, can be used as building blocks to approximate the size of an irregular three-dimensional body. The vertical attraction produced at one corner of a parallelepiped with sides x, y, and z (Fig. 7.3g) is, if we use eq. 7.11a:

$$\Delta g = G\Delta\rho \int_0^{z_1}\int_0^{y_1}\int_0^{x_1} \frac{z\,dx\,dy\,dz}{(x^2+y^2+z^2)^{3/2}} \qquad (7.38)$$

If we use the symbols:

$r_1 = (x_1^2 + y_1^2 + z_1^2)^{1/2}$

$r_x = (y_1^2 + z_1^2)^{1/2}$ $\quad r_y = (x_1^2 + z_1^2)^{1/2}$ $\quad r_z = (x_1^2 + y_1^2)^{1/2}$

$\cos l = x_1/r_1$ $\quad \cos m = y_1/r_1$ $\quad \cos n = z_1/r_1$

$\cos\alpha = z_1/r_x$ $\quad \cos\beta = z_1/r_y$ $\quad \cos\lambda = y_1/r_z$

$\sin\lambda = x_1/r_z$

the solution to eq. 7.38 becomes (e.g., Talwani, 1973):

$$\Delta g = G\Delta\rho\left\{ z\left[\frac{\pi}{2} - \sin^{-1}(\cos\beta\cos\lambda) - \sin^{-1}(\cos\alpha\sin\lambda)\right] + \frac{x}{2}\ln\left[\frac{(r_1-y_1)(r_z+y_1)}{(r_1+y_1)(r_z-y_1)}\right] + \frac{y}{2}\ln\left[\frac{(r_1-x_1)(r_z+x_1)}{(r_1+x_1)(r_z-x_1)}\right]\right\} \qquad (7.39)$$

This equation can be applied successively to compute the attraction of a parallelepiped where P is not at a corner.

The attraction of the rectangular parallelepiped CJ (opposite corners, Fig. 7.3h) is the sum and difference of eight rectangular parallelepipeds, seven of which have a corner at P (Talwani, 1973). Thus:

$$(\Delta g)_{CJ} = (\Delta g)_{PK} - (\Delta g)_{PJ} + (\Delta g)_{PG} - (\Delta g)_{PH} - (\Delta g)_{PF} + (\Delta g)_{PE} - (\Delta g)_{PC} + (\Delta g)_{PD} \qquad (7.40)$$

in which each term in the equation is of the form 7.39. Formulas for rectangular parallelepipeds based on eq. 7.40 have been published by various authors (e.g., Nagy, 1966a, and Talwani, 1973).

Vertical rectangular lamina

The vertical attraction produced by a vertical rectangular lamina (Fig. 7.3i) of thickness Δx can be obtained from eq. 7.11a. We integrate:

$$\Delta g = G\,\Delta\rho\,\Delta x \int_0^{z_1}\int_0^{y_1} \frac{z\,dy\,dz}{(x_1^2+y^2+z^2)^{3/2}} \qquad (7.41)$$

in which integration is from one corner $(x_1, 0, 0)$ to the opposite corner (x_1, y_1, z_1), and obtain (e.g., Talwani, 1973):

$$\Delta g = G\,\Delta\rho\,\Delta x \ln \frac{(x_1^2 + z_1^2)[y_1 + (x_1^2 + z_1^2)^{1/2}]}{x_1(y_1 + r_1)} \qquad (7.42)$$

in which $r_1 = (x_1^2 + y_1^2 + z_1^2)^{1/2}$. The attraction produced by a vertical lamina with opposite corners at x_1, y_1, z_1 and x_1, y_2, z_2 can be obtained, if we use eq. 7.42, as the difference between laminas with corners at $x_1, 0, 0$ and x_1, y_2, z_2 and at $x_1, 0, 0$ and x_1, y_1, z_1.

Horizontal rectangular lamina

The vertical attraction produced by horizontal lamina of thickness Δz can be obtained from eq. 7.11a. We integrate:

$$\Delta g = G\,\Delta\rho\,\Delta z \int_0^{y_1}\int_0^{x_1} \frac{h\,dx\,dy}{(x^2 + y^2 + h^2)^{3/2}} \qquad (7.43)$$

where the lamina is at depth h (Fig. 7.3j) and integration is from a corner $(0, 0, h)$ beneath P to an opposite corner (x_1, y_1, h). Integration produces:

$$\Delta g = G\,\Delta\rho\,\Delta z \left[\frac{\pi}{2} - \tan^{-1}\left(\frac{x_1 y_1}{h r_1}\right)\right] \qquad (7.44)$$

where $r_1 = (x_1^2 + y_1^2 + h^2)^{1/2}$.

Other mathematically described bodies

Similar integration procedures can be used to compute attractions by other simple-shaped bodies, such as vertical cylinders, fault (vertical or dipping) displacements, and infinite inclined dikes. Formulas for such bodies have been provided by various authors, particularly Heiland (1940, p. 151), Jakosky (1940, p. 356), and Nettleton (1976, p. 174).

In general it is difficult to approximate field structures satisfactorily with mathematically described bodies. Procedures for obtaining attractions produced by irregularly shaped bodies are described later in this chapter.

MAXIMUM DEPTH AND DEPTH-TO-TOP-OF-BODY ESTIMATES

Gravity-anomaly profiles provide useful estimates of the maximum depth to the source body. A sphere (or horizontal cylinder) produces the same gravity field as if it were a point (or a line) source. If we assume concentrated masses, the source is at a maximum depth. If we assume realistic densities, depth-to-top of the sphere or cylinder can be estimated.

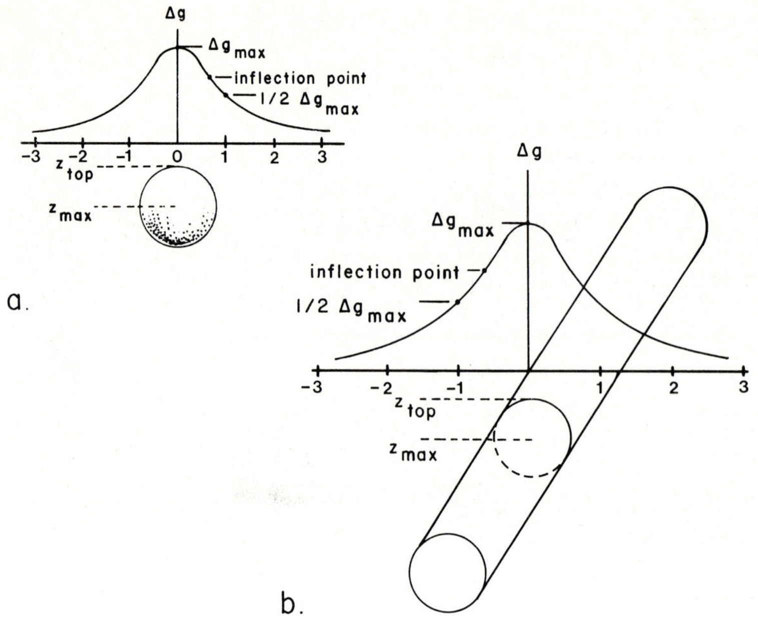

Fig. 7.4. Gravity-anomaly profiles over: a. sphere, and b. two-dimensional cylinder showing horizontal distances to the half-maximum gravity value and the inflection point on the profiles.

We can compute the maximum depth to a buried sphere (i.e., to its center) from a gravity profile across it. Eq. 7.21 gives the anomaly as horizontal distance from a point over the center (Fig. 7.4a). Directly above the sphere $(\Delta g)_{max} = GMz^{-2}$, and at x_α $\Delta g = ½(\Delta g)_{max}$. Thus:

$$\frac{(\Delta g)_{max}}{GM} = \frac{1}{z^2} = \frac{2z}{(x_\alpha^2 + z^2)^{3/2}} \tag{7.45}$$

The depth to center (maximum depth) is:

$$z = 1.3x_\alpha \tag{7.46}$$

or 1.3 times the distance to the half-maximum anomaly value.

We can also obtain the depth to the center of the sphere by twice differentiating eq. 7.21. The slope $d(\Delta g)/dx$ is zero over the sphere and is largest at the inflection point, where $d^2\Delta g/dx^2 = 0$. From the inflection point, a distance x_β from over the center of the sphere is obtained from:

$$\frac{d^2(\Delta g)}{dx^2} = 4GMz[15x_\beta^2(x_\beta^2 + z^2)^{-7/2} - 3(x_\beta^2 + z^2)^{-5/2}] \tag{7.47}$$

We obtain for the maximum depth:

$$z = 2x_\beta \tag{7.48}$$

The maximum depth is thus twice the distance to the inflection point.

For a line source (horizontal cylinder), the distance to the half-maximum value is obtained from eq. 7.26, which can be written

$$\frac{(\Delta g)_{\max}}{2GM} = \frac{1}{z} = \frac{2z}{(x_\alpha^2 + z^2)} \tag{7.49}$$

Solving for the maximum depth, we get:

$$z = x_\alpha$$

and for the inflection point at x_β the depth of a line source is:

$$z = \sqrt{3}\, x_\beta \tag{7.50}$$

These depth rules usually provide rough estimates because gravity profiles seldom correspond to true spheres or horizontal cylinders. The calculations are nevertheless useful because they indicate depths below which an anomaly cannot originate.

The above discussion indicates that anomaly profiles with high curvature correspond to shallow sources, and those with low curvature to deeper sources, provided the body has a uniform density.

TOTAL MASS DETERMINATIONS

The total mass of a buried body can be determined uniquely from its gravity field. The mass can be equated with the outward flux of gravity across an enclosing surface by applying Gauss's theorem (eq. 2.23). If we let the flux g_n in Fig. 2.7 be positive inward instead of outward across the surface of a body, eq. 2.23 becomes:

$$\int_S g_n \,\mathrm{d}S = 4\pi GM \tag{7.51}$$

in which integration is over the entire surface containing mass M. Hammer (1945) developed a procedure for determining the total mass of a buried body. All mass, including anomalous masses, are taken to lie within a hemisphere of infinite radius, as illustrated in Fig. 7.5. The solid angle subtended at a surface point P by the underlying hemisphere is 2π; no masses exist in the overlying hemisphere. We apply Gauss's theorem (e.g., eq. 2.23) and write for the downward gravitational flux over the horizontal surface (similar to eq. 2.22a):

$$\int_S g_n \,\mathrm{d}S = \int_{-\infty}^{\infty} \int_{-\infty}^{\infty} g_z \,\mathrm{d}x\,\mathrm{d}y = 2\pi GM \tag{7.52}$$

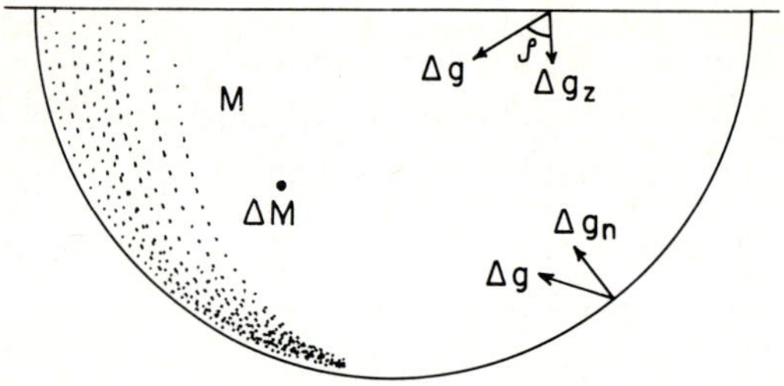

Fig. 7.5. Illustration of a semi-infinite hemisphere, where the upper surface represents the earth's surface. The total mass M consists of incremental masses ΔM, each producing an incremental attraction Δg.

in which g_z is the vertical component of gravity, x and y are horizontal coordinates, and M is the sum of all masses within the hemisphere beneath P. The total enclosed mass is then given by:

$$M = \frac{1}{2\pi G} \int_{-\infty}^{\infty} \int_{-\infty}^{\infty} g_z \, dx \, dy \tag{7.53}$$

An anomalous mass ΔM, superimposed on the mass of a country rock, can similarly be determined from the resultant gravity anomaly Δg_z:

$$\Delta M = \frac{1}{2\pi G} \int_S \Delta g_z \, dS \tag{7.54}$$

in which the integration extends over the area included by the gravity anomaly. The accuracy of the ΔM determination depends on how accurately the residual anomaly Δg_z is known.

The mass of an ore body, M_b (density ρ_b) can be determined from the anomalous mass computation by adding the mass of the displaced country rock, M_C (density ρ_c), to the value of ΔM. This requires knowledge of the density values ρ_b and ρ_c, as can be obtained from bore-hole or other data. Since $M_b/\rho_b = \Delta M/(\rho_b - \rho_c)$, the total mass of an ore body is given by:

$$M_b = \frac{1}{2\pi G} \left(\frac{\rho_b}{\rho_b - \rho_c} \right) \int_S \Delta g_z \, dS \tag{7.55}$$

LINE-INTEGRAL METHODS

Digital computers have application to the determination of attractions produced by complex structures when the body perimeter has been digitized. Line-integral techniques can be applied to two- and three-dimensional bodies.

A line-integral form for the attraction produced by a two-dimensional body is obtained by substituting into eq. 7.16b. We substitute $r \sin\theta$ for z (Fig. 7.6a):

$$\Delta g = 2G\,\Delta\rho \int_z \int_x \frac{z\,dx\,dz}{(x^2+z^2)} = 2G\,\Delta\rho \int_z \int_x \frac{\sin\theta\,dx\,dz}{r} \quad (7.56a)$$

and $d\theta$ for $\sin\theta\,dx/r$ to obtain:

$$\Delta g = 2G\,\Delta\rho \int_z \int_\theta d\theta\,dz \quad (7.56b)$$

The attraction Δg_j (Fig. 7.6b) produced by any one solenoid (Hubbert, 1948a), an infinitely long bar with a trapezoidal cross section and sides

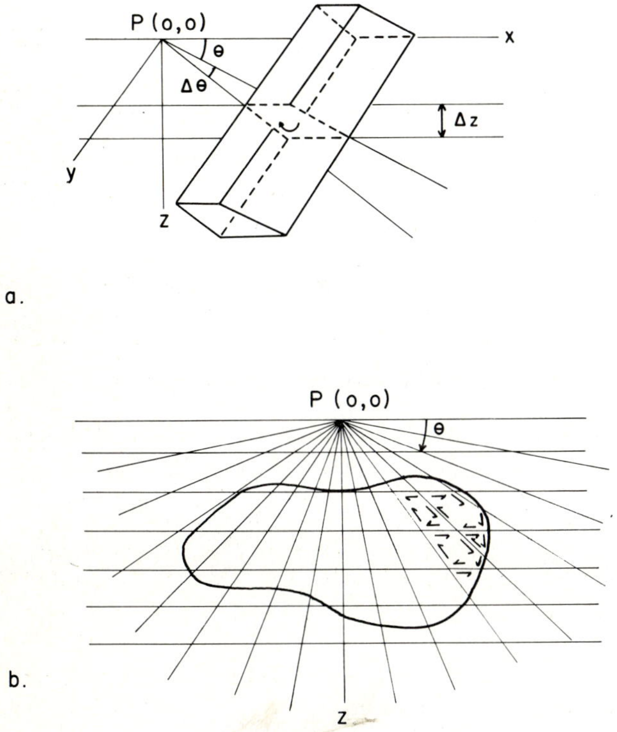

Fig. 7.6. a. Two-dimensional solenoid $\Delta\theta\ \Delta z$. b. Template of solenoids $\Delta\theta\ \Delta z$ each of which exerts an identical vertical attraction at P. The attraction of the body can be obtained by counting the number of included solenoids or by integrating around the perimeter of the body (integrations along interior sides of each solenoid cancel out, such that the line integral around the perimeter of the body describes the total attraction).

$\Delta\theta\ \Delta z$, is given by:

$$\Delta g_j = 2G\ \Delta\rho \int_z^{z+\Delta z} \int_\theta^{\theta+\Delta\theta} \mathrm{d}\theta\ \mathrm{d}z = 2G\ \Delta\rho\ \Delta\theta\ \Delta z \tag{7.57}$$

As shown below, eq. 7.57 is equivalent to the line integral around a solenoid, that is:

$$\Delta g_j = 2G\ \Delta\rho \oint z \mathrm{d}\theta = -2G\ \Delta\rho \oint \theta\ \mathrm{d}z \tag{7.58}$$

If we integrate around the perimeter $\Delta\theta\ \Delta z$:

$$\oint z\ \mathrm{d}\theta = \int_\theta^\theta z \Big|_{z=z}^{z=z+\Delta z} \mathrm{d}\theta + \int_\theta^{\theta+\Delta\theta} (z+\Delta z)\ \mathrm{d}\theta + \int_{\theta+\Delta\theta}^{\theta+\Delta\theta} z \Big|_{z=z+\Delta z}^{z=z} \mathrm{d}\theta$$

$$+ \int_{\theta+\Delta\theta}^{\theta} z\ \mathrm{d}\theta \tag{7.59a}$$

The first and third terms on the right side are zero because θ is constant along the path of integration. The second and fourth terms produce:

$$\oint z\ \mathrm{d}\theta = (z+\Delta z)\Delta\theta - (z)\Delta\theta = \Delta z\ \Delta\theta \tag{7.59b}$$

If we integrate eq. 7.59a in a counterclockwise direction, the limits of integration are reversed and the line integral in eq. 7.59b becomes $-\Delta z \Delta\theta$. Integration around the trapezoid can be performed in either direction, but the same direction must be used around all loops.

If we similarly integrate $\oint \theta\ \mathrm{d}z$ in a clockwise direction, it is found that:

$$\oint \theta\ \mathrm{d}z = -\Delta\theta\ \mathrm{d}z \tag{7.60}$$

Vertical attractions produced by any of the solenoids are thus:

$$\Delta g_j = 2G\ \Delta\rho\ \Delta z\ \Delta\theta = 2G\ \Delta\rho \oint z\ \mathrm{d}\theta = -2G\ \Delta\rho \oint \theta\ \mathrm{d}z \tag{7.61}$$

We can divide an irregularly shaped two-dimensional body (Fig. 7.6b) into solenoids of equal $\Delta\theta$ and equal Δz, each having identical vertical attractions Δg_j at a point P. The total attraction is the sum of the solenoids enclosed by the body, that is, Δg_j times the number of included solenoids. A simpler procedure than counting solenoids is to obtain a line integral around the perimeter of the entire body.

It is seen in Fig. 7.6b that along every common side between adjacent solenoids the integration paths are in opposite directions, such that the sum of integrals along common sides cancel out. It follows then, that the only lines which contribute to the line integral comprise the perimeter of the body. Thus:

$$\Delta g = \sum_j \Delta g_j = 2G\ \Delta\rho \sum_j \int_\theta \int_z \mathrm{d}z\ \mathrm{d}\theta = 2G\ \Delta\rho \left[\oint z\ \mathrm{d}\theta\right]_{\text{perimeter}} \tag{7.62}$$

which can also be written:

$$\Delta g = -2G\,\Delta\rho\,\left[\oint \theta\;\mathrm{d}z\right]_{\text{perimeter}} \tag{7.63}$$

The line-integral method can be illustrated by calculating the vertical attraction at P due to an infinite horizontal slab. We use an equation equivalent to eq. 7.63 and obtain:

$$\Delta g = 2G\,\Delta\rho \oint z\;\mathrm{d}\theta = 2G\,\Delta\rho\,\left[\int_0^{\pi} (z + \Delta z)\;\mathrm{d}\theta + \int_{\pi}^{0} z\;\mathrm{d}\theta\right] \tag{7.64}$$

which reduces to the slab formula $\Delta g = 2\pi G\,\Delta\rho\,\Delta z$ (eq. 7.28). Eq. 7.64 could also have been written:

$$\Delta g = -2G\,\Delta\rho \oint \theta\;\mathrm{d}z = -2G\rho\left[\int_z^{z+\Delta z} \tan^{-1}\left(\frac{z}{x_1}\right)\mathrm{d}z + \int_{z+\Delta z}^{z} \tan^{-1}\left(\frac{z}{x_2}\right)\mathrm{d}z\right] \tag{7.65}$$

in which x_1 and x_2 are horizontal distances from a point P to the ends of the slab (i.e., $\pm\infty$). The arc tangents then become 0 and π, and we obtain the same slab formula.

Hubbert (1948b) used line integrals to develop curves of gravity profiles corresponding to variously shaped two-dimensional bodies, some of which are shown in Fig. 6.4. These curves can be applied in the field to make rapid estimates of the terrain correction which will be associated with a two-dimensional feature and, conversely, to determine optimum locations of gravity stations which will entail minimal terrain corrections. Line-integral methods can be readily applied to compute gravity profiles corresponding to complex two-dimensional, buried cross sections. Where possible, depth control, as from seismic measurements, provides constraints. Fig. 7-7 illustrates such a section. Attractions along a profile are computed which include the effects produced by loops around each layer comprising the hypothetical model, and then summing the effects of all loops. Computer procedures are described in the next section.

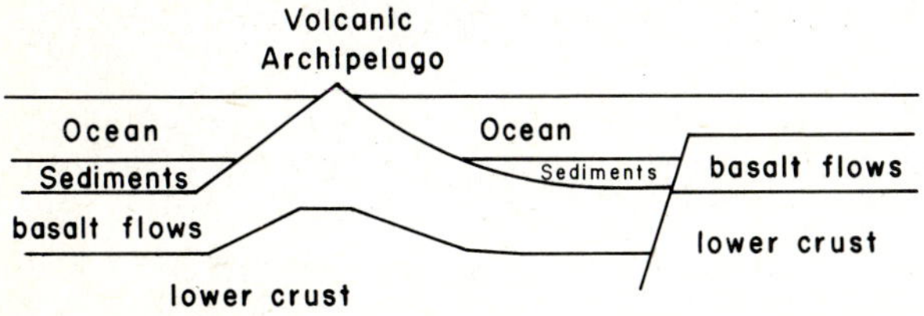

Fig. 7.7. Schematic representation of a two-dimensional cross section of an oceanic region. The gravity anomaly can be computed using the sum of attractions obtained by line integrals around each layer.

Three-dimensional bodies

Line-integral methods can also be used to analyze three-dimensional bodies. Talwani and Ewing (1960) developed an effective three-dimensional method. A body is divided into horizontal slabs, line-integrals around the perimeter of each slab are obtained, and the total attraction is the sum of the attractions of the individual slabs.

The attraction which a three-dimensional body produces can be developed from eq. 7.12b. Where P is above some part of a slab of thicknesses Δz, the attraction becomes:

$$\Delta g = G\,\Delta\rho\,\Delta z \int_0^{2\pi}\int_0^{r} \frac{zr\,\mathrm{d}r\,\mathrm{d}\theta}{(r^2+z^2)^{3/2}} \tag{7.66}$$

We integrate over r to obtain:

$$\Delta g = G\,\Delta\rho\,\Delta z \int_0^{2\pi}\left[1-\frac{z}{(r^2+z^2)^{1/2}}\right]\mathrm{d}\theta \tag{7.67}$$

and then integrate around the slab perimeter to obtain:

$$\Delta g = G\,\Delta\rho\,\Delta z\left\{\oint \mathrm{d}\theta - \oint\left[\frac{z}{(r^2+z^2)^{1/2}}\right]\mathrm{d}\theta\right\} \tag{7.68}$$

which is readily adaptable to computer programing. For an infinite slab, the second term on the right in eq. 7.68 becomes zero ($r\to\infty$), and the equation reduces to the slab formula.

If P is not directly above the slab, the calculation can be obtained in two steps, illustrated in Fig. 7.9b. The slab size is arbitrarily increased to include the projection of point P, and the attraction caused by the enlarged slab is computed. The attraction produced by the enlarged part of the slab is then subtracted, and we thus obtain the attraction of the true slab. Attractions produced by separate slabs are summed to obtain the total attraction. Computer programing for this procedure is described in the following section.

NUMERICAL METHODS

Digital computers can compute attractions produced by quite irregularly shaped two- or three-dimensional bodies. The hypothetical model must first be digitized. Various methods can be used to permit the computer to compute an anomaly. The computed anomaly is then compared with the observed and the model adjusted until a good fit is obtained.

Two-dimensional numerical methods

Irregularly shaped two-dimensional models can be readily approximated by polygons of any number of sides. We can obtain the resultant vertical

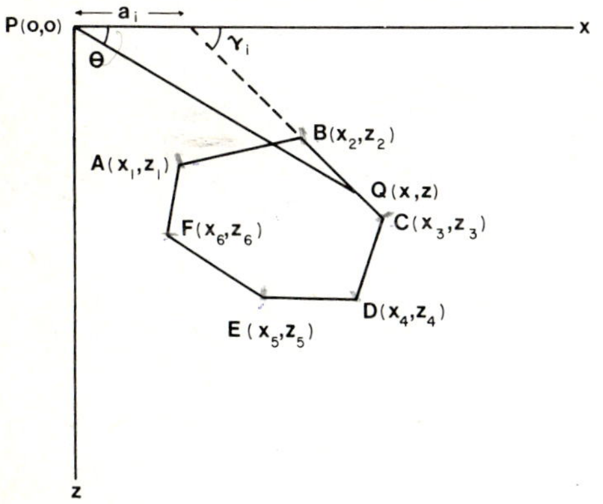

Fig. 7.8. Two-dimensional polygon *ABCDEF*; a_i is the intercept of any side along the x axis.

attraction if we use the line-integral methods described by Hubbert (1948a) and elegantly adapted for computer programing by Talwani et al. (1959; see also Talwani, 1973). Talwani's program has been widely used, where various individuals have modified it to work on the particular computers available to them.

Fig. 7.8 illustrates a constant-density, two-dimensional body approximated by a polygon. The vertices of the polygon, in this case a hexagon, can be digitized, as at points $A\ (x_1, z_1)$, $B\ (x_2, z_2) \ldots F\ (x_6, z_6)$. The attraction is obtained from the line integral around the perimeter *ABCDEFA*, such that:

$$\Delta g = 2G\ \Delta\rho \left[\int_A^B z\ d\theta + \int_B^C z\ d\theta + \ldots \int_F^A z\ d\theta\right] \tag{7.69}$$

The depth of a point $Q(x, z)$ on a side of the hexagon is $z = x \tan\theta = (x - a_1)\tan\gamma_i$, where $i = 1, 2 \ldots 6$, and a_i is a distance along the x axis (Fig. 7.8). Eliminating x gives for the depth, as a function of θ (Talwani et al., 1959):

$$z = \frac{a_i \tan\theta \tan\gamma_i}{\tan\gamma_i - \tan\theta} \tag{7.70}$$

The integral for the side *BC* becomes, in the general case:

$$\int_B^C z\ d\theta = a_i \sin\gamma_i \cos\gamma_i$$

$$\left\{(\theta_i - \theta_{i+1}) + \tan\gamma_i \ln\left[\frac{\cos\theta_i\ (\tan\theta_i - \tan\gamma_i)}{\cos\theta_{i+1}\ (\tan\theta_{i+1} - \tan\gamma_i)}\right]\right\} \tag{7.71}$$

To digitize this equation we substitute:

$$a_i = x_{i+1} + z_{i+1} \text{ cotan } \gamma_i = x_{i+1} + z_{i+1} \left(\frac{x_{i+1} - x_i}{z_{i+1} - z_i} \right)$$

$$\text{where } \gamma_i = \tan^{-1} \left(\frac{z_{i+1} - z_i}{x_{i+1} - x_i} \right) \tag{7.72}$$

and reduce terms, obtaining for the depth at Q:

$$z = \frac{x_{i+1}\,(z_{i+1} - z_i) - z_{i+1}\,(x_{i+1} - x_i)}{(z_{i+1} - z_i) \text{ cotan } \theta - (x_{i+1} - x_i)} \tag{7.73}$$

This expression is substituted into eq. 7.71 to produce for the attraction around the entire polygon (e.g., Talwani, 1973, with slight modification):

$$\Delta g = 2G\,\Delta\rho \sum_{i=1}^{6} \left[\frac{x_i z_{i+1} - z_i x_{i+1}}{(x_{i+1} - x_i)^2 + (z_{i+1} - z_i)^2} \right] \left[(x_{i+1} - x_i)(\theta_i - \theta_{i+1}) + (z_{i+1} - z_i) \ln \left(\frac{r_{i+1}}{r_i} \right) \right] \tag{7.74}$$

in which $r_i = (x_i^2 + z_i^2)^{1/2}$.

The attraction produced by a two-dimensional body consisting of a series of polygons, each with its own configuration and density, can be computed by successive application of eq. 7.74 and summing. An illustration of such a body, Fig. 7.7, is an oceanic crustal section consisting of a water layer, sediments, and underlying igneous layers. The loops extend to $\pm\infty$ in the x direction, an effect which is approximated in computers if we use very large distances, so that θ (Fig. 7.8) is made both very small and near 180°. Particularly in models that include lower crustal layers, θ must be made sufficiently small to eliminate edge effects, which is done either by integrating to very large horizontal distances or by flagging that path of integration and setting θ equal to 0 or π.

In models represented by polygons with only rectangular sides, the attraction can be represented by a function of the coordinates of the vertices and evaluated without use of a digital computer. The attraction produced by a rectangle with P (origin) at one corner is (similar to eq. 7.65):

$$\Delta g = 2G\,\Delta\rho \int_0^{z_1} \tan^{-1} \left(\frac{z}{x_1} \right) dz \tag{7.75}$$

where x_1, z_1 are the coordinates of the corner opposite to P.
Integrating this equation gives:

$$\Delta g = 2G\,\Delta\rho z_1 \left[\tan^{-1} \left(\frac{x_1}{z_1} \right) + \frac{x_1}{z_1} \ln \left(1 + \frac{z_1^2}{x_1^2} \right) \right] \tag{7.76}$$

To obtain the attraction of a rectangle which does not have a corner at P, eq. 7.76 is applied successively, as sums and differences of the rectangles that do have corners at P (analogous to the procedure in eq. 7.40).

Three-dimensional numerical methods

We can approximate three-dimensional models of irregular shape if we use variously shaped "building blocks" and summing on the blocks. Talwani and Ewing (1960) developed methods for computing attractions produced by horizontal polygonal laminas or slabs, which represent horizontal sections of the model. The total attraction is the sum of the separate slab attractions.

The Talwani polygonal slab, illustrated in Fig. 7.9, has probably had the widest application of the various building-block approaches. It is a laborious procedure for complicated models, but it can provide reliable values for the attraction. Slabs of thickness Δz are given polygonal perimeters which approximate the model at the specified depth. The attraction at P (Fig. 7.9) produced by a slab of thickness Δz is given by eq. 7.68. The formula is evaluated along lines AB, BC, . . . FA and the values summed to obtain the slab attraction.

The angle ψ in Fig. 7.9 is seen to equal:

$$\psi = \pi - \zeta_i - \phi_{i+1} + \phi \tag{7.77}$$

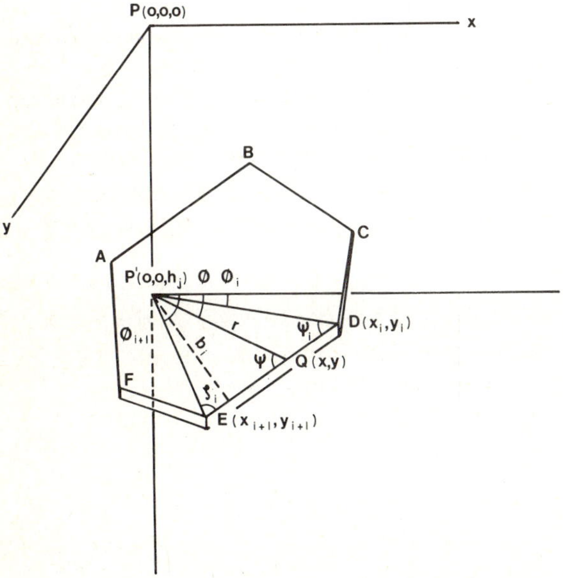

Fig. 7.9. Polygonal lamina or slab of thickness Δz at depth h_j. P' is the projection of the origin to the slab; i indicates the ith vertex, j the jth depth.

such that the radius vector r from P' can be expressed in terms of b_i, the normal to DE, by:

$$r = \frac{b_i}{\sin \psi} = \frac{b_i}{\sin (\pi - \zeta_i - \phi_{i+1} + \phi)} \tag{7.78}$$

The attraction along DE at depth h_j can be written (similar to eq. 7.68):

$$(\Delta g)_{DE} = G\, \Delta\rho\, \Delta z \left[\int_D^E d\phi - \int_D^E \frac{h_j \sin \psi \, d\phi}{(r^2 + h_j^2)^{1/2}} \right] \tag{7.79}$$

which, upon integration, becomes:

$$(\Delta g)_{DE} = G\, \Delta\rho\, \Delta z \left\{ (\phi_{i+1} - \phi_i) - \sin^{-1} \left[\frac{h_j \cos \psi_i}{(b_i^2 + h_j^2)^{1/2}} \right] \right.$$

$$\left. - \sin^{-1} \left[\frac{h_j \cos \zeta_i}{(b_i^2 + h_j^2)^{1/2}} \right] \right\} \tag{7.80}$$

To program eq. 7.80 for computer calculations, we rewrite the equations in terms of x_i, y_i, x_{i+1}, and y_{i+1}. Talwani (1973) thus obtained for the slab attraction:

$$\Delta g = G\, \Delta\rho\, \Delta z \sum_i \left\{ \Lambda \cos^{-1} \left[\frac{x_i}{r_i} \left(\frac{x_{i+1}}{r_{i+1}} \right) - \frac{y_i}{r_i} \left(\frac{y_{i+1}}{r_{i+1}} \right) \right] \right.$$

$$\left. + \sin^{-1} \left[\frac{h_j c_i \mathrm{B}}{(b_i^2 + h_j^2)^{1/2}} \right] - \sin^{-1} \left[\frac{h_j e_i \mathrm{B}}{(b_i^2 + h_j^2)^{1/2}} \right] \right\} \tag{7.81}$$

in which B = +1 if b_i is positive and −1 if b_i is negative; $\Lambda = +1$ if the term $\left[\frac{y_i x_{i+1} - x_i y_{i+1}}{r_i r_{i+1}} \right]$ is positive and −1 if the term is negative;

$$b_i = \frac{x_i (y_i - y_{i+1}) - y_i (x_i - x_{i+1})}{[(y_i - y_{i+1})^2 + (x_i - x_{i+1})^2]^{1/2}}$$

$$c_i = \frac{x_i (x_i - x_{i+1}) - y_i (y_i - y_{i+1})}{r_i \, [(y_i - y_{i+1})^2 + (x_i - x_{i+1})^2]^{1/2}}$$

$$e_i = \frac{x_{i+1} (x_i - x_{i+1}) - y_{i+1} (y_i - y_{i+1})}{r_{i+1} \, [(y_i - y_{i+1})^2 + (x_i - x_{i+1})^2]^{1/2}}$$

$$r_i = (x_i^2 + y_i^2)^{1/2} \qquad r_{i+1} = (x_{i+1}^2 + y_{i+1}^2)^{1/2}$$

The term $\Sigma_i(\phi_{i+1} - \phi)$ in eq. 7.80 equals 2π if P' (origin projected on to the slab) lies within the polygon, is zero if P' lies outside it, and equals the angle subtended if P' lies on the boundary of the polygon.

The vertical attraction resulting from a series of polygonal slabs is clearly the sum of the separate slab attractions.

Other types of building blocks can be similarly programed for computer calculations and summed to obtain Δg. Various investigators have used vertical prisms and parallelepipeds. Cordell and Henderson (1968) used a series of vertical prisms, extending up from a flat bottom, down from a flat top, or symmetrically about a horizontal plane, as a model. They used an iterative method to modify the model, based on ratios of observed to computed anomaly, until a satisfactory fit with observed gravity was obtained. Each prism was digitized and its attraction computed with the use of formulas of the types of eq. 7.39 and 7.40.

Nagy (1966a) developed a computer subroutine for a rectangular parallelepiped as a unit building block. These blocks can be assigned separate values of density and combined to represent a hypothetical structure. The equations using digitized coordinates of the building block are lengthy and cumbersome. However, the same computer subroutine is applicable to all digitized blocks.

Bhattacharyya and Navolio (1975) represented models by a finite number of rectangular parallelepipeds. They developed a novel approach for computing anomalies produced by arbitrarily shaped bodies. The anomaly is expressed as a convolution of the shape of the source body and a distance function (describing the distance from source to station). Bhattacharyya and Navolio describe the convolution method.

TEMPLATES

Templates of various types can be constructed to obtain attractions produced by irregularly shaped two- and three-dimensional models. Fig. 7.6b is one such template for two-dimensional bodies, where each solenoid exerts identical vertical attractions at P, equal to $2G\ \Delta\rho\ \Delta\theta\ \Delta z$ (eq. 7.61), and the total attraction is the product of this attraction per solenoid times the number of solenoids comprising the model. Fig. 6.2 is an example of a template used for computing terrain corrections. Templates can be similarly constructed to obtain attractions produced by various three-dimensional models, as described by Gassman (1951) or Talwani (1973). The advent and versatility of digital computers has resulted in a generally decreased emphasis on use of templates.

INVERSE METHODS FOR INTEGRATING GRAVITY ANOMALIES

Inverse methods solve integral equations which satisfy observed anomalies, and thereby determine unknown mass distributions. Solutions require that we assume either density contrasts or shape of the body. The equations to be solved are of the form of eqs. 7.10a or b for bodies with three dimensions and eqs. 7.16a or b for those with two dimensions. In many of the inverse problems, an equivalent layer or sheet at some depth h is the assumed shape of the body, such that integration over depth is simplified to produce an integral equation in the variable x (or x and y).

Inverse methods (Bott, 1973) concern either linear or nonlinear integral equations. If linear, the shape of the body is assumed, usually an equivalent layer, and density distributions are solved for in that layer which satisfy the integral equations. If nonlinear, a density contrast(s) is assumed and the shape(s) satisfying the integral equation is solved for.

Three types of solutions are used to solve linear inverse problems (Bott, 1973). One is the Fourier approach, where Fourier analyses are made of the anomaly and the equivalent layer is synthesized from each of the Fourier components; the Tsuboi (1938) method for the downward continuation of gravity to an equivalent layer applies this approach to determine variations in depths to Moho. The second is the convolution method, in which appropriate filters are applied for the downward continuation to an equivalent layer containing density distributions which satisfy the integral equation; the $(\sin x)/(x)$ method of Tomoda and Aki (1955) provides such a filter. The third applies matrix methods, in which linear algebra is used to obtain matrices for synthesizing the shape of the body. The Fourier and convolution methods can be readily applied to obtain equivalent layers of source bodies. The matrix method can be used to analyze more generally shaped bodies. Linear methods can usually provide solutions if the anomaly-wavelength components are longer than twice the station spacing and longer than about three-fourths of the depth to the equivalent layer (Bott, 1973).

Bott (1973) also discussed nonlinear inverse problems. Tanner (1967) developed a practical method in which he iterated the nonlinear problem by linear approximations until he obtained a good fit of the observed anomalies. Tanner used computer programs for this iterative procedure, which he applied successfully to specific types of problems. Nonlinear optimization techniques have been developed with the advent of digital computers. Al-Chalabi (1970, 1971) described methods to determine optimum values of parameters in a given function.

Fourier (linear) method of inversion

An observed gravity anomaly can be analyzed into individual Fourier components, where each component is interpreted in terms of a corresponding density variation on a sheet (or equivalent layer). By summing the individual Fourier components, the density distribution on the sheet (layer) which fits the observed anomaly is synthesized. The Fourier method assumes that the anomaly is periodically repeated. It is therefore desirable to surround the anomaly by a border of zero values, so as to avoid end distortions.

A two-dimensional analysis illustrates the Fourier method, where the earth's curvature is neglected. Gravity anomalies are projected downward to a plane at depth h with source masses represented by a surface-coated density expressed as a harmonic function of x; thus:

$$\Delta\sigma(x, h) = \Delta\sigma_0 \cos kx \tag{7.82}$$

in which k is the wave number $2\pi/\lambda$, and λ the wavelength of the anomaly. The gravity anomaly on a large sheet at depth h is, from eqs. 7.29 and 7.82:

$$\Delta g(x, h) = 2\pi G\, \Delta\sigma(x, h) = 2\pi G\, \Delta\sigma_0 \cos kx \tag{7.83}$$

The upward projection of this anomaly to the earth's surface decreases exponentially. At the surface the anomaly (which satisfies Laplace's equation) can thus be written:

$$\Delta g(x, 0) = e^{-kh}\, \Delta g(x, h) = 2\pi G e^{-kh}\, \Delta\sigma_0 \cos kx \tag{7.84}$$

Expressed as a Fourier series, the anomaly at the surface can also be written:

$$\Delta g(x, 0) = \tfrac{1}{2}\, a_0 + \sum_{k=1}^{n} (a_k \cos kx + b_k \sin kx) \tag{7.85}$$

in which:

$$a_0 = \frac{1}{\pi}\int_{-\pi}^{\pi} \Delta g\, dx \qquad a_k = \frac{1}{\pi}\int_{-\pi}^{\pi} \Delta g \cos kx\, dx$$

$$b_k = \frac{1}{\pi}\int_{-\pi}^{\pi} \Delta g \sin kx\, dx \tag{7.86}$$

Each Fourier component can be projected from the surface downward to depth h by multiplying it by e^{kh}. Thus:

$$\Delta g(x, h) = e^{kh}\, \Delta g(x, 0) = e^{kh}\left[\tfrac{1}{2}a_0 \sum_{k=1}^{n} (a_k \cos kx + b_k \sin kx)\right]$$

$$= 2\pi G\, \Delta\sigma_0 \cos kx \tag{7.87}$$

from which $\Delta\sigma(x, h)$ can be synthesized. The density distribution on the plane at depth h is thus obtained for an observed gravity profile.

Tsuboi (1938) developed and applied this approach to determine undulations of the Moho for the case of constant crustal and mantle densities. The surface-coated density at depth h is converted to volume density for an equivalent layer, where $\Delta\sigma(x, h) = \Delta\rho\, \Delta h$. The layer thickness Δh represents undulations of the Moho about an average depth h. For density contrasts at the Moho of $\Delta\rho = 3.30-2.85 = 0.45$ g/cm^3 we obtain:

$$\Delta\sigma = 0.45\ \Delta h \quad \Delta h = 2.2\ \Delta\sigma \tag{7.88}$$

where $\Delta\sigma$ takes on the value for each point along the profile as determined in eq. 7.87. By this procedure Moho undulations can be obtained along the observed profile.

The Fourier method can be applied to three-dimensional cases, where a grid of equidistant gravity stations is used. The procedure has been described by Kanasewich and Agarwal (1970) and summarized by Bott (1973). The Fourier method has become attractive since the development of the fast Fourier transform algorithm (Cooley and Tukey, 1965).

Convolution (linear) method of inversion

Another method for the downward continuation of gravity to an equivalent layer is based on the operation in which a convolution is obtained (the operation is called a convolution) of the anomaly with an analytically derived filter function. Formulas for obtaining the density distribution in the equivalent layer, based on the convolution theorem (which states that the Fourier transform of a convolution equals the product of the separate Fourier transforms), have been obtained by Bott (1967, 1973) for two-dimensional cases, and can be extended to three-dimensional problems.

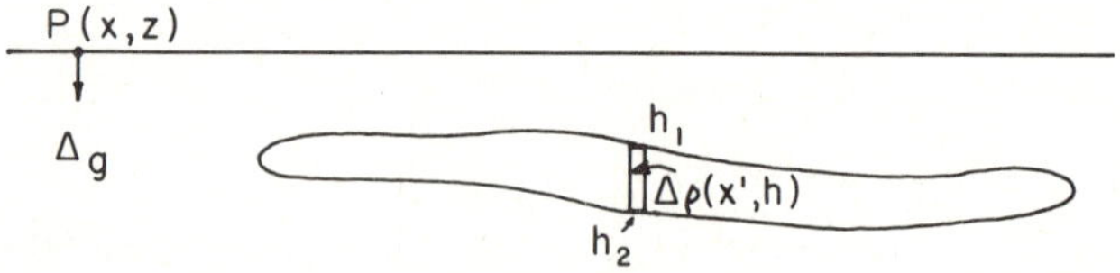

Fig. 7.10. A two-dimensional body (e.g., an equivalent layer) with density $\Delta\rho(x', h)$ producing a gravity anomaly Δg at point P.

We assume a density distribution which produces a two-dimensional gravity anomaly to be concentrated as a surface coating on a sheet of depth h or as a volume density in an equivalent layer between depths h_1 and h_2 (Fig. 7.10). The gravity anomaly along a traverse can be computed from eq. 7.16a. We integrate with respect to z and obtain an integral equation for $\Delta\rho(x', h)$ in the variable x. We first integrate eq. 7.16a, written as:

$$\Delta g(x) = 2G \int_{h_1}^{h_2} \int_{-\infty}^{\infty} \Delta\rho(x', h) \left[\frac{(z' - z)\,\mathrm{d}x'}{(x' - x)^2 + (z' - z)^2} \right] \mathrm{d}z' \tag{7.89}$$

to obtain:

$$\Delta g(x) = 2G \int_{-\infty}^{\infty} \Delta\rho(x', h) \{ \ln[(h_2 - z)^2 + (x' - x)^2]^{1/2} - \ln[(h_1 - z)^2 + (x' - x)^2] \}\ \mathrm{d}x' \tag{7.90}$$

which can be written in the form (Bott, 1973):

$$\Delta g(x) = \frac{1}{\sqrt{2\pi}} \int_{-\infty}^{\infty} \Delta\rho(x', h)\, K\,(x - x'; h_1; h_2)\, dx' \tag{7.91}$$

in which $\Delta\rho(x', h)$ is the density of the layer to be solved for and $K(x - x'; h_1; h_2)$ is the data kernel or the nucleus in the integral equation. K is characterized by a separate value for each observed value of $\Delta g\,(x)$. The formula for K which applies in eq. 7.90 is:

$$K = \sqrt{2\pi}(2G)\{\ln[(h_2 - z)^2 + (x' - x)^2]^{1/2} - \ln[(h_1 - z)^2 + (x' - x)^2]^{1/2}\} \tag{7.92}$$

Eq. 7.91 is a convolution integral which gives the solution $\Delta g\,(x)$ as the convolution (sometimes referred to as Faltung) of the two functions $\Delta\rho$ and K in the variable x over the interval $-\infty$ to $+\infty$. We define the convolution (eq. 7.91) symbolically as:

$$\Delta g(x) = \Delta\rho * K \tag{7.93}$$

The equivalent layer is usually at a finite depth. The anomaly $\Delta g\,(x)$ must then be smoothed (i.e., filtered) before it can be inverted to recover the lost variable x so as to avoid singularities in $\Delta g\,(x)$ at short wavelengths, that is, as $k \to \infty$. Singularities can be avoided by replacing $\Delta g\,(x)$ with a filtered anomaly function $\Delta f\,(x)$ (Bott, 1967, 1973). Thus:

$$\Delta f(x) = \frac{1}{\sqrt{2\pi}} \int_{-\infty}^{\infty} \Delta g(\bar{x})\, \Gamma\,(x - \bar{x})\, d\bar{x} = \Delta g * \Gamma \tag{7.94}$$

in which $\bar{x} = x/a$, a is the station interval, and $\Gamma(x - \bar{x})$ is the appropriate filter function.

We use Fourier transforms to transform solutions for Δg and Δf from the x to the k domains. The Fourier transform of $\Delta g(x)$ is $\overline{\Delta g}(k)$, here defined as:

$$\overline{\Delta g}\,(k) = \frac{1}{\sqrt{2\pi}} \int_{-\infty}^{\infty} \Delta g(x)\, e^{-ikx}\, dx \tag{7.95}$$

in which the transform operator k is taken as the wave number, $i = (-1)^{1/2}$ and

$$e^{-ikx} = \cos kx - i \sin kx$$

is the data kernel. The corresponding inverse Fourier transform is here defined as:

$$\Delta g(x) = \frac{1}{\sqrt{2\pi}} \int_{-\infty}^{\infty} \overline{\Delta g}(k)\, e^{ikx}\, dk \tag{7.96}$$

in which:

$$e^{ikx} = \cos kx + i \sin kx$$

Convolution in the x domain (i.e., $\Delta g(x) = \Delta\rho * K$) is equivalent to multiplication in the k domain (i.e., $\overline{\Delta g}\ (k) = \overline{\Delta\rho}(k)\overline{K}(k)$) (e.g., Jenkins and Watts, 1968, p. 45). The Fourier transform of an output function in the k domain is therefore equivalent to multiplying the input and filter functions at the same value of k. The transforms of eqs. 7.93 and 7.94 can thus be written as:

$$\overline{\Delta g}(k) = \overline{\Delta\rho}(k)\ \overline{K}\ (k) \tag{7.97a}$$

$$\overline{\Delta f}(k) = \overline{\Delta g}(k)\ \overline{\Gamma}\ (k) \tag{7.97b}$$

When $\overline{\Delta f}\ (k)$, the transform of the filtered anomaly, is taken equal to $\overline{\Delta g}\ (k)$, the transform of the observed anomaly, the transform of the density is:

$$\overline{\Delta\rho}(k) = \frac{\overline{\Delta g}(k)\ \overline{\Gamma}\ (k)}{\overline{K}(k)} \tag{7.98}$$

where the filter function is chosen such that $\overline{\Delta\rho}(k)$ is a Fourier transform.

To recover the density distribution in the equivalent layer it is necessary to synthesize or add up contributions for all values of k for the same value of x. This we obtain from the the inverse Fourier transform:

$$\Delta\rho(x', h) = \frac{1}{2\pi} \int_{-\infty}^{\infty} \overline{\Delta\rho}(k)\ e^{ikx'}\ dk \tag{7.99}$$

We substitute from eq. 7.98 into 7.99 to obtain:

$$\Delta\rho(x', h) = \frac{1}{2\pi} \int_{-\infty}^{\infty} \overline{\Delta g}(k) \left[\frac{\overline{\Gamma}\ (k)}{\overline{K}\ (k)}\right] e^{ikx'}\ dk \tag{7.100}$$

and substitute eq. 7.95 for $\Delta g\ (k)$ to obtain:

$$\Delta\rho(x', h) = \frac{1}{\sqrt{2\pi}} \int_{-\infty}^{\infty} \Delta g(x) \left\{ \int_{-\infty}^{\infty} e^{-ik(x-x')} \left[\frac{\overline{\Gamma}\ (k)}{\overline{K}(k)}\right] dk \right\} dx \tag{7.101}$$

Eq. 7.101 permits formal calculation of density distributions at points x' and h in the layer from the observed anomaly $\Delta g\ (x)$.

If the equivalent layer is a sheet at zero depth, the surface-density distribution is:

$$\Delta\sigma(x, y) = \frac{1}{2\pi G}\ [\Delta g(x, y)] \tag{7.102}$$

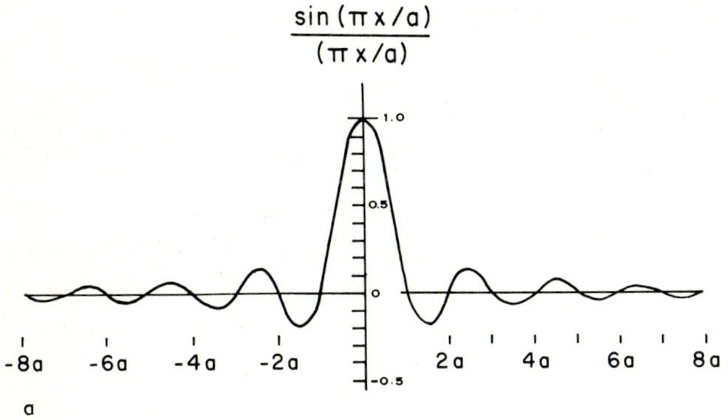

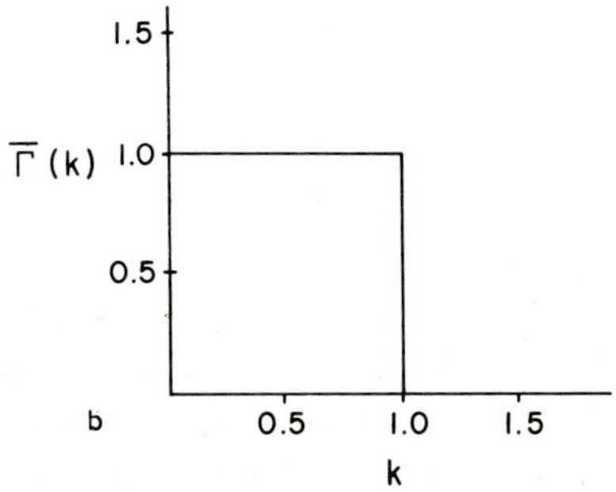

Fig. 7.11. a. Curve $\Gamma(\pi x/a) = \sin(\pi x/a)/(\pi x/a)$ as a function of horizontal distance a; b. Fourier transform $\overline{\Gamma}(k)$ of $\Gamma(\pi x/a)$ as function of wavenumber k.

The convolution method can be illustrated when we apply a filter function $\sin x/x$, which sharply filters anomalies with wavelengths shorter than twice the station interval. Tomoda and Aki (1955) applied this type of filter to obtain two-dimensional density distributions on a sheet buried at depth h.

A filter function $\Gamma(x - \bar{x})$ is determined uniquely by values distributed at equal distances along a line, according to a theorem in information theory, if no wavelengths are shorter than twice the station interval. Following Tomoda and Aki, the filter function $\sin x/x$ (where x is in radians) is known to be unity at $x = 0$ and zero at $x = n\pi$ (Fig. 7.11a). To give x in units of distance, we write for the filter:

$$\Gamma\left(\frac{\pi x}{a}\right) = \frac{\sin(\pi x/a)}{(\pi x/a)} \begin{array}{l} = 1 \text{ at } x = 0 \\ = 0 \text{ at } x = \pm a, \pm 2a, \ldots \pm na \end{array} \qquad (7.103)$$

The Fourier transform of $\Gamma(\pi x/a)$ (from the definition of eq. 7.95) is:

$$\begin{aligned}\bar{\Gamma}(k) &= \frac{1}{a}\sqrt{\frac{\pi}{2}}\int_{-\infty}^{\infty}\Gamma\left(\frac{\pi x}{a}\right)\, e^{-i(\pi kx/a)}\, dx \\ &= \frac{1}{a}\sqrt{\frac{\pi}{2}}\int_{-\infty}^{\infty}\frac{\sin(\pi x/a)}{(\pi x/a)}\, e^{-i(\pi kx/a)}\, dx \begin{array}{ll} = 1 & 0 < k \leqslant 1 \\ = 0 & k > 1 \end{array}\end{aligned} \tag{7.104}$$

The equation shows that the amplitudes of the component waves of $\Gamma(\pi x/a)$ as illustrated in Fig. 7.11b, are unity for wave numbers less than one and zero for wave numbers greater than one. The inverse transform is:

$$\Gamma\left(\frac{\pi x}{a}\right) = \frac{1}{\sqrt{2\pi}}\int_{-\infty}^{\infty}\bar{\Gamma}(k)\, e^{i(\pi kx/a)}\, dk \tag{7.105}$$

which reduces to:

$$\Gamma\left(\frac{\pi x}{a}\right) = \frac{1}{\sqrt{2\pi}}\int_{0}^{1} e^{i(\pi kx/a)}\, dk \tag{7.106}$$

and has for its real part:

$$\Gamma\left(\frac{\pi x}{a}\right) = \frac{1}{\sqrt{2\pi}}\int_{0}^{1}\cos\left(\frac{\pi kx}{a}\right) dk \tag{7.107}$$

We apply this analysis to a two-dimensional gravity anomaly having values $\Delta g(0)$, $\Delta g(a)$, $\Delta g(2a) \ldots \Delta g(na)$ at nodal points $x = 0, a, 2a, \ldots na$. A function $\Delta\gamma(x)$, which is a superposition of terms, with the origin displaced appropriately:

$$\begin{aligned}\Delta\gamma(x) &= \sum_n \Delta g(na)\left\{\frac{\sin[(\pi x/a) - n\pi]}{[(\pi x/a) - n\pi]}\right\} \\ &= \Delta g(0)\left\{\frac{\sin(\pi x/a)}{(\pi x/a)}\right\} + \Delta g(a)\left\{\frac{\sin[(\pi x/a) - \pi]}{[(\pi x/a) - \pi]}\right\} \\ &\quad + \ldots \Delta g(na)\left\{\frac{\sin[(\pi x/a) - n\pi]}{[(\pi x/a) - n\pi]}\right\}\end{aligned} \tag{7.108}$$

can represent the observed anomaly exactly at the nodal points because the terms on the right side vanish for $x = na$, except at $x = 0$ where it is $\Delta g\ (0)$.

The nth term in eq. 7.108 can be written, applying eq. 7.107:

$$\Delta\gamma_n(x) = \Delta g(na)\int_{0}^{1}\cos\left(\frac{\pi kx}{a} - n\pi k\right) dk \tag{7.109}$$

which represents the wave-number spectrum of a single term in eq. 7.108. Values of $\Delta\gamma\ (x)$ can be continued downward to depth h, the nth term having the form (e.g., Bott, 1973):

$$\Delta\gamma_n(x, h) = \Delta g(na) \int_0^1 e^{(\pi kh/a)} \cos\left(\frac{\pi kx}{a} - n\pi k\right) dk \tag{7.110}$$

When we integrate the above equation over the wave number, we obtain values at depth h produced by the nth term (Bott, 1973):

$$\Delta\gamma_n(x, h) =$$

$$\Delta g(na) \frac{a e^{(\pi h/a)} \left\{ nh \cos\left[\left(\frac{\pi x}{a}\right) - n\pi\right] + (nx - n\pi a) \sin\left[\left(\frac{nx}{a}\right) - n\pi\right]\right\} - \pi ha}{[(\pi x - n\pi a)^2 + (nh)^2]} \tag{7.111}$$

The value of $\Delta\gamma_n(x, h)$ at the jth field point ($x = ja$) is:

$$\Delta\gamma_n(ja, h) = \Delta g(na) \frac{ah[(-1)^{j-n} e^{(\pi h/a)} - 1]}{\pi[a^2(j-n)^2 + h^2]} \tag{7.112}$$

We sum all terms $\Delta\gamma_n$ to obtain the value of $\Delta\gamma$ at depth h beneath the jth field point; thus:

$$\Delta\gamma(ja, h) = \sum_n \Delta\gamma_n(ja, h) \tag{7.113}$$

The downward continued value of gravity at every jth field point can be converted to a surface density on the equivalent layer (depth h) by dividing by $2\pi G$ (eq. 7.102); thus:

$$\sigma_j(ja, h) = \frac{1}{2\pi G} \sum_n \Delta\gamma_n(ja, h) \tag{7.114}$$

We can convert the surface density distributions along an equivalent layer to density distributions along other-shaped bodies.

Tomoda and Aki (1955) applied this method to obtain variations in depth to Moho, making assumptions about a crustal-mantle density contrast.

Oldham (1967) has extended the sin x/x method for three-dimensional anomalies.

Matrix (linear) method of inversion

Methods of linear algebra have been recently applied to the solution of the linear inverse problem in gravity; they have been discussed by Bott (1973). The geometry of the anomaly-producing body must first be specified, and this can be three-dimensional in contrast to a simple equivalent layer. The source body can be divided into a finite number of volume elements, such as two-dimensional prisms, sheets, or columns, each having a uniform density.

In practice, the number of observed anomaly values must at least equal, and preferably exceed, the number of prisms (sheets, or columns).

We can approximate the integral equation 7.93 by (Bott, 1973; Tanner, 1967):

$$\Delta g_i = K_{il} \Delta \rho_l \tag{7.115}$$

in which $i = 1 \ldots n$ (observed stations), $l = 1 \ldots m$ (prisms), and the repeated subscript l is summed over all values for every value of i. The kernel function K_{il} is the contribution which the volume element makes to the ith anomaly value. We can obtain formulas for the kernel function by direct integration of bodies. The kernel function for a rectangular two-dimensional prism is $2G\,\Delta\theta\,\Delta z$ in eq. 7.32.

Eq. 7.115 consists of a system of n equations in m unknowns. Where the number of equations exceeds the number of unknowns, solutions are obtained by the method of least squares. In matrix notation (e.g., Arfken, 1970, p. 166):

$$\Delta \rho = K^{-1} \Delta g \tag{7.116}$$

where K^{-1} is the inverse kernel array and Δg the array of observed anomalies. The solution of eq. 7.116 specifies a system of prisms of variable density $\Delta \rho_l$ or a system of horizontal two-dimensional sheets of variable mass per unit area $\Delta \sigma_l$, which satisfy the observed anomalies.

Bott (1973) discussed the theory and provides applications of the matrix method to both gravity and magnetic problems.

Nonlinear methods of inversion

Nonlinear methods provide an approach to the most commonly encountered problem in gravity interpretation, the determination of the shape of a body producing observed anomalies. The problem is nonlinear because the unknowns which define the shape occur in the kernel function in eq. 7.91 (or its three-dimensional equivalent). Some assumptions must be made about the shape of the body before solutions can be attempted. The nonuniqueness of nonlinear methods has been discussed by Smith (1961), Tanner (1967), Al-Chalabi (1971), and Bott (1973).

One method for determining the shape satisfying observed gravity anomalies is by trial and error, as can be done by computer. Linear solutions can provide successive approximations when the density and one surface of the body are known, and iterations are used to obtain the shapes of the unknown surfaces. The method was first applied by Bott (1960) to determine the shape of the bottom of a two-dimensional sedimentary basin from observed anomalies. Tanner (1967) developed a more rapidly converging method using two-dimensional rectangular prisms to compute an adjusted density. He applied these methods to determine the shape of a sedimentary basin, where the density contrast and upper surface are given, and the shape of a granitic body of given density with part of the upper surface specified and the lower surface at a constant but unknown depth.

Tanner (1967) considered the iteration of the nonlinear problem by linear approximations, making assumptions about one of the surfaces of the body. He used two-dimensional rectangular prisms which produce attractions as given in eqs. 7.30 and 7.32. These formulas provide the kernel in the integral equation (eq. 7.115) used in the matrix equation 7.116. The solution to this equation specifies a system of horizontal prisms of variable density $\Delta\rho_l$. We can approximate distribution of prisms of uniform $\Delta\rho$ by taking one surface (e.g., upper or lower) as fixed and transforming such that the total mass of the prisms is held constant. However, because the attraction of each prism is nonlinear with depth, the transformed prisms will usually not satisfy the observed anomaly. The transformed prisms are therefore adjusted; this can be accomplished by iterations in which linear approximations are used (and care is taken to avoid instabilities in the process).

In the iterative method, the mass producing the anomaly is first taken to be concentrated in a thin horizontal sheet along the top of the structure. Solutions to eq. 7.102 then provide an equivalent sheet with variable $\Delta\sigma$. Horizontal prisms are selected next (with prism widths greater than depth to the upper surface to avoid instability in the solutions), which have uniform density contrasts. Tanner's sedimentary basin program assumes a known depth h_1 to the upper surface and estimates a depth h_2 to the lower surface. The depth h_2 can be estimated from:

$$h_{2i} = h_{1i} + \left(\frac{\Delta\sigma_i}{\Delta\rho}\right) \tag{7.117}$$

A more stable but less rapid computer program uses the entire transformed prism obtained from the first estimate to adjust the model. After each adjustment a set of new transformed prisms is obtained from the formula:

$$h_{2i} = h_{1i} + (z_{2i} - h_{1i})\left(\frac{\Delta\rho_l}{\Delta\rho}\right) \tag{7.118}$$

in which z_{2i} is the previously estimated depth of the lower surface. A solution is obtained when $\Delta\rho_l$ nearly equals the assumed value $\Delta\rho$.

We can solve three-dimensional structural models with similar iterative methods. Cordell and Henderson (1968) applied automatic successive approximation methods to models consisting of vertical prisms, specifying either a flat top, base or midsection in the model. A first approximation of the structure can be obtained if we use the Bouguer slab relationship. The gravity field of the first model is calculated and the ratio of observed to computed anomaly at each station is used to modify the model. The process is iterated until a satisfactory fit between observed and computed anomalies is obtained.

A somewhat more versatile nonlinear method uses optimization techniques (described, for example, in Luenberger, 1969), whereby digital computers search for optimum values (e.g., a maximum of minimum) of one or more unknown variable parameters of a function. The parameters can include the coordinates of an entire body or those of specified prisms or blocks, of variable size and shape, which comprise the body, and also the density contrast of the body or variable contrasts of its constituent parts. We can optimize for functions chosen to be the minimum sum of the squares of residuals of the observed minus computed anomaly, or other optimization procedures. Al-Chalabi (1970, 1971) published optimization methods applied to two-dimensional gravity and magnetic anomalies. He applied a number of suitable objective functions to assumed initial models, finding that direct search methods provide a good approach in the early search stages and gradient methods in the later stages. The behavior of the functions can be quite complex, so that it is important to program carefully and use all available model constraints.

Several optimization subroutines are available. One effective subroutine (Bott, 1973) is in MINUIT, described by James and Roos (1969), which consists of three subroutines: (1) Monte Carlo SEEK, which searches for a fit when no starting point is apparent; (2) TAUROS, a fast stepping method of Rosenbrock; and (3) MIGRAD, a rapidly converging subroutine (based on a method by Davidson) in the vicinity of a minimum.

Optimization methods can solve for a larger number of variables in the solution sought than can iterative methods; however, iterative methods use less computer time when problems are well defined.

Chapter 8

MARINE GRAVITY STUDIES

Vening Meinesz-type, three-pendulum apparatuses were the first instruments used routinely to measure gravity at sea. Measurements were made in submerged submarines because pendulums could not operate reliably on surface ships, even in calm seas. Most sea measurements were made by gravimeters operating on surface ships because the cost of operation is relatively low and, in recent years, the accuracy of measurement became quite high. Observed perturbations in paths of artificial earth satellites made it possible to compute free-air anomalies over the earth with wavelengths greater than about 4,000 km; more recently, satellites have made it possible to determine variations in sea level over oceans quite accurately, and therefore geoidal configurations.

From 1923 to 1929 pendulum measurements were made in most of the world oceans; these measurements were unaffected by first-order horizontal accelerations experienced by submarines, but they were modified by second-order (Browne) effects (eq. 6.45b). The anomalies so obtained could be in error by as much as tens of milligals. Pendulum measurements made since 1938 are generally free of second-order effects, such that these anomaly determinations are generally reliable and have provided check points for later gravimeter surveys. Since about 1959, pendulums have not been in general use at sea, however, because the continuous-reading and cheaper-operating gravimeters were then showing promise in their initial tests aboard surface ships. Gravity anomalies obtained from sea-pendulum measurements have been published by Vening Meinesz (1929, 1948) and Worzel (1965).

Sea gravimeters were developed initially for use in submarines in the 1950's. Measurements were attempted on several occasions when such a submarine surfaced in a calm sea, with sufficiently good results to indicate that these meters could be modified to operate on surface ships. The first satisfactorily designed surface-ship gravimeters were the Graf-Askania meter, mounted on a stabilized platform (Worzel, 1958), and the LaCoste and Romberg meter, suspended in a gimbal system (LaCoste, 1959).

Measurements made by first-generation surface-ship gravimeters were commonly affected by ship-induced accelerations; reliable measurements could thus be obtained only in low sea states. Second-generation meters were designed to operate successfully in moderate sea states. In the Graf-Askania meter (circa 1962) damping of the sensing element was increased substantially; the LaCoste and Romberg meter (circa 1965) was redesigned to operate on a stabilized platform.

Ship positioning and consequent Eötvös corrections (eq. 6.42 or 6.43) applied in the early sea measurements were based on star fixes, dead reckoning, and fixes obtained with long-range electronic receivers. The resultant uncertainties in navigation meant that even accurate gravity measurements can result in anomaly uncertainties up to about 10 mgal. By 1966 accurate ship positioning could be obtained with artificial satellite receivers; the effect was to obtain considerably more accurate deep-ocean anomaly values.

Most ocean gravity measurements were obtained by the Lamont-Doherty Geological Observatory of Columbia University, which has been routinely making measurements aboard its two research vessels (R/V) *Vema* and *Robert D. Conrad* on their cruises in the world oceans, and also aboard the *USNS Eltanin* when this ship was operating in the southern oceans (during the 1960's). These measurements were made by Graf-Askania Gss-2 gravimeters; measurements made since about 1966 have estimated uncertainties of only a few milligals.

Graf-Askania Gss-2 meters have also been used successfully for many years by the Bedford Institute of Oceanography (Nova Scotia, Canada), particularly in the North Atlantic Ocean, and by the Istituto Osservatorio Geofizico (Trieste, Italy), mostly in the Mediterranean Sea. Several organizations in Germany have been using the Gss-2 and also the Gss-3 successfully in the North Atlantic and adjacent seas.

Extensive gravity measurements have been made routinely aboard several Woods Hole Oceanographic Institution (WHOI) ships. Initially, a LaCoste and Romberg gimbal-suspended meter was used. Since the successful testing of the Massachusetts Institute of Technology (MIT) vibrating-string meter aboard its ships, WHOI has been using the MIT meter for routine measurements, the anomaly uncertainties being a few milligals. WHOI researchers have also obtained measurements by using a LaCoste and Romberg stable-platform system.

LaCoste and Romberg gimbal-suspended meters were used at sea until the latter part of the 1960's, particularly by the U.S. Coast and Geodetic Survey (1961—1965), the Environmental Science Services Administration (1965—1970), and the National Ocean Surveys of the National Oceanic and Atmospheric Administration (U.S. Department of Commerce) since 1970 in the North Pacific Ocean; by Texas A & M University in the Gulf of Mexico; by Oregon State University in the northeast and north-central Pacific (including the Inside Passage of British Columbia and Alaska); and by the University of Hawaii in the central and western Pacific. The U.S. Naval Oceanographic Office made many measurements with gimbal-suspended meters, but anomalies obtained were generally not published in the open literature.

The gimbal-suspended system was designed to measure gravity when vertical and horizontal accelerations do not exceed 50 gal. The open ocean usually imposes larger ship accelerations, however, so that this meter measured accurately in gentle seas only (for uncertainties obtained at sea with this type of meter see Allan et al., 1962; Dehlinger and Yungul, 1962; Dehlinger, 1964; and Dehlinger et al., 1966). To improve measurements at larger accelerations, LaCoste and Romberg modified its sea meter to operate on a stabilized platform. These platform-mounted meters operated successfully at sea, providing anomalies which are accurate to within a few milligals (see e.g., Chiburis and Dehlinger, 1974, for reliabilities obtained in the Beaufort Sea).

LaCoste and Romberg platform-mounted meters have been used by Oregon State University along the west coasts of North, Central, and South America, by the University of Hawaii in the central and southern Pacific, by the University of Connecticut and the U.S. Geological Survey in the Beaufort and Chukchi seas, by the National Ocean Surveys in the North Pacific and North Atlantic, and by the U.S. Naval Oceanographic Office at numerous unspecified locations. The meter has also been used as an oil-prospecting tool on continental margins.

Japanese investigators have designed and constructed their own vibrating-string types of meters which they have used to measure gravity in the northwest Pacific. Investigators in the USSR have also designed and operated vibrating-string meters.

GRAVITY ANOMALIES AND STRUCTURAL SECTIONS

This section includes selected free-air anomaly maps and profiles representative of different types of oceanic provinces and, where available, associated crustal-subcrustal sections that conform with the observed anomalies.

Gravity anomalies over ocean ridges

Gravity profiles have been obtained at many ship crossings of the mid-ocean ridge system. These profiles have similar characteristic shapes, although they exhibit variations due to local structural conditions. Talwani (1970) published twelve profiles across various ridges. Fig. 8.1 shows a rather typical free-air anomaly profile across the north Mid-Atlantic Ridge, and also the corresponding bathymetry and a hypothetical crustal section. The gravity measurements were obtained on R/V *Vema* cruise 17 (Talwani et al., 1965). Fig. 8.2 shows a similar typical profile across the East Pacific Rise, in which measurements were obtained on R/V *Vema* cruise 19. A profile across the Juan de Fuca Ridge, west of the coasts of Washington and

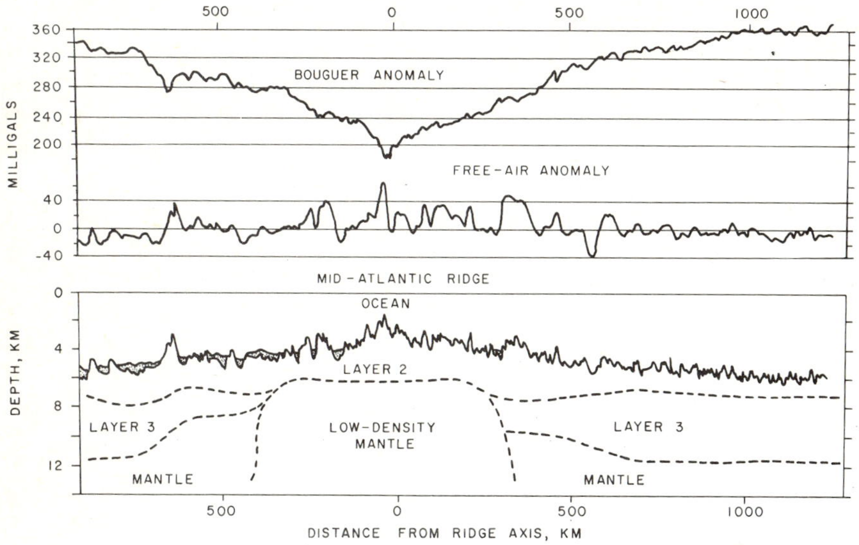

Fig. 8.1. Gravity anomalies and seismically determined structures across the Mid-Atlantic Ridge obtained on R/V *Vema* cruise 17. The left side of the figure is near 36°N 49°W, and the right side near 25°N 29°W. Bouguer anomaly computations are for two-dimensionality and a basement-rock density of 2.60 g cm^{-3}. Oceanic layer 2 is basement rock, characterized by P-wave velocities of 4.5—5.8 km sec^{-1}; oceanic layer 3 is the basal crust, characterized by velocities of 6.5—7.0 km sec^{-1}; mantle rocks have velocities of 7.9—8.4 km sec^{-1}, except beneath the ridge axis, where they are lower. The stipled layer represents sediments. (After Talwani et al., 1965.)

Oregon, appears at the top of Fig. 8.21 (Dehlinger et al., 1970). The figure also shows the corresponding bathymetry and a crustal section consistent with both the seismic refractions obtained by Shor et al. (1968) and the gravity measurements by Dehlinger et al. A free-air anomaly map over the Juan de Fuca Ridge is shown in the left part of Fig. 8.16, and a map over the Gorda Ridge, which is an offset of the Juan de Fuca at its southern end, in the northeast part of Fig. 8.15. Fig. 8.3 shows a detailed free-air anomaly map across the Mid-Atlantic Ridge near latitude 45°N (Woodside, 1972). In contrast to the Juan de Fuca Ridge, which has low bathymetric relief, the Gorda and Mid-Atlantic ridges have characteristic median valleys flanked by ridges of considerable height. Free-air anomalies across the central valleys are negative and those over the flanking ridges positive, in conformance with topography.

Average values of free-air anomalies across mid-ocean ridges are about 20—30 mgal larger than those over the adjacent ocean floors. Characteristically the anomalies are nearly independent of ridge height (exept locally over seamounts or islands in the ridge systems). The near-zero average free-air values over the typically high topographic ridge systems means that

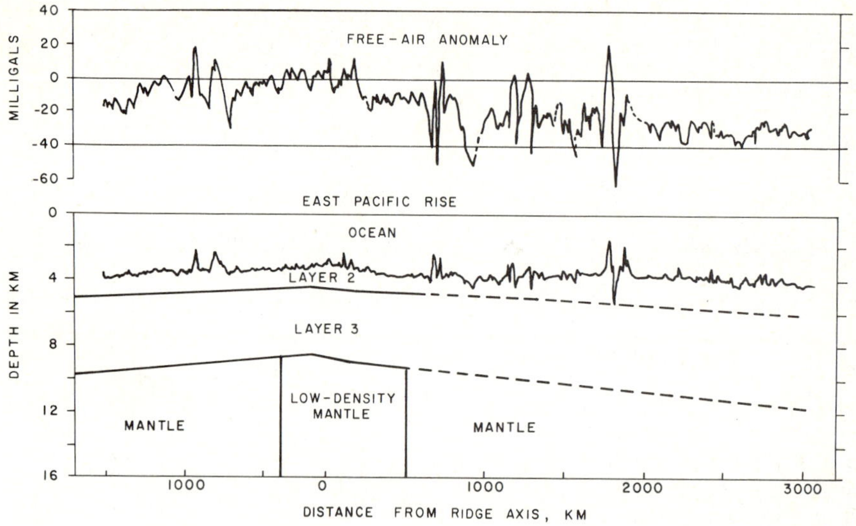

Fig. 8.2. Free-air gravity anomalies across the East Pacific Rise obtained on R/V *Vema* cruise 19. The left side of the figure is near 16°S 138°W, and the right side near 12°S 77°W. See Fig. 8.1 caption for descriptions of layers 2 and 3. (After Talwani et al., 1965, who used Raitt's, 1956, and Menard's, 1960, data to construct the seismically determined structures.)

the ridges are nearly in isostatic balance, requiring compensating low densities (mass deficiencies) beneath the ridges.

Near ridge crests ocean depths are a minimum (usually 2 to 3 km), the crust is thinnest, and Moho depths are commonly about 7 km. Compensation along the ridges must occur primarily in the uppermost mantle, which requires relatively low mantle densities beneath the ridges. We do not know the precise shapes of these low-density bodies, however. Fowler (1976) recently showed that over at least one part of the crest of the Mid-Atlantic Ridge, sub-Moho seismic velocities are relatively low directly beneath the median valley but that they have typical values (near 8.1 km sec^{-1}) beyond the edges of the median valley. Whereas velocities in sub-Moho materials appear to be low only beneath the median valley, to account for the observed free-air anomalies the materials must clearly have relatively low densities across most if not the entire ridge. Materials with typical velocities but with low densities may be in accord with the Sclater et al. (1970) thermal model of a ridge, in which elevations along the flanks of a ridge system decrease with increasing age.

Cochran and Talwani (1977) developed a method for estimating attractions produced by an ocean ridge, in effect determining a regional gravity value that can be subtracted from the observed anomaly to obtain residual patterns across the ridge. As indicated in Chapter 7, geologic ages for 1° x 1° squares across a ridge are obtained from magnetic isochron maps; these ages,

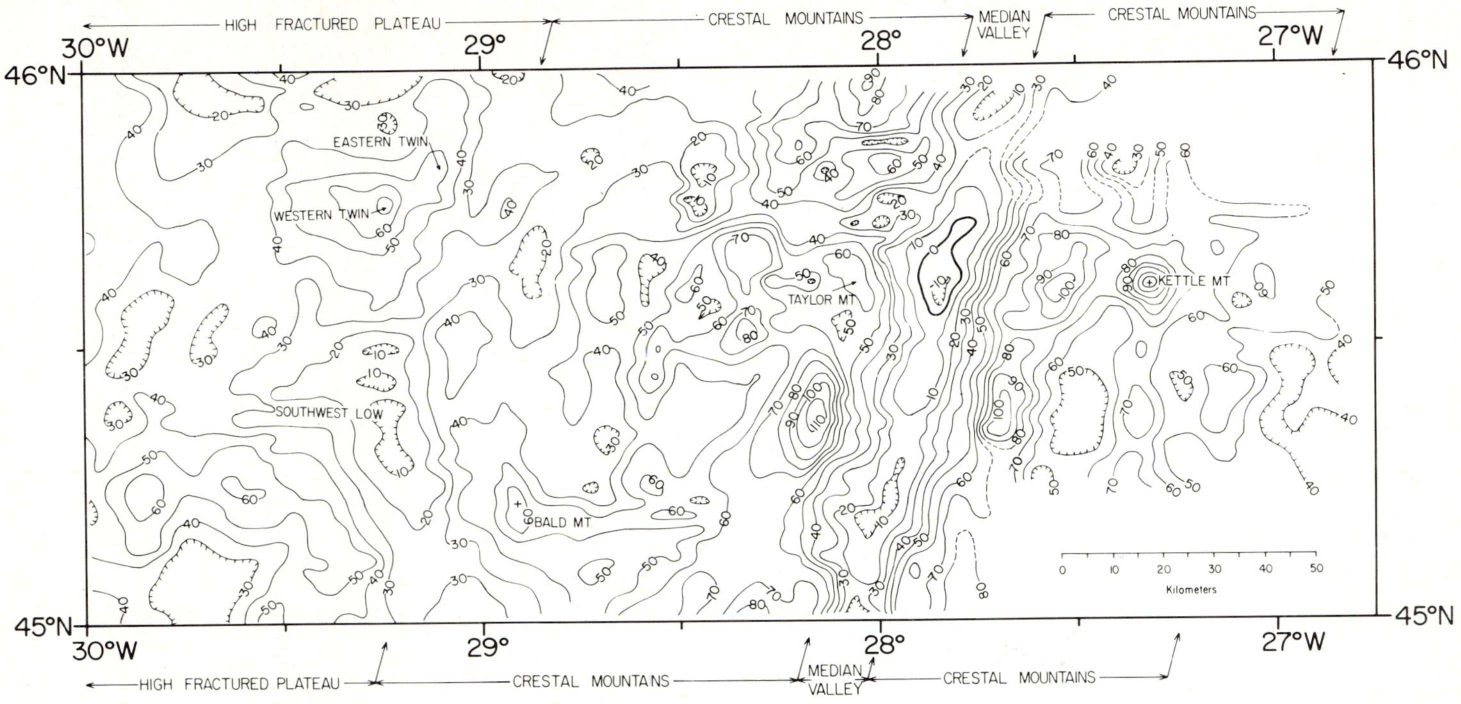

Fig. 8.3. **Free-air anomaly map over part of the Mid-Atlantic Ridge. The median valley along the ridge axis is characterized by a small negative anomaly, which is flanked by positive anomalies corresponding to the adjacent ridge escarpments. (Reproduced from Woodside, 1972, with permission.)**

when plotted as a function of the average free-air anomaly for each square, produce an empirical gravity—age relationship. Cochran and Talwani used such residual anomalies, in effect having eliminated the masking effect of the ocean ridge, to evaluate thermal models of the crust and subcrust.

Gravity anomalies over ocean trenches

Anomalies across ocean trenches exhibit characteristically large negative amplitudes over trench axes and positive values over the landward island arc, as first observed by Vening Meinesz (1941b; or see Heiskanen and Vening Meinesz, 1958, p. 388) across the Sunda Arc (south of Java). Numerous pendulum measurements made across various arcs since then showed similar anomaly patterns.

More recently, many surface-ship gravity measurements have been made across trenches; Talwani (1970) published ten such free-air gravity profiles over different parts of the same or across different trenches. When these profiles are superimposed with respect to the trench axis, they show strikingly characteristic patterns. The anomaly minimum occurs near the trench axis, with values between —200 and —350 mgal; the anomaly maximum occurs over the adjacent volcanic island arc, with values ranging to +300 mgal or more. There can be little doubt that these anomalies are created by the same types of structural features (as a subduction zone with an adjacent volcanic lineation).

Fig. 8.4 (Talwani, 1970) shows two representative trench-anomaly profiles. Minimum free-air amplitudes lie near the trench axis; a second low occurs on the landward wall of the trench. Talwani and Hayes (1967)

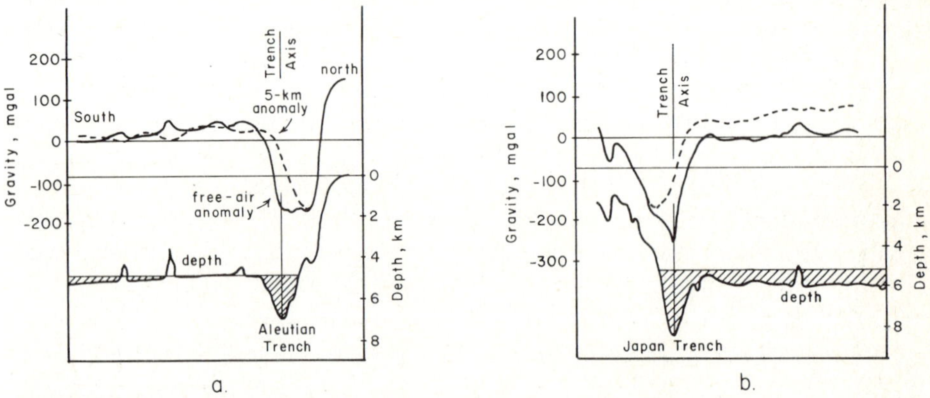

Fig. 8.4. Free-air anomaly profiles (solid lines) and associated bathymetry, and corresponding 5-km anomaly profiles (dashed lines) and the ocean filled beneath a 5-km depth across: a. the Aleutian Trench, along the Aleutian Islands of Alaska; and b. the Japan Trench, east of the Japanese Islands. (After Talwani, 1970.)

attempted to separate the effects of topography from structure in producing the free-air anomalies by filling the trench and adjacent ocean with rock ($\rho = 2.60$ g cm^{-3}) to depths of 5 km and, similarly, seaward of the inner wall (as illustrated in Fig. 8.4), removing rock at depths less than 5 km. The resultant computed anomaly is essentially independent of topographic effects; they called this the 5-km anomaly. Fig. 8.4 shows that the minimum value of the 5-km anomaly lies directly above the inner wall (see also Talwani and Hayes, 1967, or Talwani, 1970), and 10—50 km toward the island arc from the free-air minimum. The 5-km anomaly indicates that there is a mass deficiency beneath the landward wall which is not readily apparent from the free-air anomalies (nor from the Bouguer anomalies, as noted by Talwani, 1970). This offset in the minima of the 5-km relative to the free-air anomalies can be attributed to low-density sediments, possibly of great thickness, and/or to lower-density crustal materials in the upper part of a subducted plate as it bends downward near the trench axis (see, e.g., Fig. 4.6).

A further characteristic of trench anomalies is long-wavelength positive amplitudes seaward of the outer trench wall, first observed by Vening Meinesz (e.g., Heiskanen and Vening Meinesz, 1958, p. 388). Fig. 8.4 shows typical anomaly profiles (Talwani, 1970, p. 292); wavelengths are several hundred kilometers and amplitudes range about 50 mgal greater than those over the adjacent abyssal plain. Satellite-based free-air anomaly maps (e.g., Fig. 2.23) show generally positive anomalies associated with the circum-Pacific belt of trench axes. These anomalies are positive because the combined effect of the positive amplitudes on both sides of a trench axis is greater than that of the shorter-wavelength, larger-amplitude negative values along the trench axis.

Free-air anomaly maps of parts of the Aleutian Trench and adjacent regions are illustrated in Figs. 8.5 and 8.6. These maps are part of a larger map of the Aleutian Trench and Bering Sea which Watts (1975b) constructed, based on gravimetric measurements that numerous institutions obtained (most were obtained by the Lamont-Doherty Geological Observatory and by the U.S. Coast and Geodetic Survey, the Environmental Science Services Administration, and the National Oceanic and Atmospheric Administration). Figs. 8.5 and 8.6 show the characteristic large-amplitude negative anomaly lineation along the trench axis, reaching minimum values of —200 mgal. Watts and Talwani (1974) point out that the minimum values are generally displaced several tens of kilometers landward of the trench axis, toward the Aleutian Terrace on the north wall of the trench. The amplitude of the negative trench anomaly diminishes east of longitude 160°W (Fig. 8.5), where the trench contains greater thicknesses of sediments.

The Aleutian Islands are characterized by a large-amplitude, positive-anomaly lineation which extends over the sea areas between the islands. The highest amplitudes reach +200 mgal, occurring over the active volcanoes near

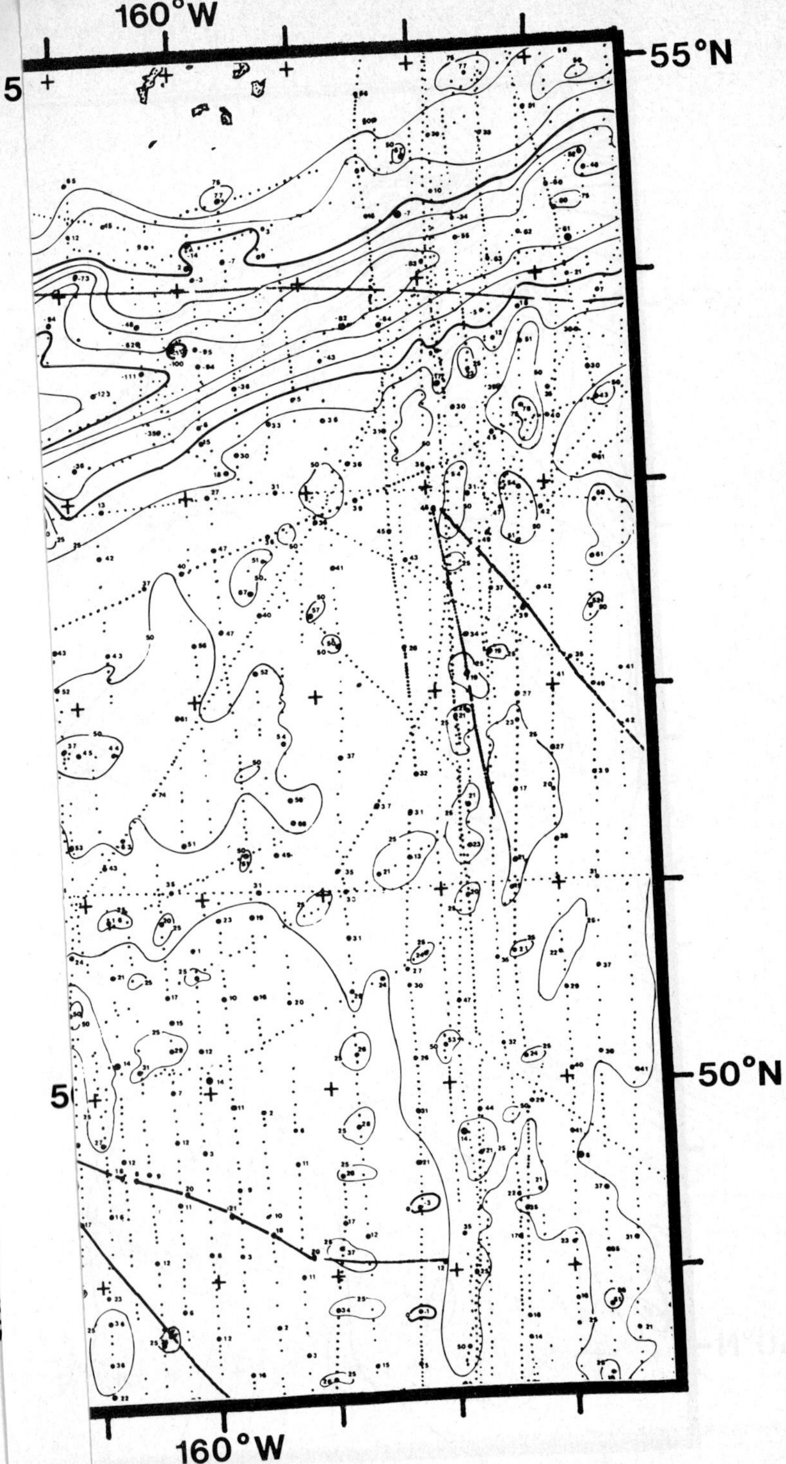

Geological Society of America.)

Rat Island (Fig. 8.6), as indicated by Watts (1975b). The arcuate bow of high-amplitude positive anomalies near 179°W lies over the Bowers Ridge, which separates the Bowers Basin to the south, characterized by low-amplitude anomalies, and the Aleutian Basin to the north.

Seaward of the trench anomaly is a characteristic zone of long-wavelength, low-amplitude positive anomalies which reach values of +50 to +80 mgal. This positive belt correlates with a regional topographic rise of several hundred meters (Watts and Talwani, 1974). The anomaly extends along the entire trench—arc system and appears to be characteristic of lithospheric plates as they bend to descend down at a trench (e.g., Fig. 4.6).

A free-air anomaly map of the Philippine Island Arc—Trench system is shown in Fig. 8.7. This map is part of a larger gravity map of the Philippine Sea compiled by Watts (1976), which is based on gravity measurements obtained by various institutions (the marine measurements were obtained principally by the Lamont—Doherty Geological Observatory of Columbia University and by the Ocean Research Institute in Tokyo). The Philippine Arc—Trench is part of a system that extends southwesterly from southern Japan to south of the Philippine Islands.

Fig. 8.7 shows several major anomaly belts. The most prominent one is the large-amplitude negative lineation over the trench, where minimum values range to —250 mgal. Major positive belts extend over the islands, where Bouguer-anomaly values range up to +300 mgal. These positive anomalies are generally continuous between the islands where the water is relatively shallow; however, where waters become deep between the islands, the anomalies are generally strongly negative.

The low-amplitude, long-wavelength positive anomaly belt that characteristically occurs seaward of a trench is observed east of the trench; it has typical values of +50 mgal (and a maximum of +95 mgal).

A free-air anomaly map of the Bonin Trench system, east of southern Japan and east of the Iwo Jima and Bonin Ridges is shown in Fig. 8.8. This map is also part of the larger above-mentioned map of the Philippine Sea compiled by Watts (1976), which contains a major trench in the island arc—trench system that extends from the Kurile Trench in the north to the Marianna Trench in the South.

The Bonin Arc—Trench exhibits the large-amplitude anomaly belts characteristic of major subduction zones. The negative lineation along the trench reaches values of —350 mgal. Anomalies are positive on the adjacent islands, the larger values occurring on the active volcanoes on the ridges (Bouguer values range up to +386 mgal on the Bonin Ridge). Seaward of the trench the characteristic long-wavelength, low-amplitude positive anomaly belt is observed, with typical values of about +50 mgal.

Gravity anomalies over the Hawaiian Archipelago

The Hawaiian Ridge provides an outstanding example of a large two-dimensional topographic load in the interior part of a lithospheric plate and associated lithospheric flexuring.

Before 1954, 31 sea-pendulum measurements were made in the Hawaiian area (Vening Meinesz, 1941b, 1948; Worzel, 1965); since 1959 a great many surface-ship measurements were made, principally by the Lamont-Doherty Geological Observatory, the University of Hawaii, the U.S. Coast and Geodetic Survey, the Environmental Science Services Administration, and the National Oceanic and Atmospheric Administration.

Watts (1975a) compiled and contoured available free-air anomaly values in the Hawaiian area, providing a regional map (Fig. 8.9) which shows three broad belts of west-northwest positive and negative anomaly trends. The center belt consists of large-amplitude positive anomalies which extend along the crest of the Hawaiian Ridge that consists of extensive volcanic rocks extruded over a period of many millions of years. The anomaly values generally exceed +100 mgal and range to +700 mgal, the maximum values

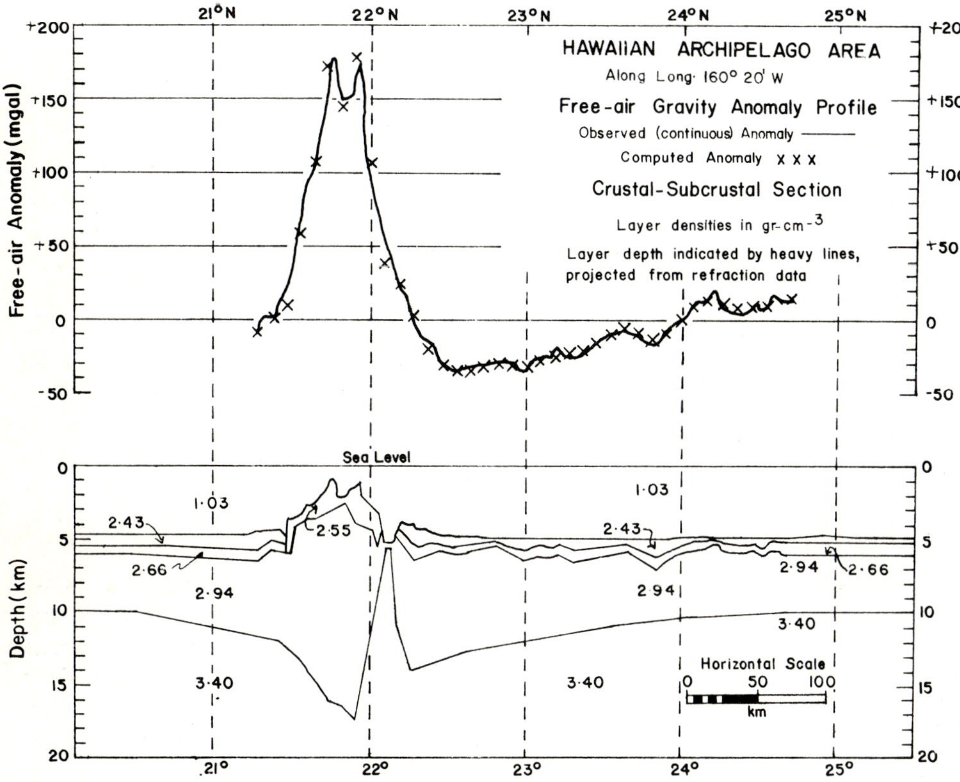

Fig. 8.10. Hypothetical crustal-subcrustal section across the Hawaiian Archipelago at 161°W which conforms with observed free-air anomalies. (Reproduced from Dehlinger, 1969, with permission.)

occurring on volcanic craters (both active and extinct). The largest anomalies occur at the southeastern edge of the ridge, where volcanic rocks are youngest. The age of rocks in the northwest end of the ridge is approximately 70 million years (m.y.) and those in the southeast end about 3 m.y.

Flanking the Hawaiian Ridge is a nearly continuous belt of negative free-air anomalies, with values less than —100 mgal in the southeastern end. These negative values correlate with a topographic depression or moat, referred to as the Hawaiian deep by Dietz and Menard (1953) and the Hawaiian trench by Malahoff and Woollard (1970).

A broad belt of positive free-air anomalies, approximately 250 to 300 km wide and with values up to +55 mgal, borders the belt of negative anomalies. A regional gravity maximum northeast of the Hawaiian Ridge correlates with

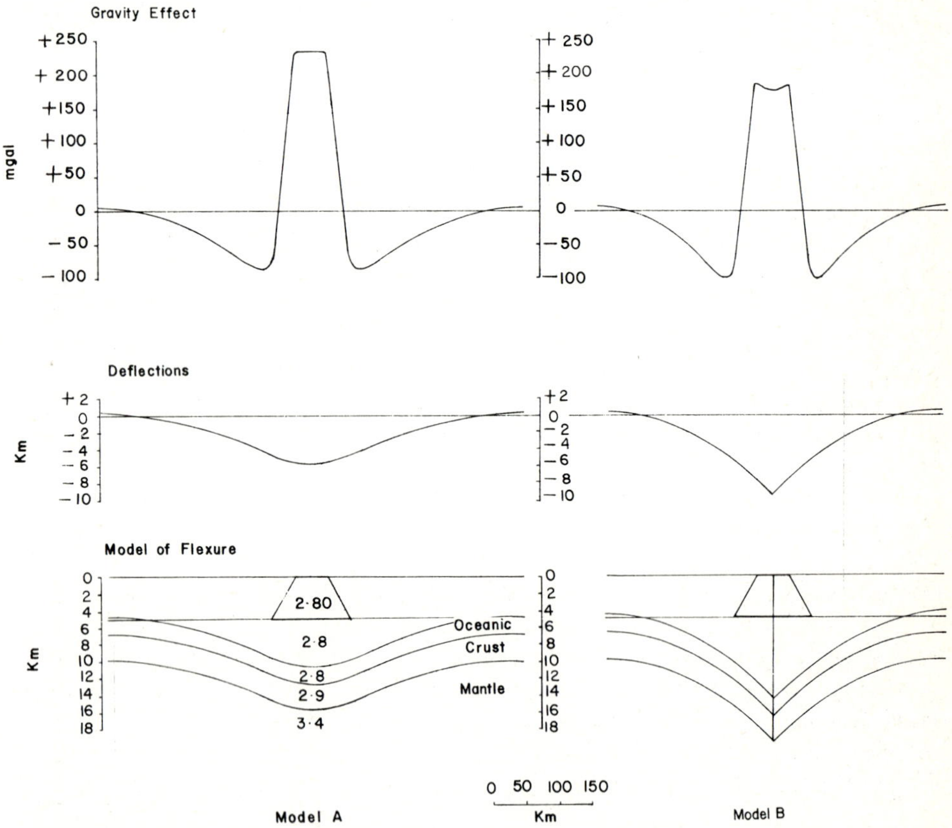

Fig. 8.11. Lithospheric deflections across the Hawaiian Archipelago, modeled as one continuous elastic sheet (left) and as two discontinuous down-bending sheets (right). (Reproduced from Watts and Cochran, 1974, with permission.)

a regional topographic rise, the Hawaiian arch of Dietz and Menard (1952). Depths of the Moho along this rise have been determined at only a few localities; along the northwestern part of the rise the depth may be as shallow as 9 km (Shor, 1960), and north of the island of Oahu it was reported to be about 10 km (Malahoff and Woollard, 1970).

Fig. 8.10, from Dehlinger (1969), shows a crustal-subscrustal section across the archipelago along 161°W, which is consistent with free-air anomalies and seismic control, based on Shor's (1960) refraction analysis projected southeasterly along the trend of the arc. The figure shows a Moho depth of 18 km beneath the ridge, which is approximately the thickness that Shor obtained to the northwest. The gravity profile indicates the existence of a mass excess beneath the topographic depression (moat) north of the ridge, a mass which can be manifested in various shapes. The magnitude of the excess is large enough to be the equivalent of the mantle extending almost to the surface. Hence, to indicate the size of the mass, rather than to indicate true structures across the moat, the figure shows the mantle coming to the ocean bottom.

The sequence of gravity-anomaly belts in the Hawaiian area suggests a flexuring of the lithosphere in response to the load along the ridge, produced by the basalt outpourings. Accordingly, Vening Meinesz (1931, 1941b) developed his concept of isostatic reductions, which incorporate the effects of lithospheric flexuring, as across the entire Hawaiian area. Walcott (1970c) later analyzed lithospheric flexuring in the Hawaiian area as consistent with observed free-air anomalies, concluding that the lithospheric plate there has viscoclastic properties. Watts and Cochran (1974) also analyzed flexuring of the Hawaiian area, using the method described in Chapter 4. They showed that if the lithosphere is modeled as one continuous elastic sheet (Fig. 8.11), the effective flexural rigidity (ERF) is about $5 \cdot 10^{29}$ dyne cm, and if modeled as two discontinuous down-bending sheets (see Fig. 8.11), the ERF is about $2 \cdot 10^{30}$ dyne cm. Watts and Cochran thus conclude that the lithosphere responds as an essentially rigid plate, that is, an elastic body, to the long duration of ridge formation. They further showed that the lithospheric deflection is about 6 km (Fig. 8.11). If we use eq. 4.21, the computed thickness of the elastic part of the lithosphere is about 28 km.

The free-air anomalies over the Hawaiian Ridge indicate that these huge topographic loads have produced lithospheric flexures which are not simply vertically compensated. This ridge and adjacent structures, and the associated free-air anomalies, are rather characteristic of topographic loads in the interior of a lithospheric plate. These structures and anomalies are markedly different from those along island arcs, where plates subduct into the mantle.

Gravity anomalies over the Great Meteor Seamount

The Great Meteor Seamount in the eastern North Atlantic (30°N 28°W) is

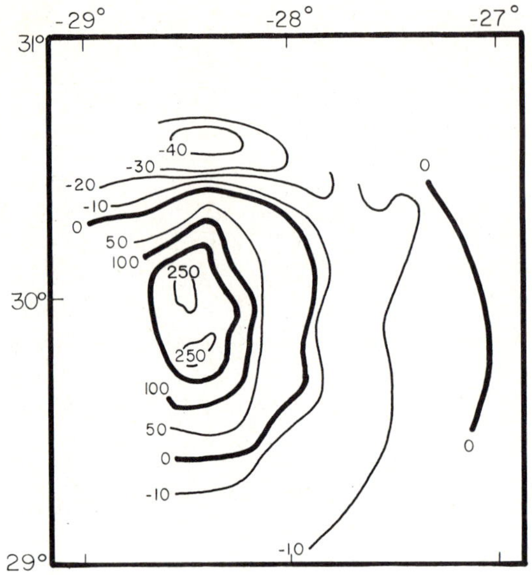

Fig. 8.12. Free-air anomaly map of the Great Meteor Seamount (North Atlantic Ocean), which extends to within 250 m of the surface. Contour values are in milligals. (After Watts et al., 1975.)

a large, three-dimensional topographic load in an interior part of a lithospheric plate. The seamount has a diameter of about 200 km at the base and extends from an abyssal plain at a depth of about 4,800 m to within 250 m of the surface. Its age is greater than 7 m.y., as indicated by foraminiferal limestones on the top of the seamount (Pratt, 1963), and less than 81 m.y., because it is situated near the 81 m.y. isochron of Pitman and Talwani (1972).

Watts et al. (1975) constructed a free-air anomaly map over the seamount (Fig. 8.12), showing that anomaly values exceed +250 mgal at the crest and that they are negative, low-amplitude at the border of the seamount, as may be anticipated for such a topographic load. Watts et al. used a three-dimensional approach to determine an EFR of $6 \cdot 10^{29}$ dyne cm, corresponding to a lithospheric flexure of nearly 3 km, illustrated in the cross section in Fig. 8.13. Because the seamount is millions of years old, this ERF value is interpreted as an elastic lithospheric response. If we use eq. 4.21 and reasonable values of constants, we find that the elastic part of the lithosphere has a thickness of approximately 19 km.

Gravity anomalies west of the United States and Canada

Continental margins are at or near to the transition zone between continental and oceanic crusts. Passive margins are commonly in isostatic

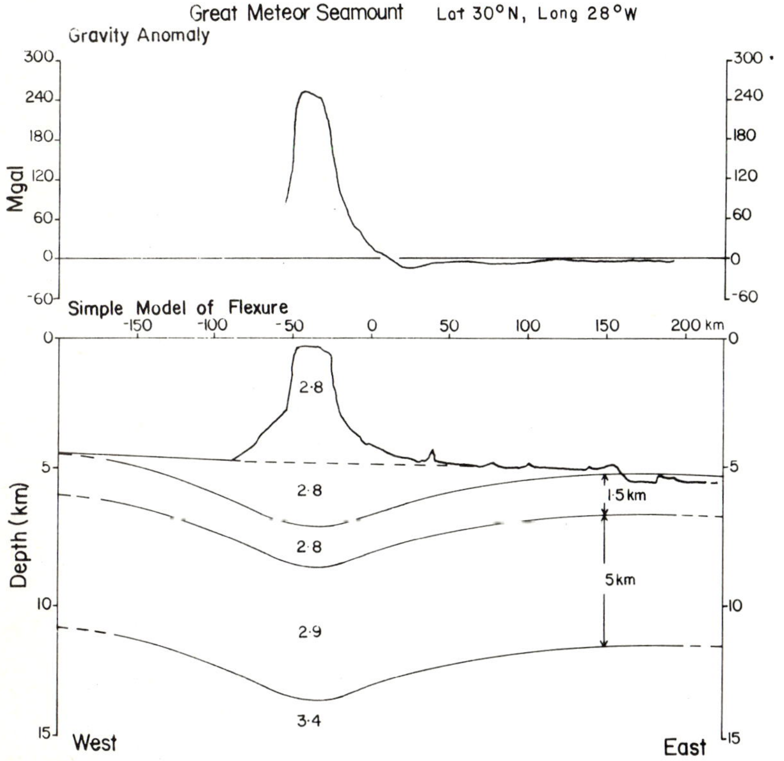

Fig. 8.13. Lithospheric deflection across the Great Meteor Seamount modeled as one continuous sheet. Layer densities are in g cm^{-3}. (After Watts et al., 1975.)

equilibrium, although the transition is characterized by edge effects between the two types of crusts, which produce a gravity high near the edge of the continental shelf and a low along the base of the continental slope. The maximum and minimum amplitudes are nearly symmetrical for a wide range of dips of the Moho beneath the continent, as seen in Fig. 8.14(a) (based on two-dimensional computations), although the amplitudes diminish in value and the distance between the peaks increases as the dip of the Moho decreases. In actual field observations, the positive amplitude is commonly smaller than the negative value, which appears to be caused by the presence of thick sediments on the continental slope or the continental shelf, as illustrated in Fig. 8.14(b, c) (also based on two-dimensional computations).

Figs. 8.15 to 8.20 show free-air anomaly maps along the continental margins and adjacent abyssal plains, and across a fracture zone (Fig. 8.15), off the west coasts of the United States, Canada, and southeastern Alaska, as published by Couch (1969), Dehlinger et al. (1970), and Couch and

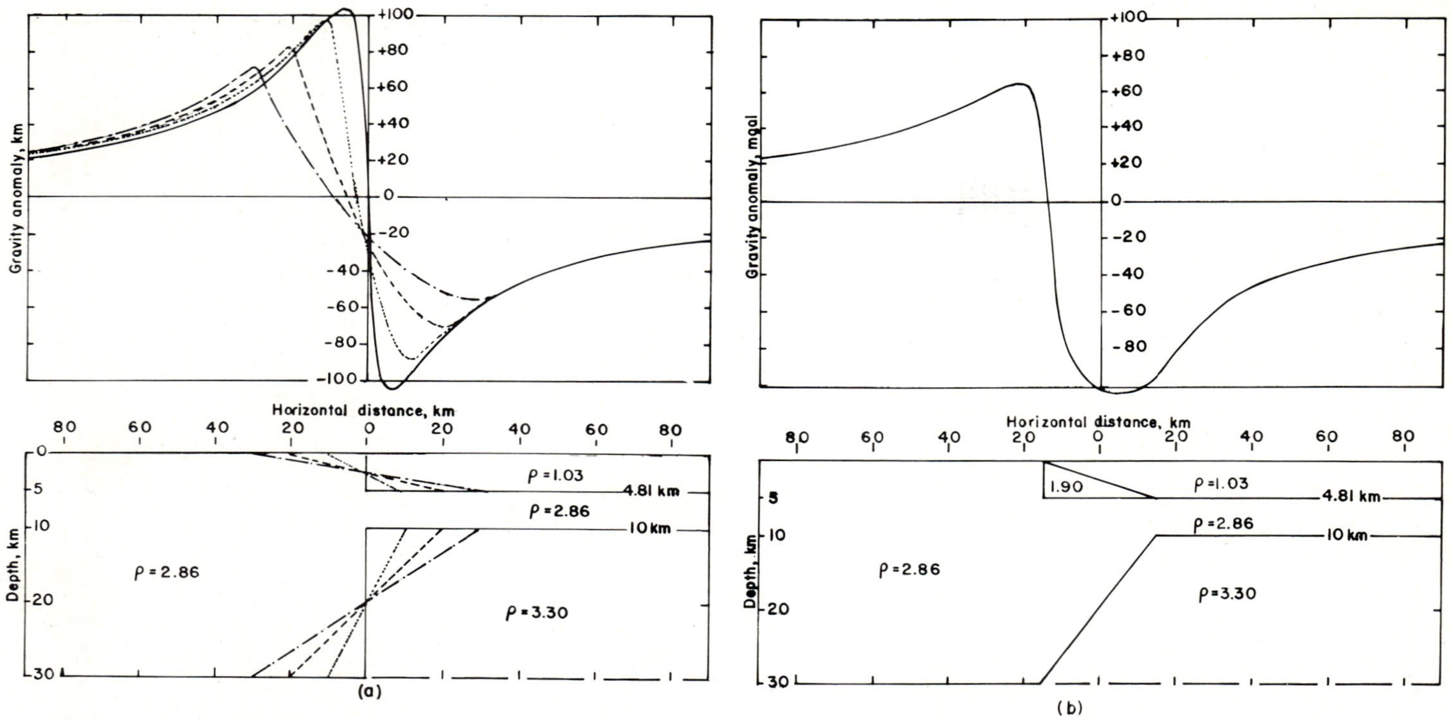

Fig. 8.14. (a) Hypothetical crustal sections across continental margins and corresponding computed two-dimensional gravity anomaly profiles. The profiles exhibit nearly symmetrical positive and negative amplitudes for a wide range of Moho dips, with smaller amplitudes for shallower Moho dips. (b) Continental margin with thick sediments at the continental slope. The effect of the sediments is to reduce the positive (computed) and increase the negative amplitudes.

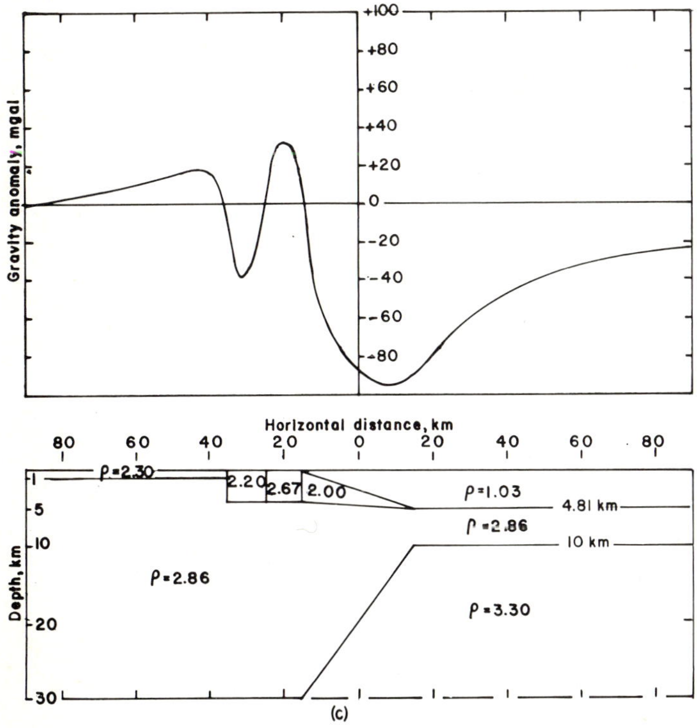

Fig. 8.14. (c) Continental margin with a deep sediment-filled trough along the outer continental shelf, an outer-shelf ridge, and thick sediments along the continental slope. The computed profile, exhibiting a negative amplitude over the shelf, a weak positive amplitude at the shelf edge, and a stronger negative amplitude near the base of the slope, is rather typical of profiles across continental margins.

Gemperle (1977). A LaCoste and Romberg gimbal-suspended meter obtained the measurements prior to 1970 and a stabilized-platform unit since then. A mean anomaly error in these maps is approximately 5 mgal, as obtained from trackline intersections (e.g., Dehlinger, 1964; Couch, 1969), resulting from errors in gravity measurement and in navigation positioning.

The maps show free-air anomaly lineations characteristic of ocean—continent transition zones: steep gradients occur on the continental slopes, and generally negative anomalies along the base of the slope, which have larger amplitudes than the positive anomalies at the shelf edge. An exception occurs off the Washington coast (Fig. 8.17), where amplitudes are small because the slope is exceptionally gentle and probably also the dip of the Moho. Along the southeast side of Alaska (Fig. 8.20) the continental margin has a positive anomaly amplitude (+120 mgal), which is considerably larger than the adjacent negative amplitude (—60 mgal). A basement high near the shelf edge is the likely cause for the large positive anomaly. The continental margin trending northwesterly along the coast of Canada and southern

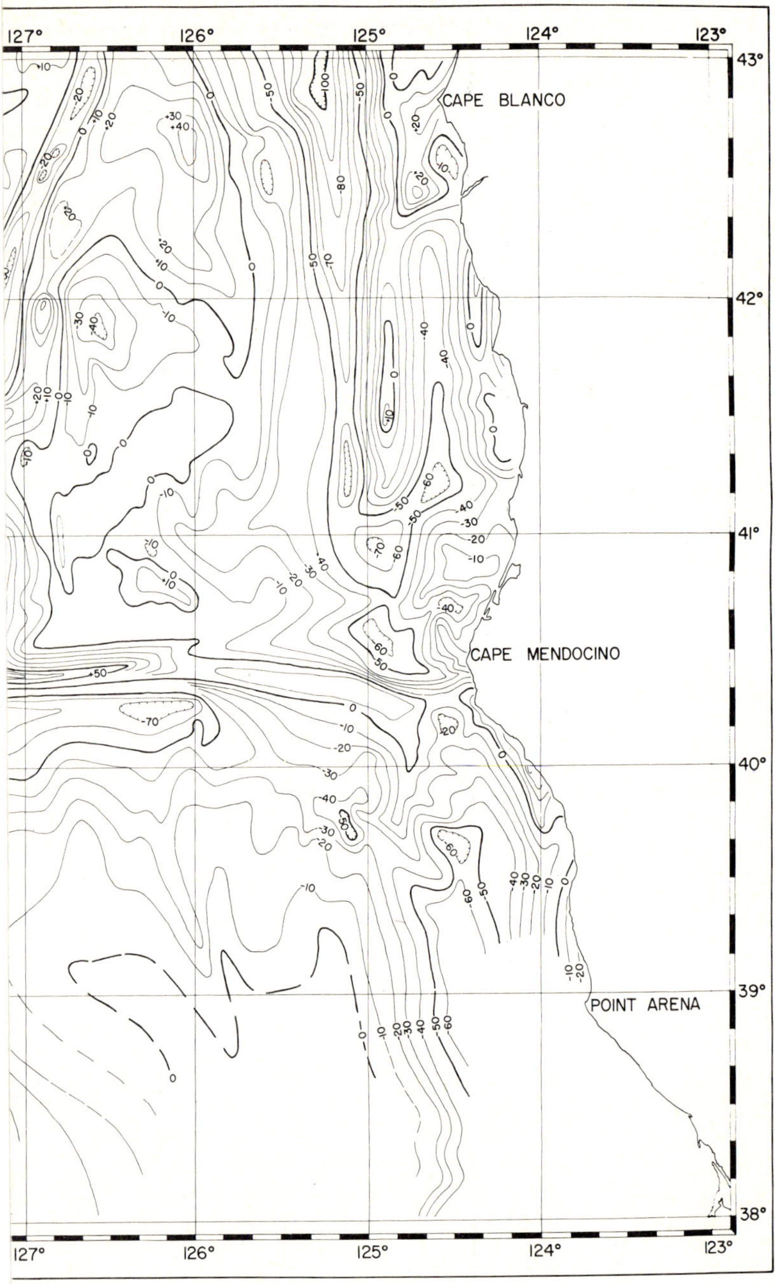

endocino Escarpment near 127°30′W.

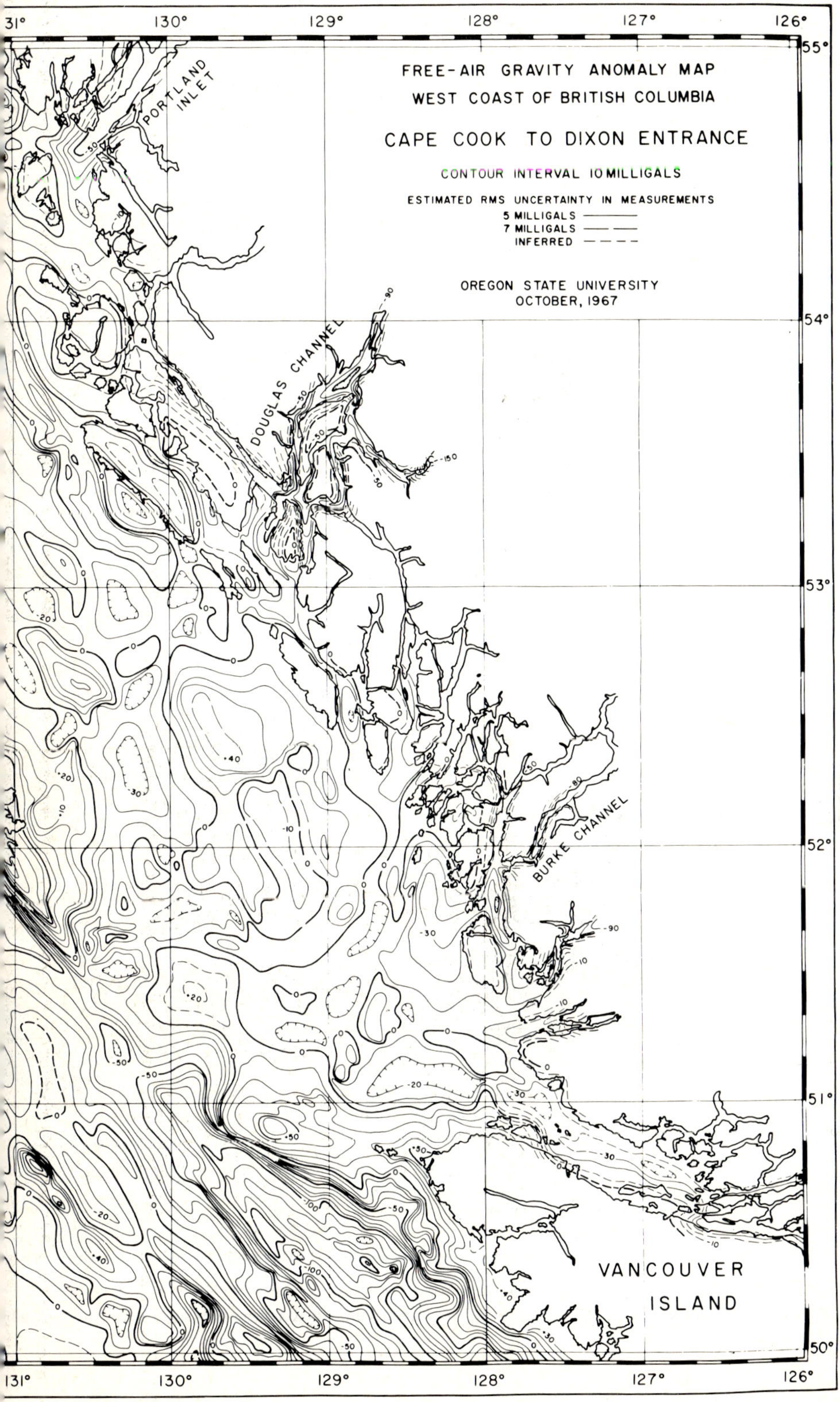

mainland) and over the Queen Charlotte fracture zone (characterized by the large

Alaska (Fig. 8.20) appears to be continuous with the eastern end of the northeasterly trending Aleutian Trench.

A negative free-air anomaly lineation is observed between the coast and shelf edge along most of the continental shelf (Figs. 8.15 to 8.19) from Alaska to California, except for periodically interrupting headland structures which extend seaward from the continent. Along Oregon this shelf lineation is known to coincide with a thick sedimentary section. It appears likely that these negative shelf lineations along most of the west coast are similarly produced by periodically interrupted narrow, elongated sedimentary basins.

Over most of the abyssal plains shown in the figures anomaly amplitudes are generally small, indicating that the areas are essentially in isostatic equilibrium. Numerous local seamounts exist in the abyssal plain west of Washington and British Columbia; these are characterized by local positive anomalies. West of Vancouver Island (Fig. 8.17) sharp bathymetric variations are associated with troughs and ridges (Dehlinger et al., 1970). West of Graham Island (Fig. 8.18) a large northwest-trending anomaly gradient extends along the Queen Charlotte fracture zone, a major transform fault characterized by frequent and also large-magnitude earthquakes.

The Juan de Fuca Ridge, a mid-ocean ridge with low topographic relief, extends northwesterly in the northwest part of Fig. 8.16 where several large gravity highs occur over seamounts. The ridge is characterized by an average free-air anomaly approximately 20 mgal greater than the average values over the adjacent plains. The anomaly variation over the ridge is thus comparable to that over other mid-ocean ridges, even though the topographic relief is substantially smaller.

The depth to Moho beneath the ridge was shown to be about 7 km (Shor et al., 1968). Since gravity anomalies indicate that the ridge is in near-isostatic equilibrium, densities of mantle rocks beneath the ridge must be relatively low. This is seen in Fig. 8.21, which shows a crustal-subcrustal section of Dehlinger et al. (1968) along latitude 44°45′N from Oregon westward across the continental margin, the sediment-filled Cascadia Basin, and the Juan de Fuca Ridge. The figure is based on two-dimensional numerical computations (eq. 7.74) for each successive layer, where the heavy lines in the figure indicate layer control points obtained from seismic refraction data (Shor et al., 1968). If we use typical mantle densities of 3.30 g cm^{-3} beneath the ridge, the computed anomaly is about +150 mgal, instead of the observed near-zero value. The required mass deficiency can hardly occur within the thin crust at the ridge; hence, it must be situated in the underlying mantle.

The Cascadia Basin, between the ridge and the continental slope (Figs. 8.16 and 8.21) has been shown by Shor et al. to contain 2—3 km thicknesses of sediments (which were brought in by the Columbia and other rivers). The near-zero anomaly values over the basin shows that it, too, is essentially in isostatic equilibrium. The continental margin also appears to be in equilibrium, the anomalies being produced by the transition from continental to oceanic crusts.

The Gorda Ridge (northeast part of Fig. 8.15) is an offset of the southward continuation of the Juan de Fuca Ridge; the offset is along the northwest-trending Blanco transform fault or fracture zone (in the southwest part of Fig. 8.16). Free-air anomalies across the Gorda Ridge are approximately −30 mgal over the median valley and +40 mgal over the flanking ridges (approximately zero average values), showing that the Gorda Ridge is in near-isostatic equilibrium. The presence of low-density mantle materials beneath this ridge is indicated in Fig. 8.22.

Geomagnetic-reversal lineations associated with the Juan de Fuca and Gorda ridges have led to the postulation (McKenzie and Morgan, 1969; Atwater, 1970) that a now-inactive subduction zone exists along the Oregon—Washington coasts. Volcanism along the Cascade Mountain Range of Washington and Oregon is consistent with such a concept. However, the relatively small gravity anomalies along the continental margin and the absence of medium to deep earthquake focal depths in the area indicate that any such subduction zone is no longer active. Hence, if an inactive subduction plate exists, it is under greatly reduced stress and is at least partially dissipated.

The Mendocino fracture zone is the east—west feature in Fig. 8.15, characterized by positive anomaly amplitudes along its north and negative amplitudes along its south sides. These anomaly values are produced by the edge effects due to juxtaposed crustal sections along the fracture zone. The anomaly gradients are largest near an east—west topographic ridge, the Mendocino Ridge, which is at the southern end of the Gorda Ridge where it is terminated by the fracture zone. Dehlinger et al. (1970) constructed three hypothetical north—south, crustal-subcrustal sections across the Mendocino fracture zone which conform with the gravity anomalies and the observed refraction data of Shor et al. (1968). Their section along longitude 127°30′ W is shown in Fig. 8.22. The section was computed using two-dimensional numerical methods (eq. 7.74) to obtain attractions produced by each layer and then summing. Densities of the several crustal layers were obtained from measured seismic velocities, by use of the Nafe and Drake (1963) empirical velocity—density relationships.

North of the fracture zone, beneath the Gorda Ridge, the refraction results of Shor et al. show the crust to be thin. If we assume a normal mantle density of 3.3—3.4 g cm^{-3} beneath the ridge, free-air anomalies will be about +150 mgal (Dehlinger et al., 1967) instead of the observed near-zero values. Thus, relatively low-density subcrustal materials are required. The shapes and densities of mantle blocks in the figure, which produce the required mass deficiency, are assumed; subcrustal densities are not derived from the Nafe-Drake velocity—density relationships, because these are based on marine sediments and are generally less applicable to higher-density,

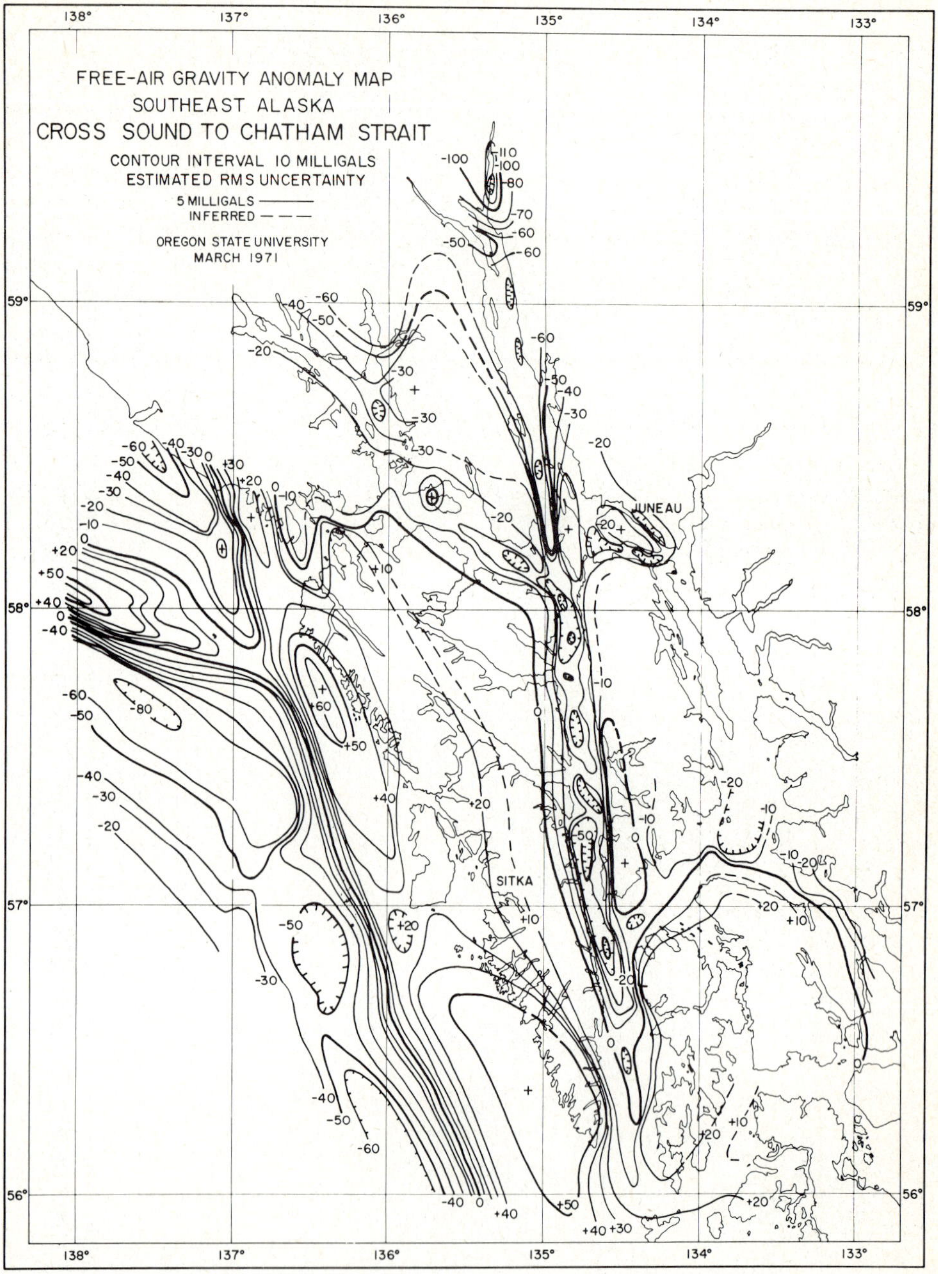

Fig. 8.19. Free-air anomaly map west of the coast of the Alaska panhandle and along the Inside Passage of southeast Alaska. (Courtesy of R.W. Couch, 1977.)

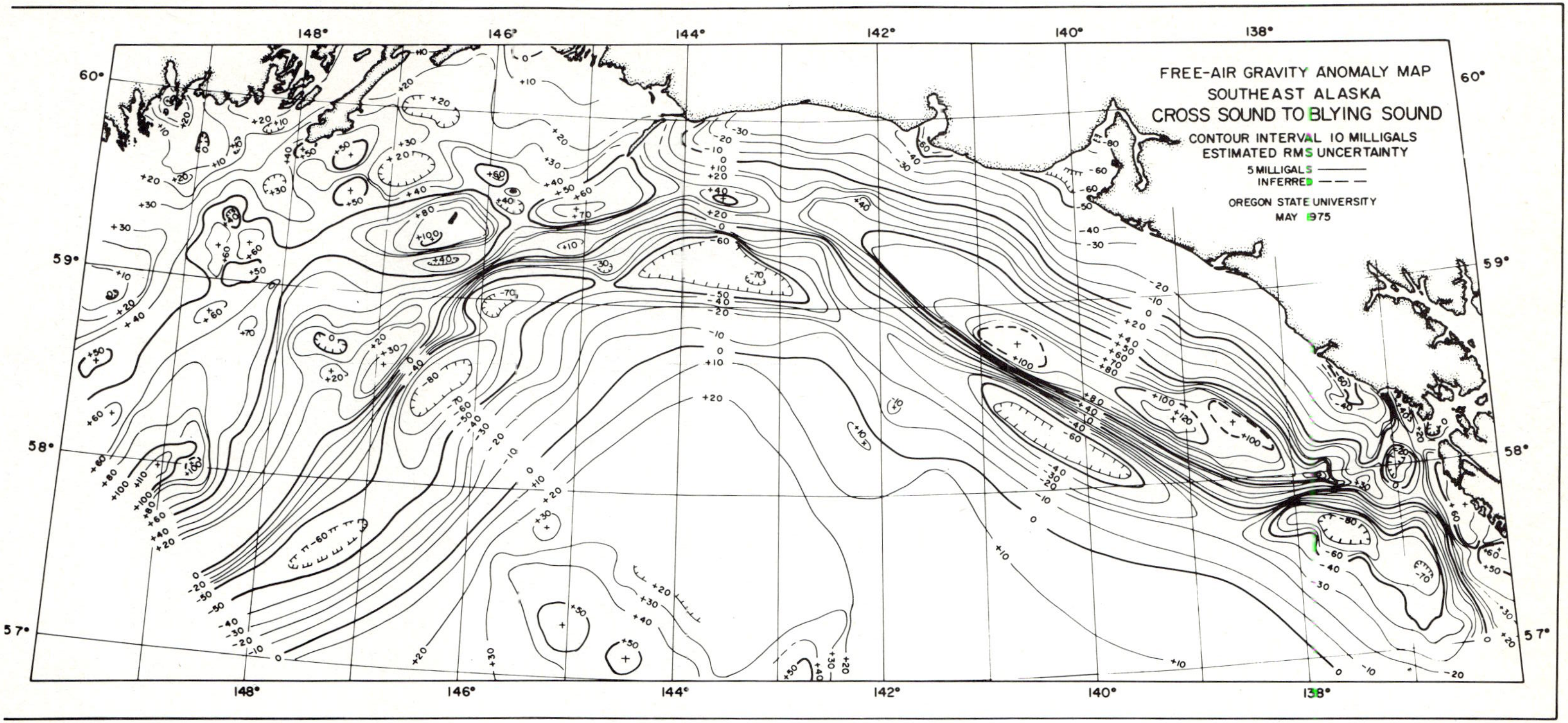

Fig. 8.20. Free-air anomaly map south of the coast of southeast Alaska, showing the junction of the Aleutian Trench to the southwest and the southeast-trending continental margin off the Alaskan panhandle. (Courtesy of R.W. Couch, 1977.)

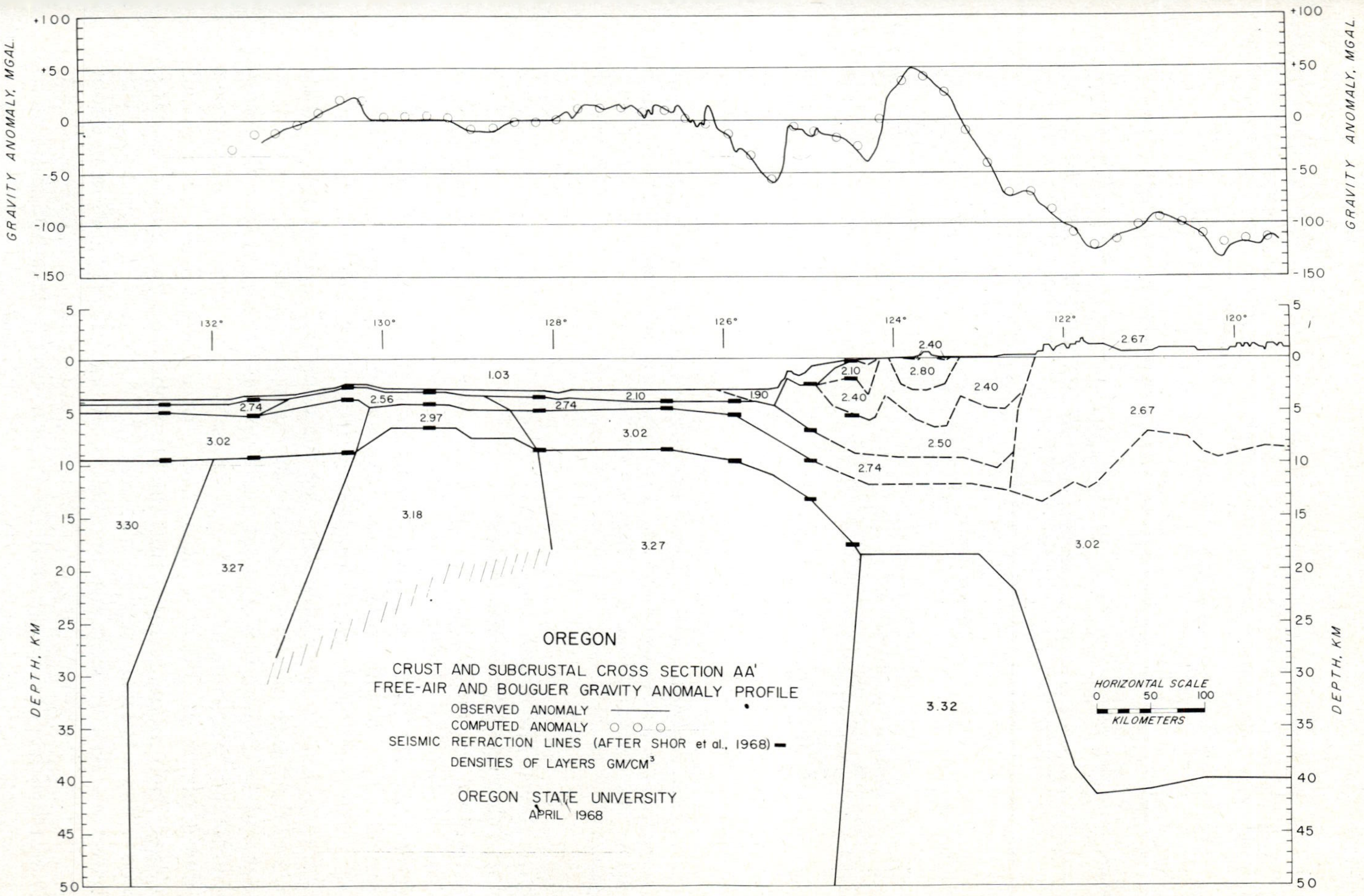

Fig. 8.21 Crustal—subcrustal section from the Oregon coast westward to the Juan de Fuca Ridge along 44°45′N; the section is consistent with seismic refraction depths (short heavy lines) and observed free-air anomalies. Dashed lines indicate questionable boundaries. (Reproduced from Dehlinger et al., 1970, with permission.)

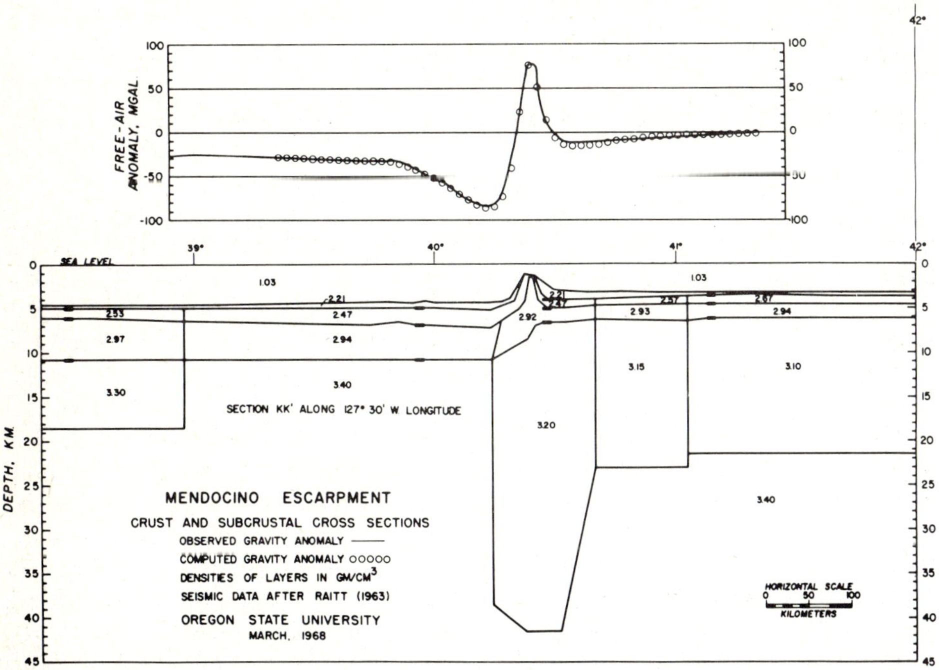

Fig. 8.22. Crustal—subcrustal section across the Mendocino Escarpment and the Gorda Ridge along 127°30′W; the section is consistent with seismic refraction depths (short heavy lines) and free-air anomalies. (Reproduced from Dehlinger et al., 1970, with permission.)

ultra-basic rocks. The mantle blocks and densities in the figure are reasonable for an oceanic ridge (the Gorda Ridge).

To simplify computations in Fig. 8.22, we assume two-dimensional structures across the fracture zone. At least near the Gorda Ridge this assumption is not fully justified; structures there are not truly two-dimensional. The actual ridge structures which produce the observed anomalies will have larger mass deficiencies than those shown in the figure; hence, mantle blocks are either larger or the density contrasts greater than shown, and the figure presents minimal subcrustal contrasts.

Sibuet and Le Pichon (1974) analyzed the free-air anomaly profile across the Mendocino Escarpment in Fig. 8.22 as if produced by juxtaposed lithospheric plates of different ages, hence of different temperatures and thicknesses. Their model assumes that the lithosphere on the south side of the escarpment has a typical oceanic thickness of 75 km and that on the north side, based on ages of magnetic anomalies, an age of 30 to 40 m.y. From their analysis of the gravity profile, they compute a plate thickness of 55 km on the north side. The Sibuet and Le Pichon model does not attempt to reproduce the observed gravity anomaly exactly; density contrasts in their model are generally consistent with mass distributions shown in Fig. 8.22.

Gravity anomalies southwest of Mexico and west of Central America

Free-air anomaly maps have been constructed of the continental margins and adjacent deeps off the southwest coast of Mexico and southeasterly from there to Panama. These regions extend along most of the length of the Middle America Trench and include a number of tectonic features in the Cocos Lithospheric Plate, which lies between the trench and the East Pacific Rise to the west. The gravity measurements were obtained by a stabilized platform-mounted LaCoste and Romberg meter.

Fig. 8.23 (from Gumma, 1974) shows free-air anomalies near the northern end of the trench and over the southeast-trending Rivera fracture zone, which is part of the East Pacific Rise. The large-amplitude negative-anomaly lineation in the western part of the figure identifies the fracture zone. To the southeast of this figure lies the northeast trending Orozco fracture zone (Fig. 8.24, from Lynn, 1975), which exhibits low-amplitude anomalies. The adjacent continental margin is characterized by large-amplitude linear anomalies on the shelf, some of which are strongly negative. The trench anomaly in Figs. 8.23 and 8.24 is strongly negative, as is characteristic of an active subducted plate, although the region seaward of the trench does not exhibit the broad positive anomaly zone commonly associated with subduction (e.g., off the Aleutian Trench and off Guatemala, as seen in the next figure).

Fig. 8.25 (Woodcock, 1975; Couch and Woodcock, 1977) shows a southeastward continuation of the high-amplitude trench anomaly and linear gravity anomalies over the northeast-trending Tehuantepec Ridge (center of the figure) and large amplitude, positive anomaly lineations on the adjacent continental shelves. Couch and Woodcock conclude that the Tehuantepec Ridge marks the boundary between two subduction provinces of different ages; also, the Moho northwest of the ridge is about 4 km shallower than that to the southeast. Watts and Talwani (1974) interpret the difference in observed amplitude and wavelength of the positive anomalies seaward of the trench on both sides of the ridge, as produced by differential lithospheric flexuring prior to subduction beneath the trench. Large linear positive anomalies trend along the continental margin to the east of the ridge and abruptly extend landward in line with the Tehuantepec Ridge. Couch and Woodcock interpret these anomalies on the shelf, which are independent of topography, as indicative of differential vertical displacements.

Fig. 8.26 shows a Couch and Woodcock crustal section across the continental margin off Guatemala, near San Jose (see Fig. 8.25), based on gravity anomalies, seismic refraction data, and a magnetic profile. Couch and Woodcock interpret the abrupt change in Moho depth near the trench axis as caused by crustal imbrication which produces seaward growth of the continent.

Fig. 8.27 (from Victor, 1975) shows free-air anomalies off the coasts of Nicaragua, Costa Rica, and northern Panama. The large-amplitude lineation along the Middle American Trench extends southeast from Nicaragua and terminates abruptly against a northeast-striking feature, the Cocos Ridge, which trends northeast toward northern Panama. In contrast to the high-amplitude anomalies which characterize the Middle America Trench, the Cocos Ridge exhibits low-amplitude, short-wavelength anomalies. Two tectonically different provinces are clearly juxtaposed in this region.

Gravity anomalies west of parts of Peru and Chile

Free-air anomaly maps have been constructed over parts of the region between the coast of South America and the East Pacific Rise (the Nazca Lithospheric Plate). The measurements were obtained by a platform-mounted LaCoste and Romberg gravimeter.

Fig. 8.28 (from Whitsett, 1975) shows anomalies west of southern Peru. The large-amplitude negative-anomaly lineation marks the trench axis, where the Nazca Plate is being subducted underneath the continent. The broad positive anomaly zone seaward of the trench has rather low amplitudes, suggesting that lithospheric flexuring prior to subduction is less pronounced here than in typical areas (e.g., the previously discussed Aleutian Trench, the Bonin Trench, the Philippine Trench, and parts of the Middle America Trench). On the continental shelf, Fig. 8.28 also shows a discontinuous, relatively large-amplitude, negative-anomaly lineation. This trend is similar to those in Figs. 8.15 to 8.18 along the west coasts of North America and also off parts of Mexico and Central America; the negative anomalies suggest elongated sedimentary basins of considerable thicknesses.

Figs. 8.29 and 8.30 (from Couch and Huehn, 1977) show free-air anomalies across part of the continental margin and trench axis off Chile. The large-amplitude, negative-anomaly lineation off northern Chile (Fig. 8.29) correlates with the deepest part of the Chilean trench and is characteristic of active subduction zones. On the shelf there is a notable absence of the negative-anomaly lineation commonly observed on other shelves. Instead, the shelf has positive-amplitude lineations, as does the shelf off Guatemala (Fig. 8.25). Off the coast of southern Chile (Fig. 8.30), where the trench is less deep, the trench-anomaly amplitudes are correspondingly lower. As the trench dies out toward the south, the shelf widens; the wider shelf is evidenced by the broader zone of low-amplitude anomalies in the southern part of the figure.

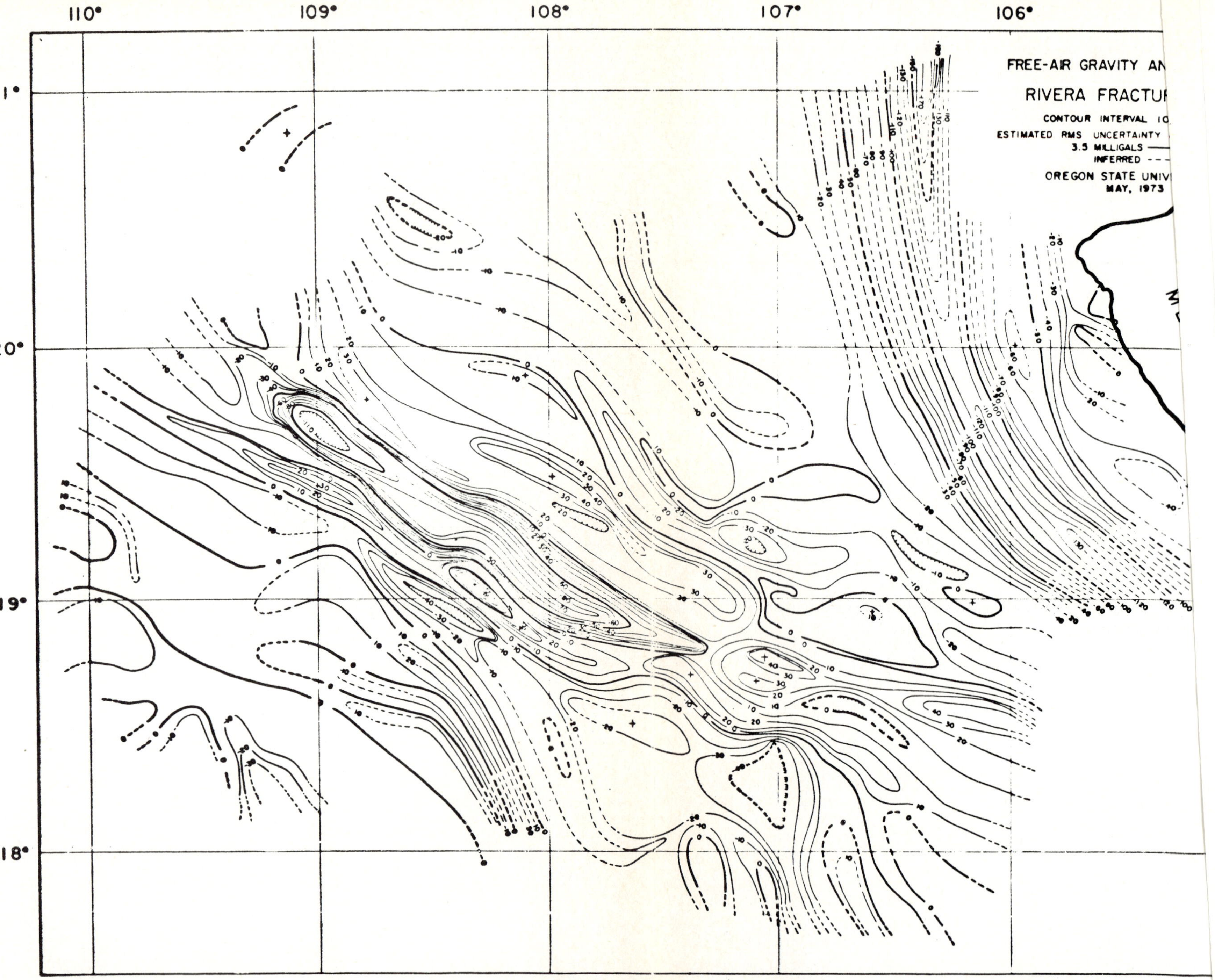

Fig. 8.23. Free-air anomaly map west of the coast of Mexico extending over the northwest-trending Rivera fracture zone (centered near 19°N 108°W). (Reproduced from G with permission.)

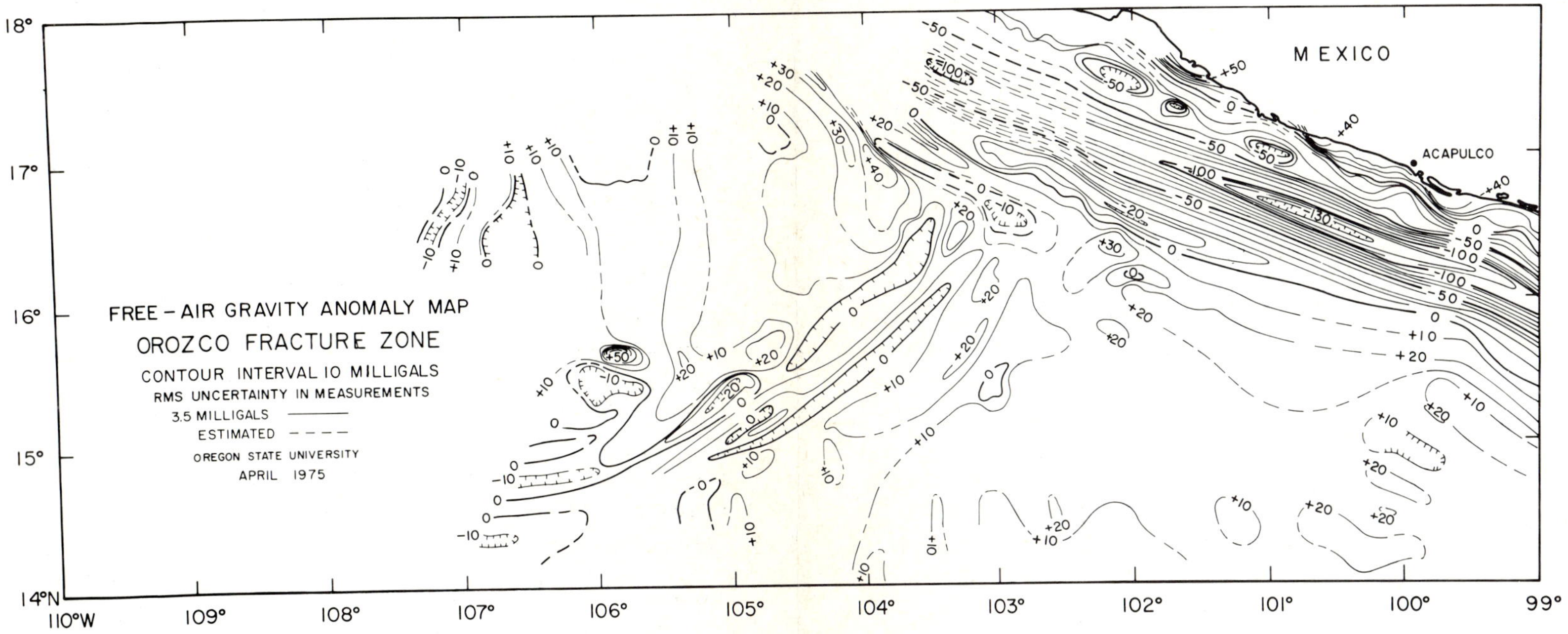

Fig. 8.24. Free-air anomaly map west of Acapulco, Mexico and over the northeast-trending Orozco fracture zone. (Reproduced from Lynn, 1975, with permission.)

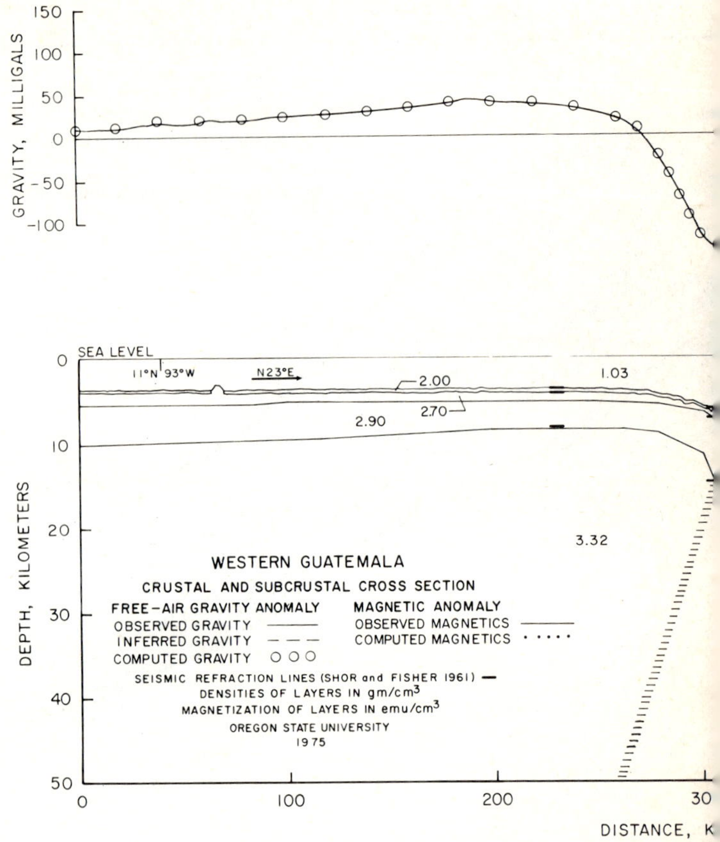

Fig. 8.26. Crustal—subcrustal section across the continental marg and magnetic data. (Courtesy of R.W. Couch, 1977.)

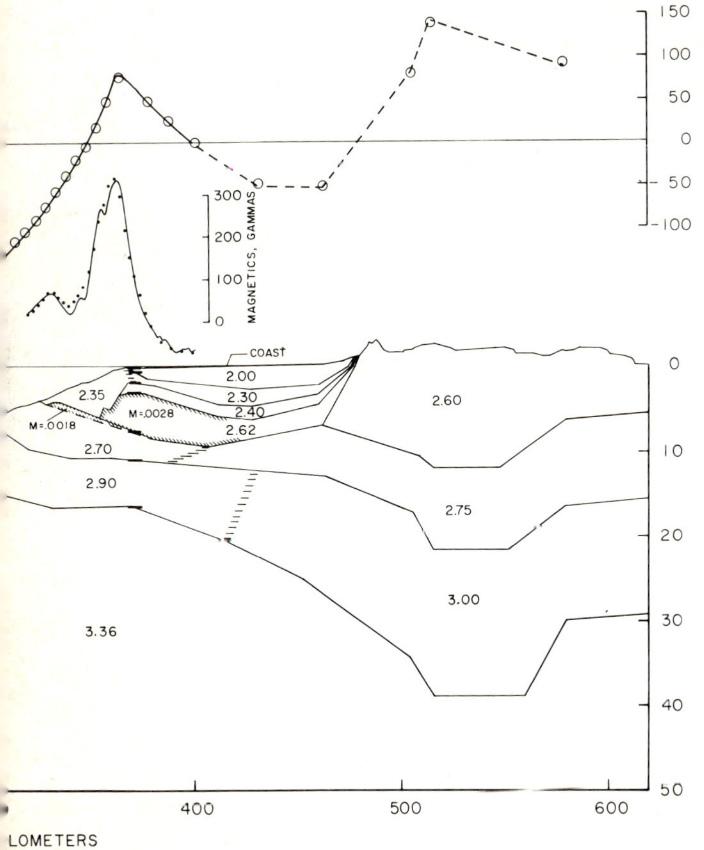

in off Guatemala, near San Jose, based on gravity, seismic refraction,

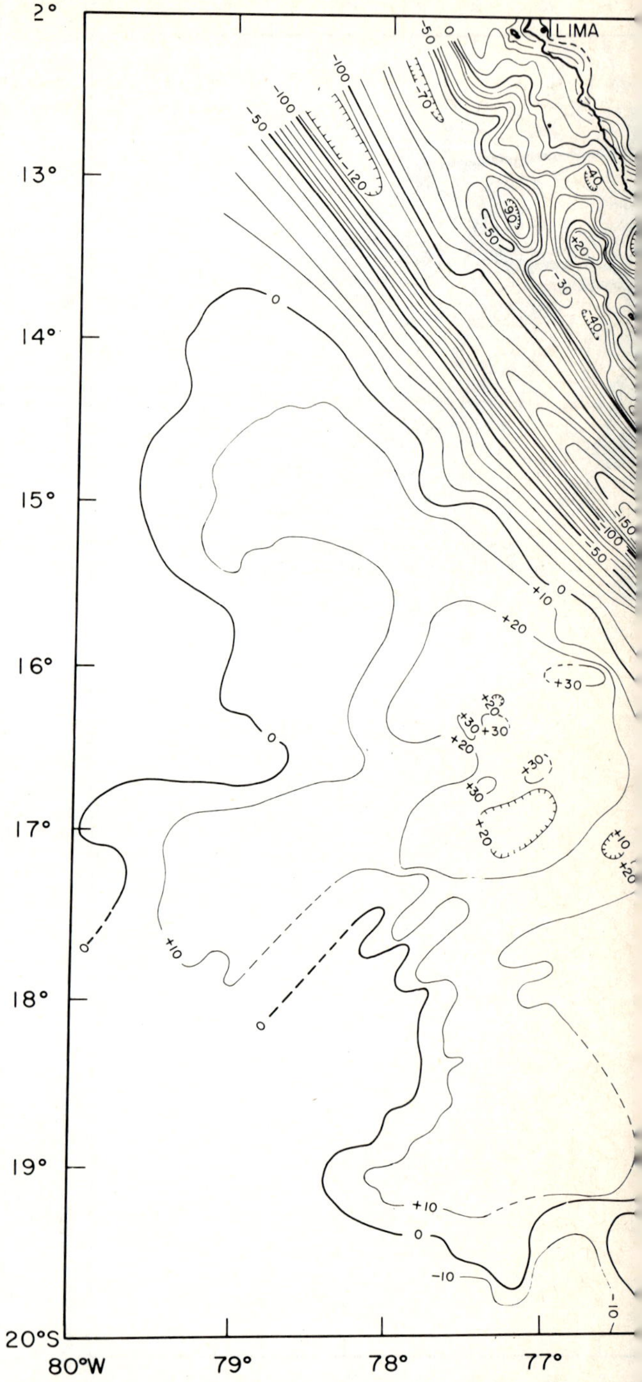

Fig. 8.28. Free-air anomaly map west of southern Peru, wh
from Whitsett, 1975, with permission.)

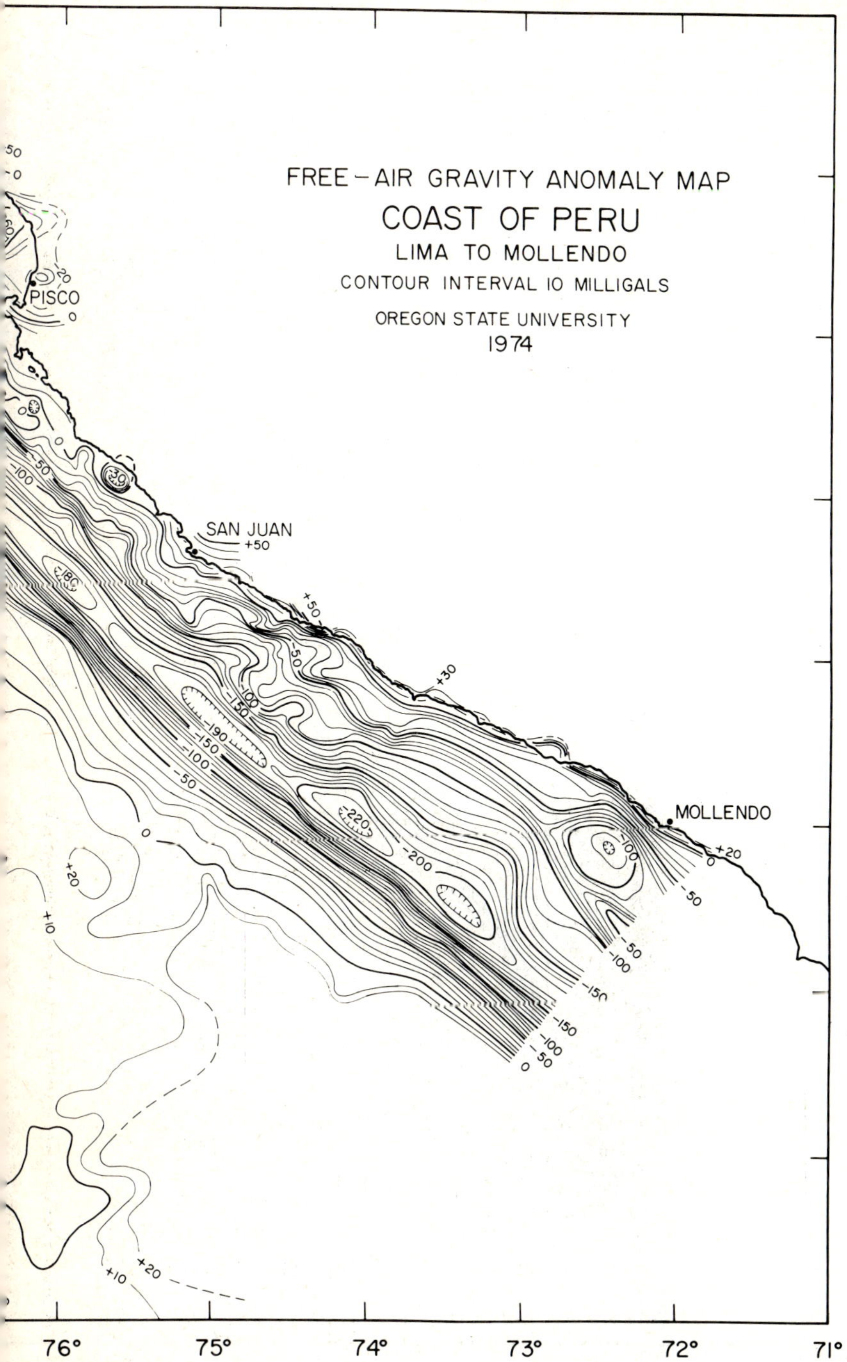

ere the Nazca Plate is subducted beneath the South America Plate. (Reproduced

Gravity anomaly map of the southern Beaufort Sea, north of Alaska

Gravity was measured aboard U.S. Coast Guard cutters (ice breakers) in the southern Beaufort Sea, north of the north slope of Alaska, from Canada westward to Barrow and out to the summer ice pack. Dehlinger and Chiburis (in prep.) obtained approximately 3,000 gravity measurements, with a mean anomaly uncertainty of 1.1 mgal (based on 166 trackline intersections), using LaCoste and Romberg platform-mounted meters. Fig. 8.31 shows a free-air anomaly map of the continental margin of northern Alaska, and Fig. 8.32 the bathymetry of the area. Additional gravity measurements in this area have also been obtained by the U.S. Geological Survey (Boucher et al., 1977).

The free-air values along the north slope of Alaska are near zero, as seen in Fig. 8.31, indicating that the coastal region of the gently sloping continental shelf (Fig. 8.32), which is a seaward continuation of the flat coastal plain of the north slope of Alaska, is essentially in isostatic equilibrium. On the continental shelf the anomaly values become increasingly more positive, reaching maximum amplitudes of +30 to +50 mgal on the shelf side of the shelf edge (the 200-m isobath in Fig. 8.32). The anomaly values decrease characteristically downward along the continental slope (between about the 200-m and 2,000-m isobaths), except in the eastern part of the area where an anomalously large linear positive anomaly (+100 mgal amplitude) parallels the slope. No bathymetric feature (Fig. 8.32) corresponds to the gravity high, such that an elongated three-dimensional mass excess must be beneath the slope. This gravity high does not extend north of the slope, as shown by the previous ice-station gravity measurements obtained by Wold and Ostenso (1971), which indicate generally negative anomaly values in the abyssal plain, known as the Canada Basin, north of the slope. The gravity high does not extend eastward, as evidenced from the free-air map of Boucher et al. (1977), which rather shows a gravity minimum along the slope off the mouth of the Mackenzie River.

The Dehlinger and Chiburis measurements in the Beaufort Sea were obtained to provide information on how regional stresses between the southward moving Arctic Ocean, spreading away from the Nansen Ridge (the Arctic continuation of the Mid-Atlantic Ridge) and the Alaskan mainland may be relieved. Does the spreading create compressional stresses at the continental margin, which are possibly no longer active, or is the compression taken up along mountain ranges to the south? Or does a transform fault characterize the north coast of Alaska, as had been suggested by Herron et al. (1974)?

Fig. 8.33 shows a free-air and a two-dimensional isostatic profile along line *A* in Fig. 8.31. In the isostatic profile, the upper 3.8 km of the section was assumed to consist either of sedimentary rock ($\rho = 2.33$ g cm^{-3}) or of ocean water, or a combination of the two. The large negative isostatic anomaly near the base of the continental slope suggests that a substantial thickness of low-density sediments exists along the slope and at its base. The Wold and Ostenso (1971) free-air determinations in the abyssal plain to the north show generally negative free-air values (—20 to —25 mgals), which imply a regional mass deficiency, as would result from a general sinking of the lithosphere which, at the same time, is receiving sediments from the Mackenzie and other rivers. Seismic refraction measurements are needed in this ice-ridden area to resolve the regional crustal—subcrustal features.

Isostatic anomalies at ocean—continent boundaries at passive continental margins

Basement rocks at an ocean—continent boundary along a passive continental margin are assumed to be products of ocean-crust material initially injected at a continental rift zone on a continental-type crust. The age of the ocean crust at or near this boundary can be obtained from the known age of the adjacent magnetic anomaly (i.e., the identified anomaly farthest from its spreading ridge).

In their analysis of gravity profiles across the Vøring Plateau margin off Norway, Talwani and Eldholm (1972) showed that the ocean—continent boundary is identifiable from two-dimensional, Airy-isostatic anomaly profiles (described in Chapter 6) across the margin. Rabinowitz (1974, 1976) similarly applied 2-D isostatic analysis to numerous gravity profiles obtained on both sides of the South Atlantic. He found that the ocean—continent boundary, as identified from observed and computed magnetic anomalies for such a boundary, is usually characterized by a diagnostic anomaly pattern which is independent of the location of the boundary with regard to the shelf edge.

Fig. 8.34 shows characteristic isostatic profiles which Rabinowitz and Labrecque (1977) obtained across the Argentine and southern African continental margins. The coast line is on the left side of each profile and the ocean on the right. The profiles are aligned with regard to the presumed ocean—continent boundary, and arrows show the position of the shelf edge. The characteristic anomaly is asymmetrical, has a wavelength of about 150 km, a positive amplitude of 30—70 mgal, and a steep slope is toward the coast and gentle slope toward the ocean. Rabinowitz and Labrecque further showed that this anomaly pattern is independent of whether thick sediments cover only the continental or only the oceanic basement.

Fig. 8.29. Free-air anomaly map west of northern Chile, where the Nazca Plate is subducted beneath the South America Plate. (Courtesy of R.W. Couch, 1977.)

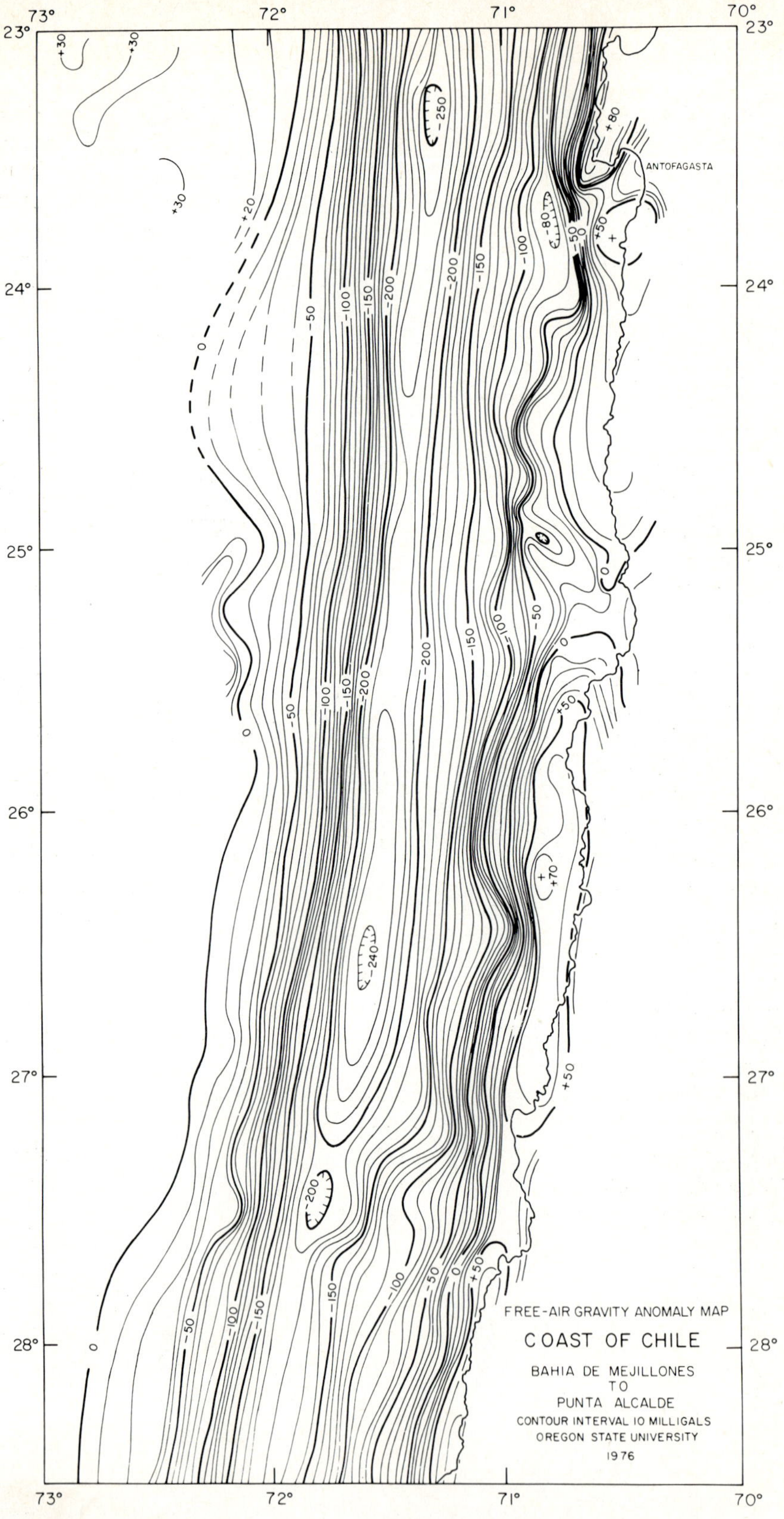
73°
72°
71°
70°
23°
24°
25°
26°
27°
28°
ANTOFAGASTA
+30
+20
+50
+80
+70
0
-50
-80
-100
-150
-200
-240
-250
FREE-AIR GRAVITY ANOMALY MAP
COAST OF CHILE
BAHIA DE MEJILLONES
TO
PUNTA ALCALDE
CONTOUR INTERVAL 10 MILLIGALS
OREGON STATE UNIVERSITY
1976

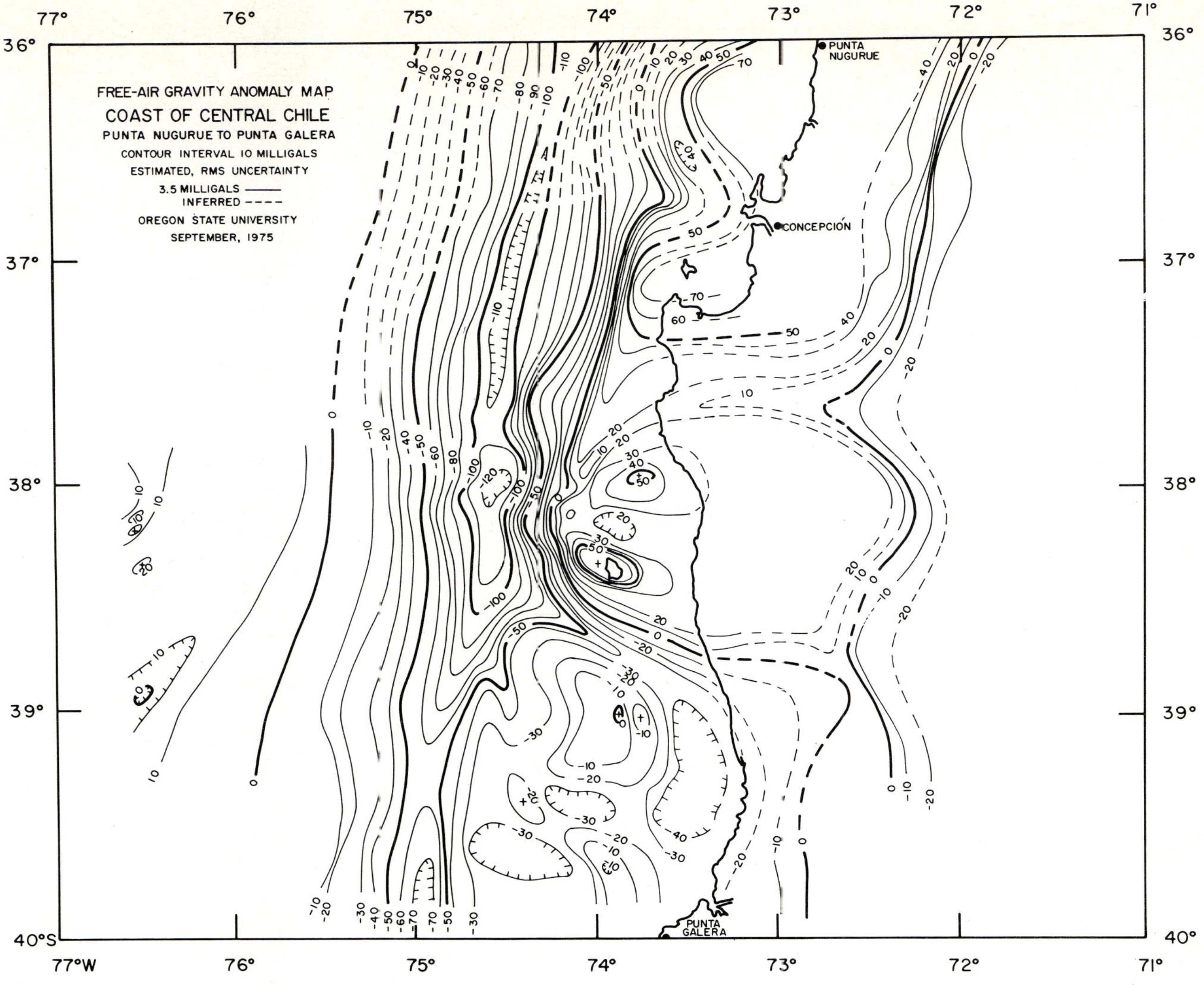

Fig. 8.30. Free-air anomaly map west of southern Chile near the southern limit of the South America Trench.

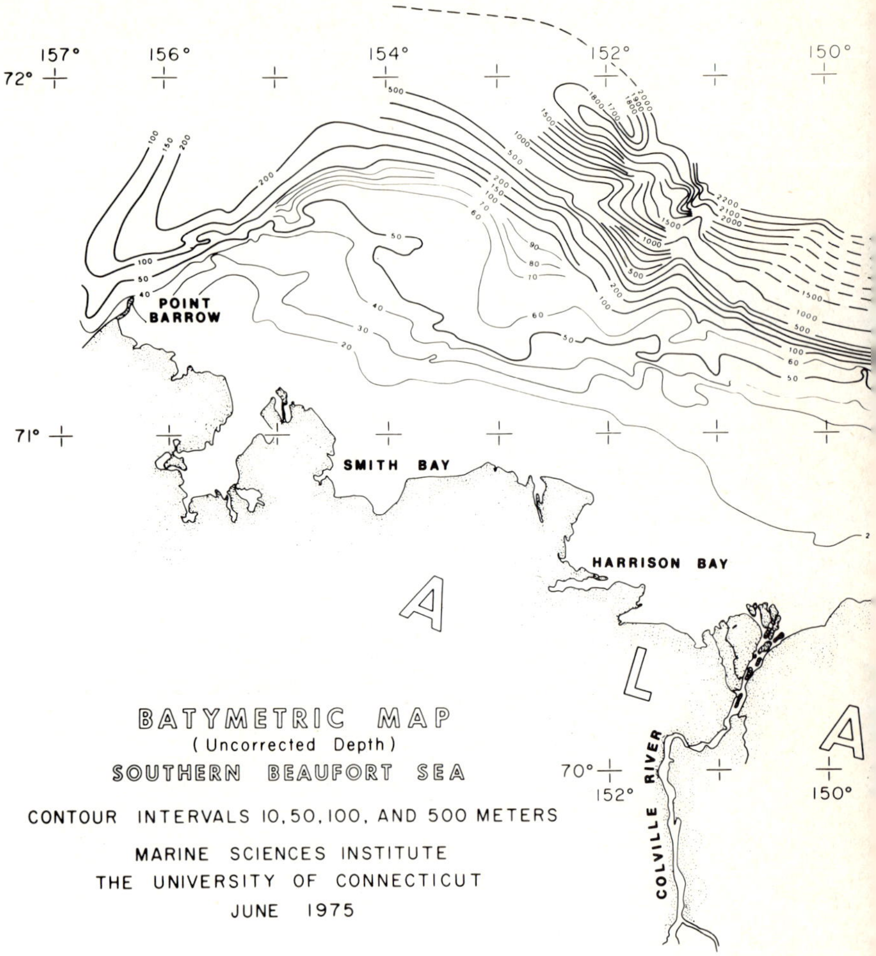

Fig. 8.32. Bathymetric map of the southern Beaufort Sea. (From Dehlinger and C

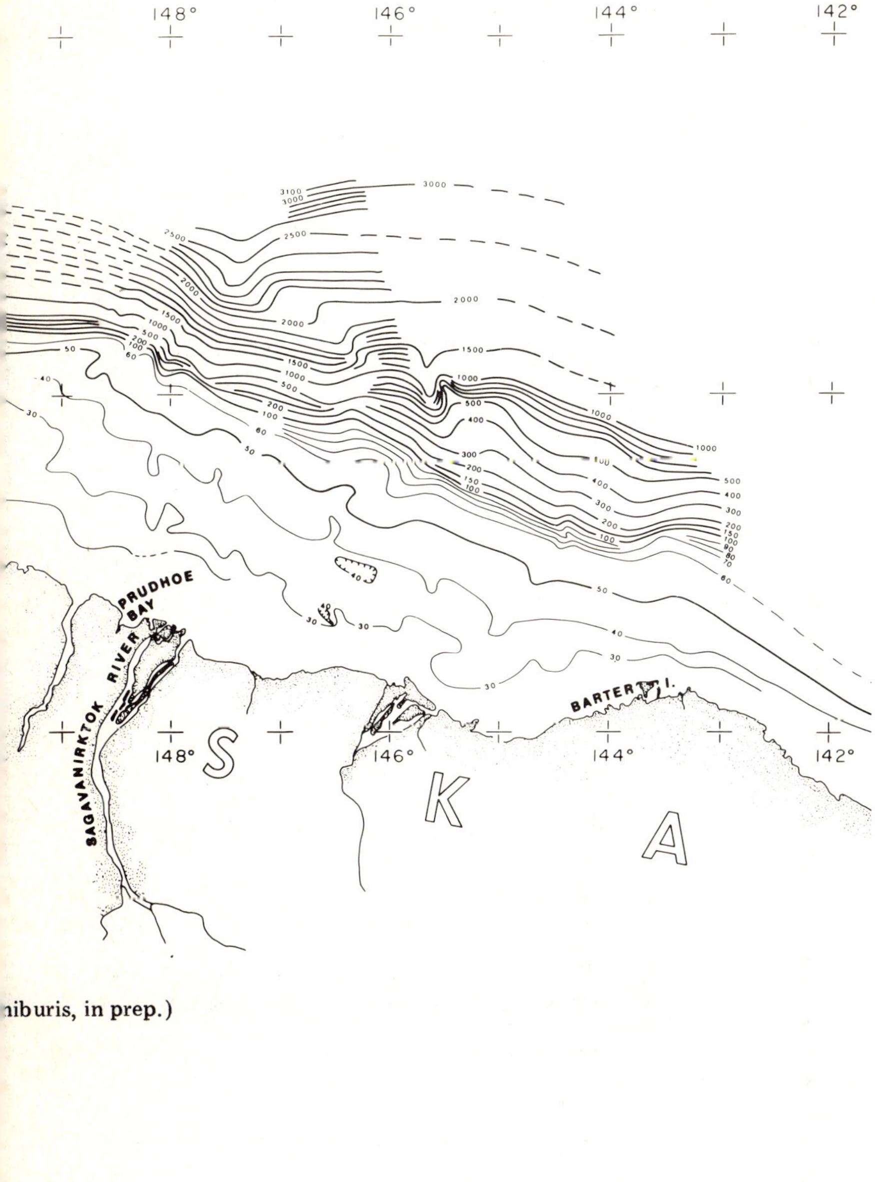

ıiburis, in prep.)

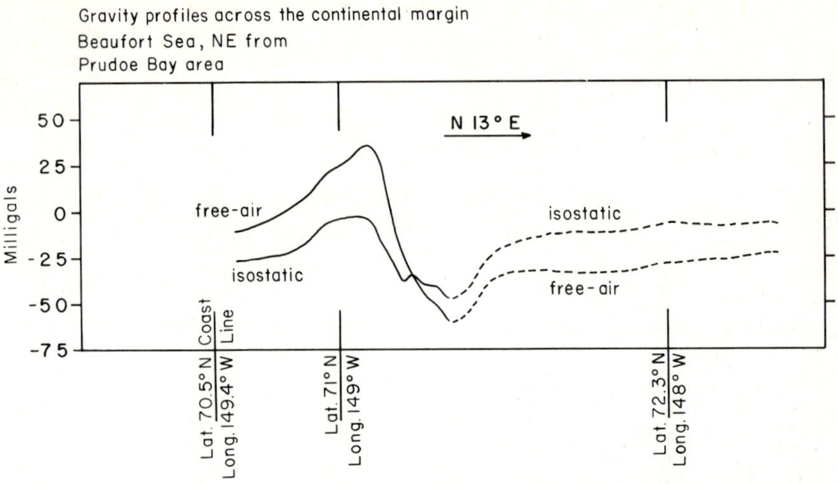

Fig. 8.33. Comparison of computed two-dimensional isostatic and observed free-air anomalies along profile *A* in Fig. 8.31. (From Dehlinger and Chiburis, in prep.)

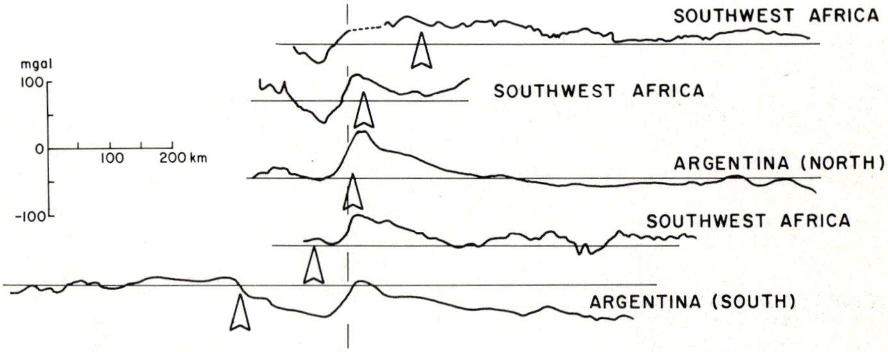

Fig. 8.34. Examples of two-dimensional isostatic anomaly profiles across continental margins. The profiles are aligned at the presumed ocean—continent boundary, with the continent on the left side of each profile and arrows pointing to the shelf edge. The characteristic anomaly extends about 150 km seaward of the boundary, with maximum amplitudes from +30 to +70 mgal. (Reproduced from Rabinowitz and Labrecque, 1977, with permission.)

The fact that a positive anomaly extends over the oceanic side of the boundary means that the ocean basement there is likely to be at a higher elevation than at mid-ocean, and also that the ocean basement is uncompensated at the boundary. Rabinowitz and Labrecque propose that the high elevation results from early rifting in a narrow basin; volcanic material similar in composition to oceanic crust will be injected at higher elevations, on a continental crust, before the basin widens sufficiently for it to develop into an ocean.

Fig. 8.35. Bathymetric map of the central and eastern Mediterranean Sea. (Reproduced from Rabinowitz and Bryan, 1970, with permission.)

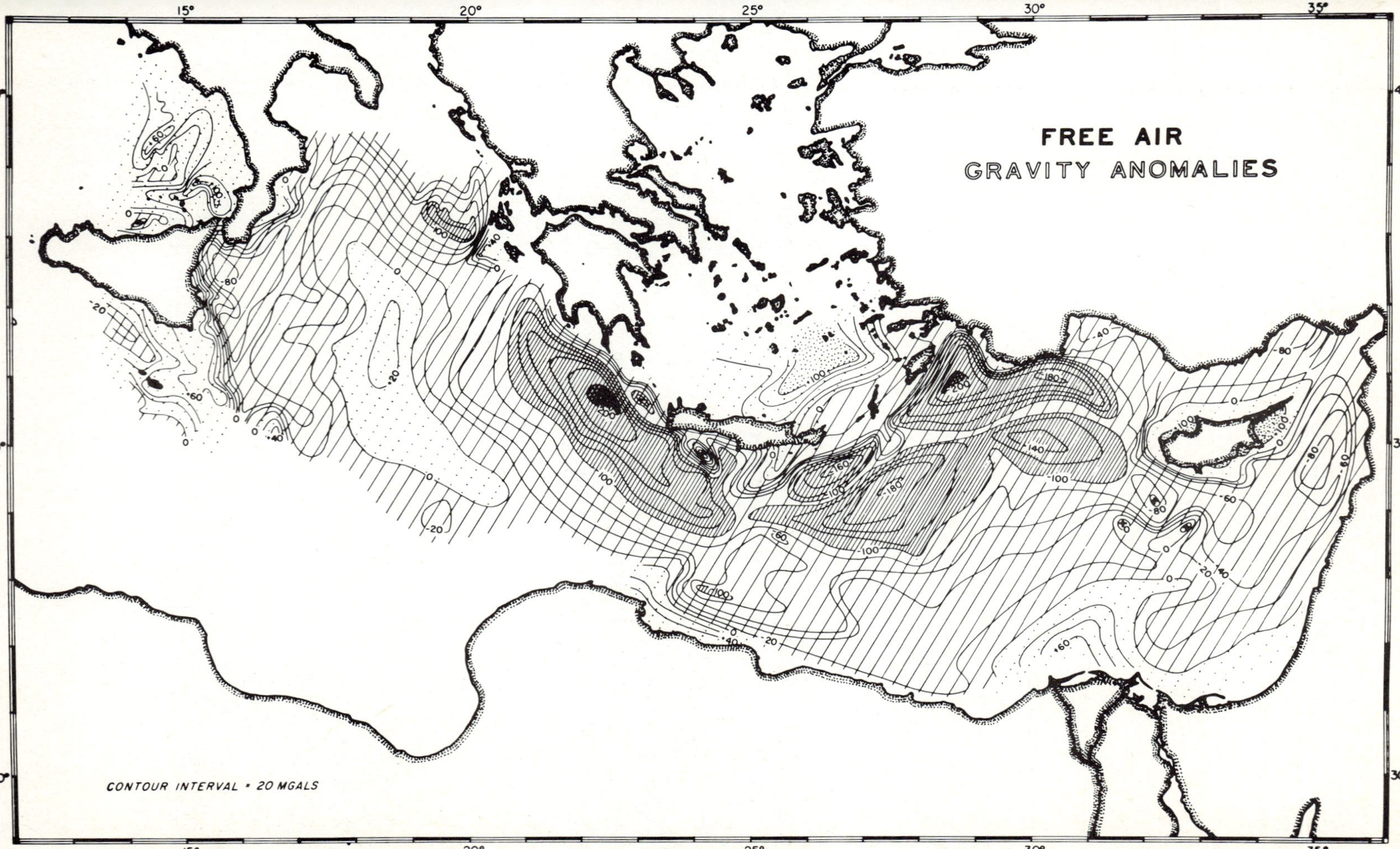

Fig. 8.36. Free-air anomaly map of the central and eastern Mediterranean Sea. (Reproduced from Rabinowitz and Bryan, 1970, with permission.)

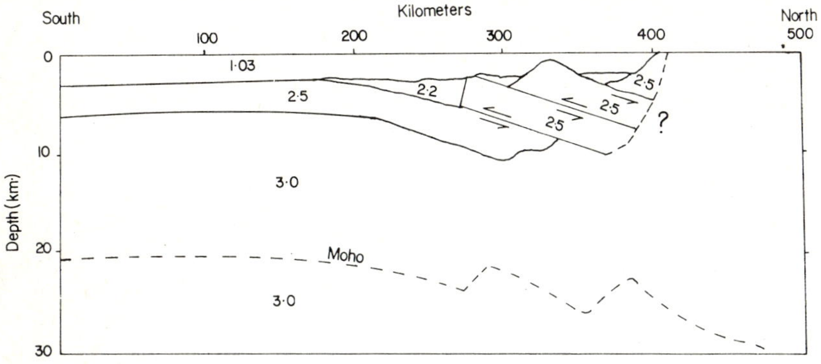

Fig. 8.37. Crustal section illustrating northward underthrusting which results in crustal shortening in the eastern Mediterranean Sea. Layer densities are in g cm^{-3}. (After Rabinowitz and Bryan, 1970.)

Gravity anomalies in the eastern Mediterranean Sea

The Mediterranean is an inland sea which has been undergoing considerable deformation, especially along two island-arc type structures. The Calabrian Arc in the western Mediterranean was recognized by Peterschmitt (1956) to extend along the southern margin of the Tyrrhenian Sea (southern Italy and westward) from distributions of earthquake foci, some exceeding depths of 300 km.

The Hellenic Arc in the eastern Mediterranean, extending around Greece from the eastern Ionian Sea (between Italy and Greece) to southern Turkey, was identified from the bathymetric contours (Fig. 8.35) of Goncharov and Mikhailov (1963). This arc, which Goncharov and Mikhailov called the Hellenic Trough, consists of two parallel trenches and an interior volcanic arc. The arc is surrounded by a broad seaward swell, the Mediterranean Ridge, which extends through the middle of the Mediterranean from the southern Ionian Sea to Cyprus. The Hellenic Trough is characterized by earthquake foci of depths less than 100 km (Rabinowitz and Ryan, 1970). Both the trough and the Mediterranean Ridge exhibit low heat-flow values (Langseth et al., 1966), characteristic of oceanic trenches, and both the trough and ridge exhibit few detectable magnetic anomalies (Allan et al., 1964).

Gravity in the eastern Mediterranean was measured by the Lamont-Doherty Geological Observatory (L-DGO) with a Graf-Askania Gss-2 gravimeter. Navigation positions were obtained with satellite fixes; resultant anomalies are reported to have estimated uncertainties of 5 mgal. Fig. 8.36 shows a free-air anomaly map, based on these L-DGO measurements and on measurements obtained by Worzel (1959) and Fleischer (1964), as well as previous sea-pendulum measurements (Rabinowitz and Ryan, 1970); how-

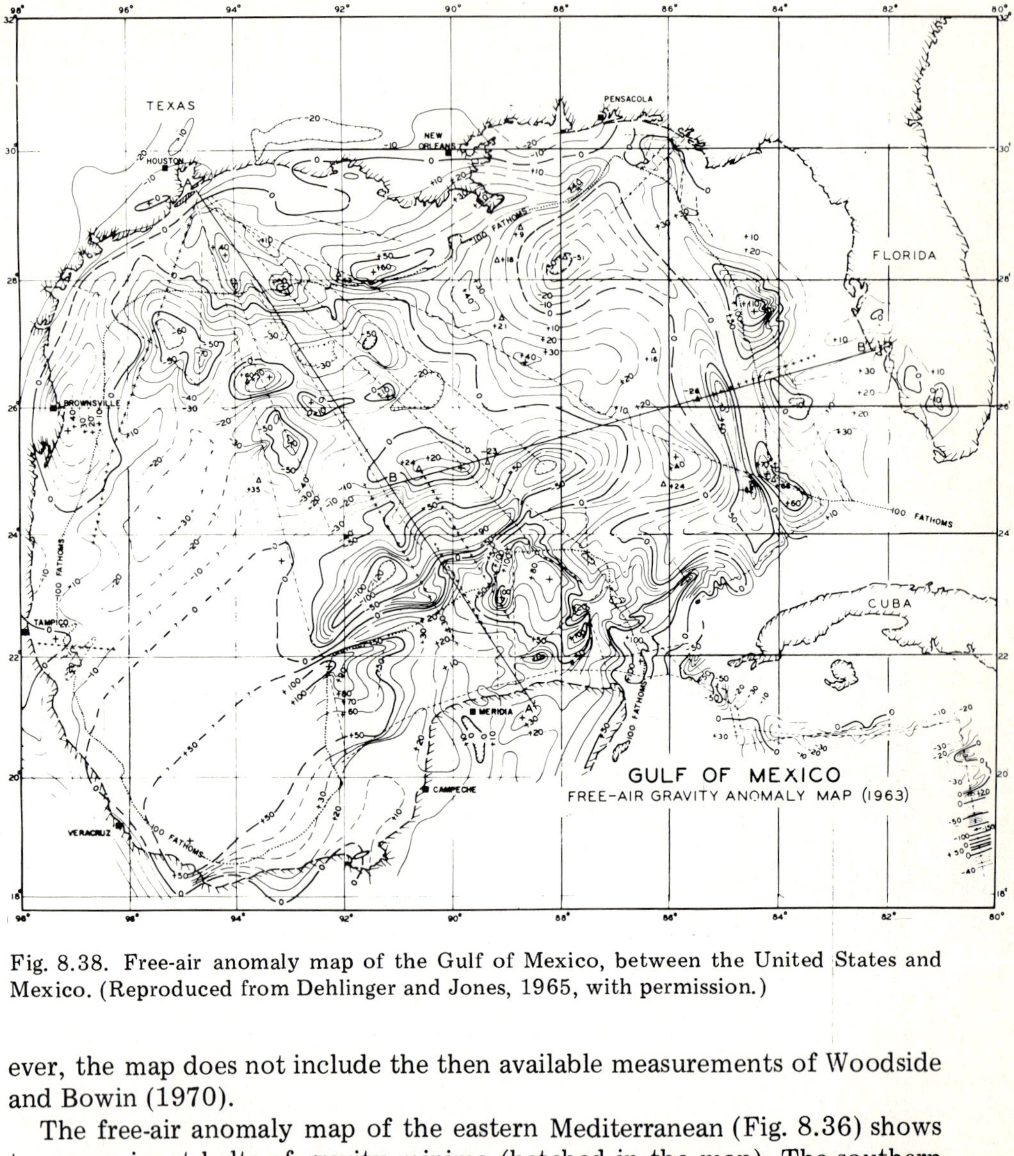

Fig. 8.38. Free-air anomaly map of the Gulf of Mexico, between the United States and Mexico. (Reproduced from Dehlinger and Jones, 1965, with permission.)

ever, the map does not include the then available measurements of Woodside and Bowin (1970).

The free-air anomaly map of the eastern Mediterranean (Fig. 8.36) shows two prominent belts of gravity minima (hatched in the map). The southern of these belts overlies the Mediterranean Ridge and follows the general curvature of the Hellenic Arc, terminating west of Cyprus as does the ridge. Anomaly values reach −150 mgal along the eastern and −250 mgal in the western parts of the belt. These low anomalies over this topographic ridge indicate considerable thicknesses of low-density materials.

The northern belt of free-air anomalies extends along the Hellenic Trough and also terminates west of Cyprus. The belt is interrupted by gravity

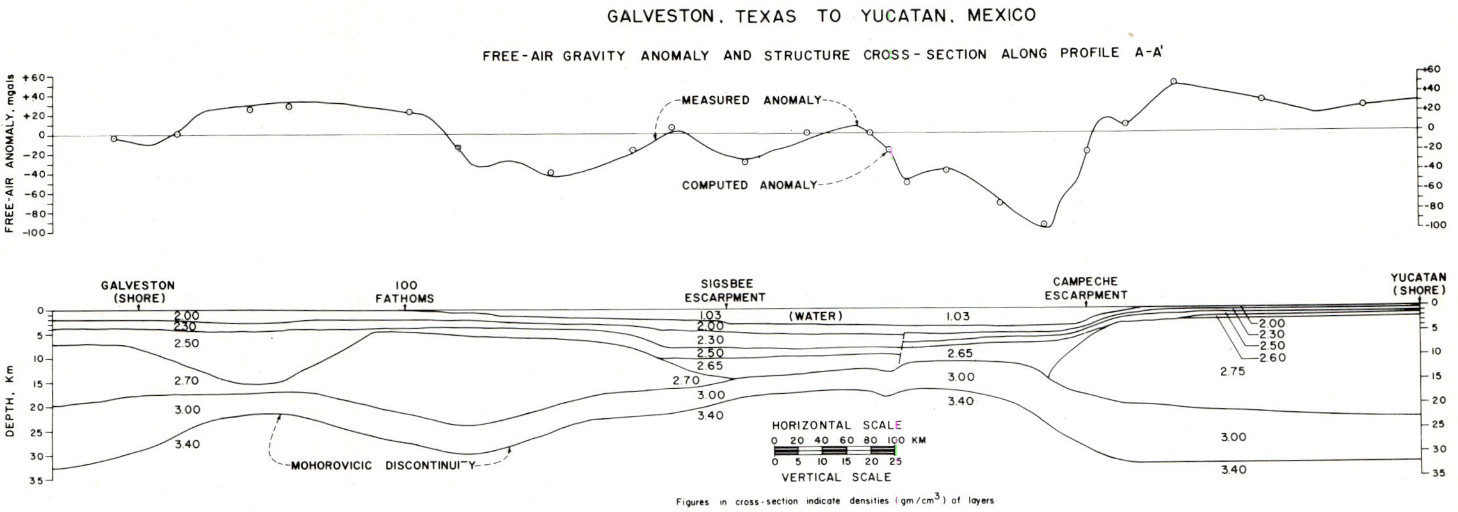

Fig. 8.39. Crustal—subcrustal section along line AA' in Fig. 8.38, from Galveston, Texas, to Yucatan, Mexico. The section is consistent with seismic refractions and gravity data. (Reproduced from Dehlinger and Jones, 1965, with permission.)

maxima produced by topographic highs (including islands). Anomaly values reach −230 mgal in the eastern and −120 mgal in the western parts of the belt.

The two belts of minima are separated by a relative high which is produced by a series of topographic highs (along the top of a feature known as the Mediterranean Wall). South of the Mediterranean Ridge, along the southern part of the Mediterranean Sea, there exists a belt of small positive anomalies. Large positive anomalies occur over Cyprus and in the Aegean Sea (between Greece and Turkey), with values exceeding +100 mgal.

Rabinowitz and Ryan conclude from analyses of their gravity data, combined with other geophysical results, that the island-arc-type structures in the eastern Mediterranean Sea are a result of underthrusting and crustal shortening, as illustrated in Fig. 8.37, produced by Africa moving north toward Europe. The Mediterranean Ridge is interpreted to consist of a thick series of deformed sediments, producing the southern of the two gravity-minima belts (the Mediterranean Ridge). Cyprus and adjacent massives are interpreted to be "upper mantle" nappes, as a result of crustal shortening, as first proposed by Gass and Masson-Smith (1963) and by Gass (1968), thus producing the gravity highs. Rabinowitz and Ryan conclude that the eastern Mediterranean "has been swallowed in the northernmost Alpine compressional belt" and that the sea "is now shrinking, and the loss of surface area is being compensated by underthrusting beneath the Hellenic and Calabrian arcs and crustal shortening beneath the Mediterranean Ridge and Hellenic Trough".

Gravity anomalies in the Gulf of Mexico

The Gulf of Mexico is a partly landlocked sea. The Sigsbee Deep in the middle of the Gulf overlies an essentially oceanic crust (Ewing et al., 1955, 1960). Fig. 8.38 is a free-air anomaly map of most of the Gulf, obtained by Dehlinger and Jones (1965) using a LaCoste and Romberg gimbal-suspended meter. The anomalies have an estimated uncertainty of 7 mgal. Fig. 8.39 shows a northwest—southeast crustal section across the Gulf from Galveston, Texas to Yucatan, Mexico (line AA' in Fig. 8.38), which conforms with observed gravity and seismic refraction data.

A deep geosyncline, the Gulf Coast Geosyncline, parallels the north coast of the Gulf. Anomalies over the geosyncline are approximately +50 mgal (Figs. 8.38 and 8.39); the deep sedimentary section in the syncline appears to be compensated isostatically by a shallow Moho, as indicated in the cross section. South of the syncline a broad platform overlies a continental crust.

The Sigsbee Deep (Fig. 8.39) was shown (Ewing et al., 1955) to consist of an oceanic crust overlain by 6—7 km of sediments. The near-zero, free-air anomaly values (Fig. 8.38) indicate that this deep is essentially in isostatic

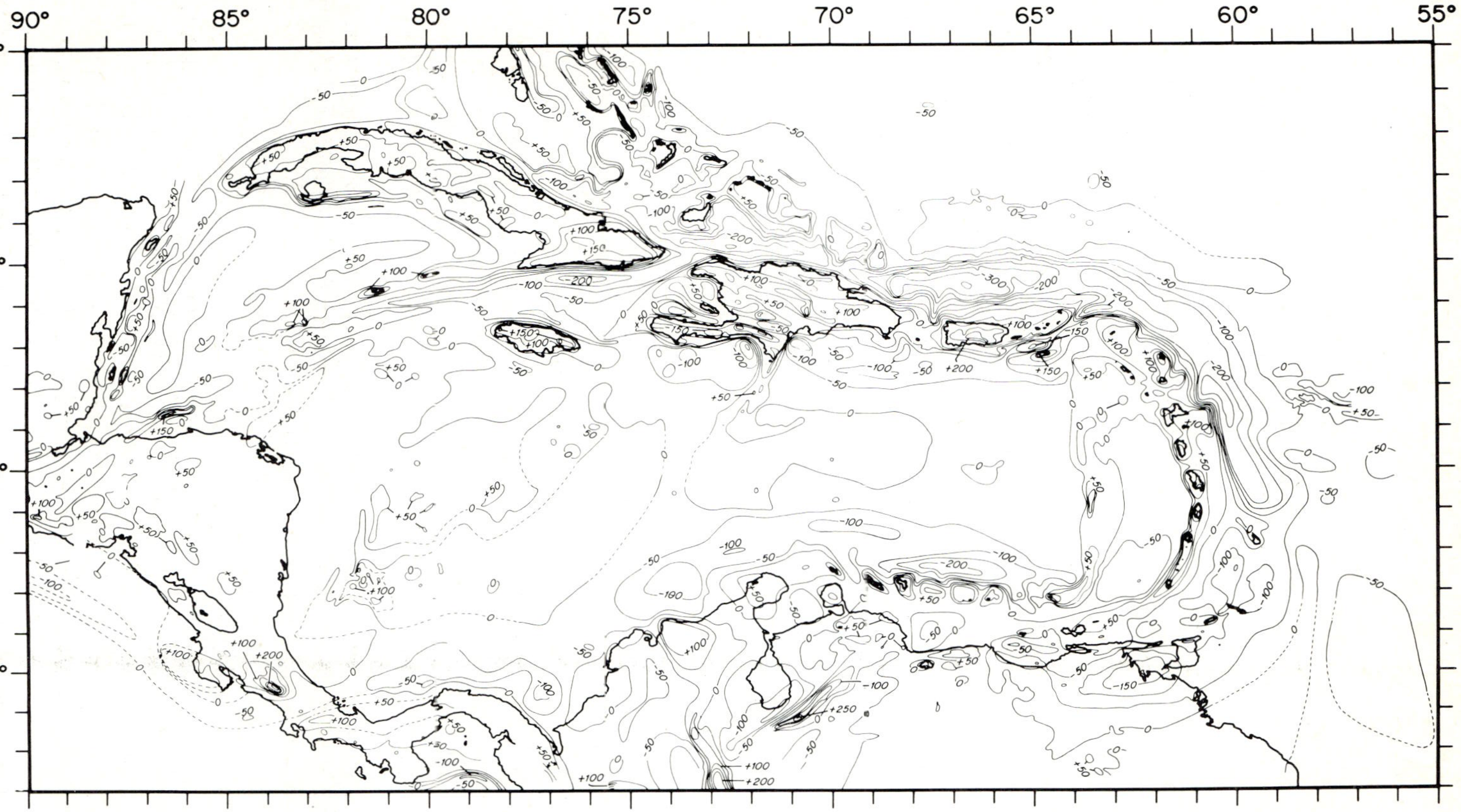

Fig. 8.40. Free-air gravity anomaly map of the Caribbean Sea and surrounding land areas. (Courtesy of C.O. Bowin, 1977.)

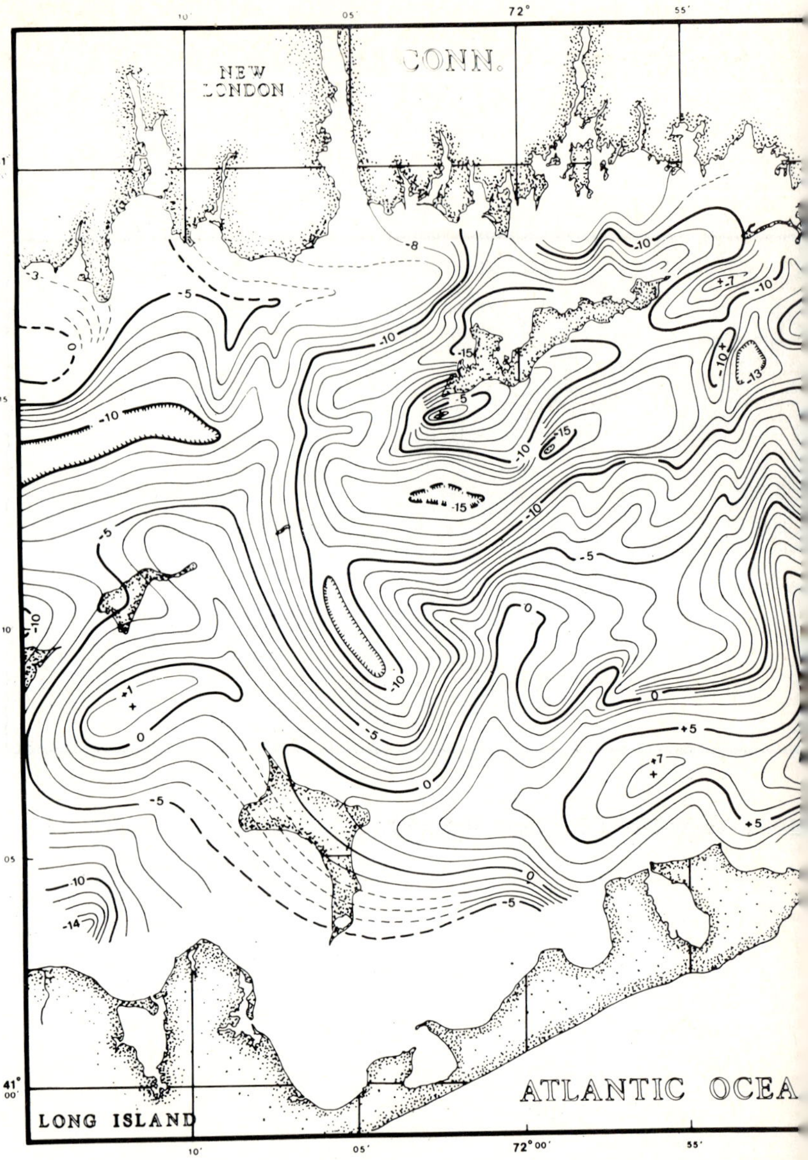

Fig. 8.42. Free-air anomaly map of Block Island Sound, between Rhode Islan

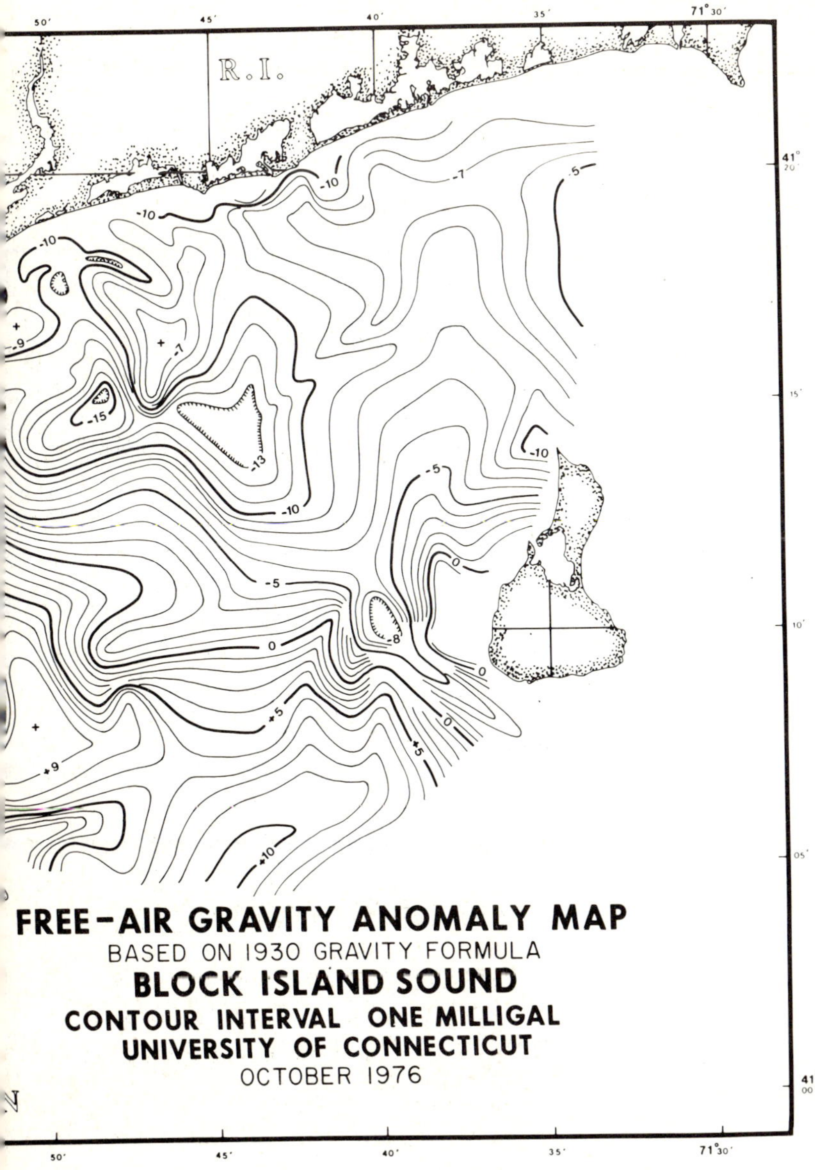

d and New York. (From Dehlinger, in prep.)

equilibrium. Since the sediment load in the deep is brought in largely by the Mississippi River, it appears that the deep has been adjusting itself to the increased sediment load, so as to remain in near isostatic equilibrium.

Along the Yucatan Peninsula (Mexico), the cross section shows a transition from oceanic to continental crust. Platform carbonate rocks occur along the Bank of Campeche, where free-air anomalies have values up to +100 mgal. Seaward of the escarpment, anomaly values range from —90 to —120 mgal. The positive and negative anomalies across the Campeche Escarpment are produced by edge effects due to the transition from oceanic to continental crust. Similarly, along the West Florida escarpment, edge effects produce positive anomalies up to +100 mgal on the platform and negative anomalies seaward of the slope.

Gravity anomalies in the Caribbean Sea

Bowin (1976) presented a detailed discussion and analysis of the gravity field of the Caribbean Sea and its tectonic implications. The gravity data are based on shipboard measurements; most data were obtained on ten cruises on Woods Hole Oceanographic Institution R/V *Chain* and R/V *Atlantis II*, but also include measurements obtained by other organizations.

Fig. 8.40 shows a regional free-air anomaly map of the Caribbean area. The large amplitudes of the anomalies, contoured on a 50-mgal interval, indicate that the borders of the area are tectonically active. The low anomaly values in the central part of the Caribbean Sea indicate that this region is in near-isostatic equilibrium.

Anomaly amplitudes are generally positive (+50 to +200 mgal) on the island chains of the Greater and Lesser Antilles and show strong negative lineations along the Puerto Rico Trench (to —350 mgal), north of Puerto Rico, and east of the Lesser Antilles island arc (to —200 mgal). The Puerto Rico Trench negative lineation continues southwestward along the Cayman Trough, between Cuba and Jamaica, where anomalies exceed —200 mgal.

Bowin concludes that the south-trending part of the negative lineation east of the Lesser Antilles island arc is produced by underthrusting of the Atlantic Plate beneath the island arc, and that the east-trending lineations along the Puerto Rico Trench and the Cayman Trough are a result of associated transform faulting. Seismic first-motions have indicated predominantly strike-slip displacements along this negative anomaly belt. Bowin interprets the positive anomalies over the islands north and east of the Caribbean Sea, and also in Central America, as regions probably being uplifted. Compressive forces, which may result from differential motion between the North and South American plates, may cause a down warping of the crust in the Trinidad and eastern Venezuelan areas (Bowin, 1976, p. 69), producing negative anomalies there.

The large-amplitude gravity anomalies surrounding the Caribbean Sea appear to result from differential motions between the Caribbean, North American, and South American plates.

Gravity anomalies in Long Island and Block Island sounds

Long Island Sound is an inland sea located between New York and Connecticut; it opens to Block Island Sound to the east, which, in turn, opens to the Atlantic Ocean. The sounds are on the eastern seaboard of the northeastern United States and are underlain by a continental crust. The region has been tectonically inactive since the Triassic Basin of central Connecticut was formed along a zone of local weakness in the crust between two semi-parallel chains of older gneiss domes (Eaton and Rosenfeld, 1960).

Gravity was measured along numerous tracklines in Long Island and Block Island sounds by a LaCoste and Romberg platform-mounted gravimeter (Dehlinger, in prep.) aboard a 65-ft. research vessel, the University of Connecticut *T-441*. Measurements were made in calm seas and navigation positions were obtained by radar fixes. The mean uncertainty of approximately 650 anomaly determinations is 1.4 mgal, based on misties at 60 trackline intersections.

This survey shows that in calm seas accurate gravity anomalies can be obtained from measurements in relatively small boats, although these measurements could not have been obtained in even moderate seas. Long Island Sound has sizable and continuously changing tidal currents. These currents caused errors in the computed Eötvös corrections, resulting in anomaly uncertainties which could have been reduced if positioning had been obtained by electronic navigation methods.

Figs. 8.41 and 8.42 show free-air anomaly maps of the sounds with a 1-mgal contour interval. Anomalies in central Long Island Sound, east of about 73°W (east of New Haven), have longer wavelengths than those to the west. The Connecticut Triassic Basin is known to extend southward in Connecticut to New Haven, but it is not known whether the basin continues into the sound. The gravity anomalies themselves do not completely resolve this question, but the change in anomaly wavelength indicates a juxtaposition of two different regional structures.

Regional anomaly trends in Long Island Sound have been observed to continue landward both in Connecticut and on Long Island, as seen on Bouguer anomaly maps of these adjacent areas (Urban et al., 1972). A north-trending gravity high of +45 mgal in the western part of Long Island Sound (73°25′W) continues northward into Connecticut west of the Triassic

Fig. 8.43. Maps showing 5°-average free-air anomalies over the northern and southern Atlantic oceans. Shadings indicate bathymetric depths; stipled = less than 2,000 fathoms; clear = 2,000—2,500 fathoms; vertical lines = >2,500 fathoms. (Reproduced from Talwani and Le Pichon, 1969, with permission.)

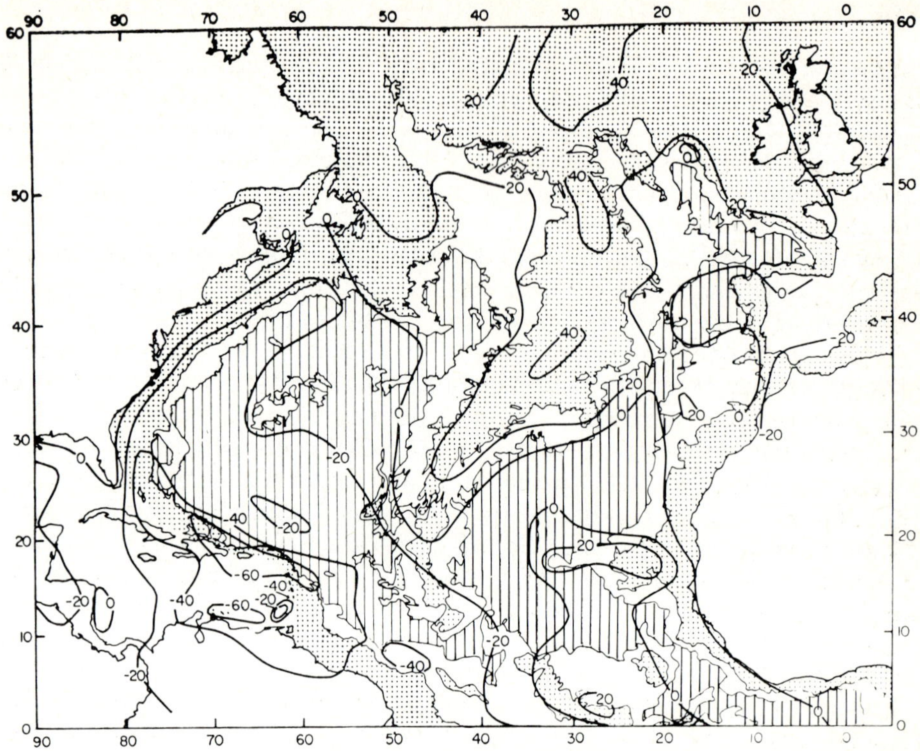
90
80
70
60
50
40
30
20
10
0
60
50
40
30
20
10
0

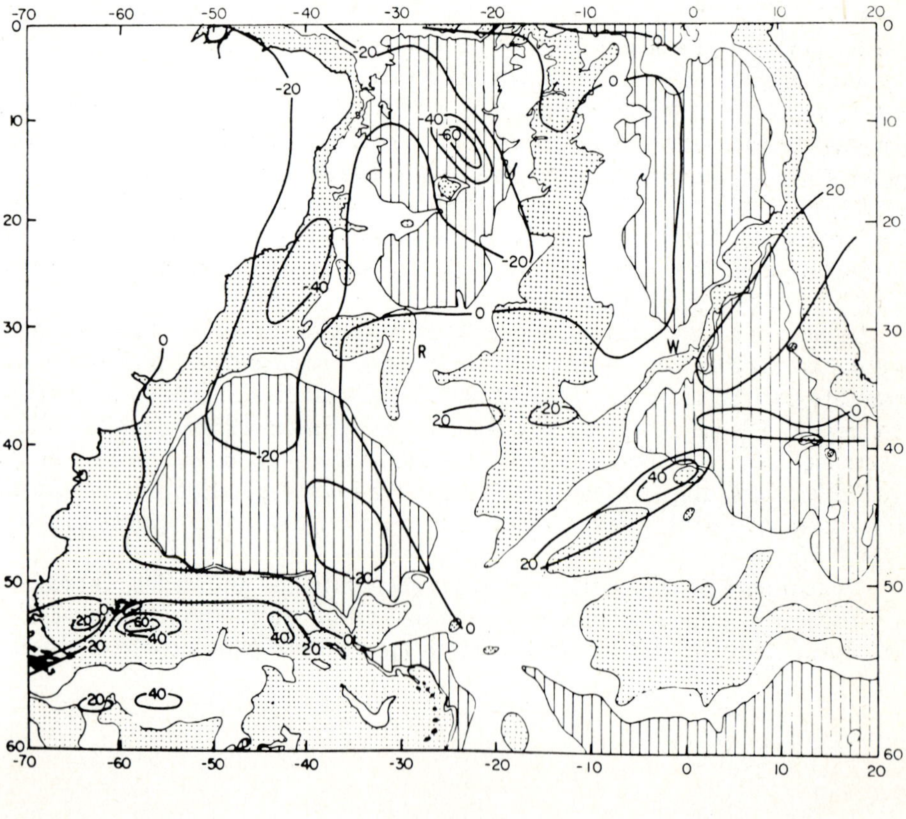
-70
-60
-50
-40
-30
-20
-10
0
10
20
0
10
20
30
40
50
60
R
W

Basin and southward across Long Island. Along several profiles south of Long Island Cochran and Talwani (1976) obtained a gravity high which appears to be a continuation of the same feature. Similarly, the long-wavelength anomalies in central Long Island Sound continue northward into Connecticut, although they tend to disappear in the southern part of the sound and are not observed on Long Island. Also, a northeast-trending sequence of a weak gravity high between two longer-wavelength lows in eastern Long Island Sound is observed to continue as two broad lows in Connecticut to the north and Long Island to the south.

Approximately 3 km south of the Connecticut coast a free-air anomaly high (about 3-mgal residual value) parallels the coast between New Haven and New London, being terminated by the boundary of the Triassic Basin at New Haven. This anomaly could be produced by an east—west trending, near-surface basement structure along the northern coast of the sound.

Block Island Sound exhibits anomaly trends which are different from those of eastern Long Island Sound. The Block Island Sound anomalies generally trend north—south and the wavelengths are relatively shorter. The largest trend extends north from the east tip of Long Island to the Connecticut—Rhode Island border (71°53′W), with the east side having 5—10 mgal larger values than the west side. The anomaly suggests a north—south striking fault in the basement, with the west side downthrown.

Regional gravity field over the Atlantic Ocean

Talwani and Le Pichon (1969) constructed a regional free-air anomaly map of the Atlantic Ocean which they compared with topography and with satellite-determined anomalies. The regional anomalies were obtained by averaging observed free-air anomalies over 5° squares. Fig. 8.43 shows the 5°-average free-air anomaly maps and generalized bathymetric for the North and South Atlantic. A general correlation of low positive anomalies with bathymetric highs and low negative anomalies with bathymetric lows indicates that the areas are isostatically compensated for the 1,000- to 5,000-km wavelengths considered.

Talwani and Le Pichon also show comparisons of these observed free-air anomaly profiles across the North Atlantic and the South Atlantic with satellite-derived anomalies, based on tesseral spherical-harmonic coefficients to degree 12, as computed by Gaposchkin (Kaula, 1966). Profiles with 5° averages of the free-air anomalies have wavelengths of approximately 4,000 km and amplitudes of ±25 mgal. The corresponding satellite profiles have amplitudes of about ±5 mgal and wavelengths which appear to exceed 4,000 km.

To compare further surface with satellite-obtained anomalies, Talwani and Le Pichon averaged surface-ship and pendulum data over 20° squares; the resultant anomaly wavelengths correspond to tesseral-harmonic coefficients of degree and order 9. The agreement between these surface-based and the satellite-based anomalies was found to be good in the North Atlantic, and somewhat less good in the South Atlantic (see Talwani and Le Pichon, 1969, pp. 346 and 347 for comparisons).

Clearly, surface-based free-air anomalies, particularly the observed values, have very much shorter wavelengths and higher amplitudes than those obtained from satellite data. It is apparent that while satellite perturbations provide very long-wavelength anomalies (over 4,000 km), they are not capable of providing the usually important, short-wavelength information obtainable from surface measurements.

APPENDIX A — CONVERSION OF UNITS

From cgs system to SI (rationalized mks) system

Mass	1 g = 10^{-3} kg (kilogram)
Length	1 cm = 10^{-2} m (meter)
Time	= sec (second)
Density	1 g cm^{-3} = 10^{3} kg m^{-3}
Velocity	1 cm sec^{-1} = 10^{-2} m sec^{-1}
Gravity, acceleration	1 gal = 1 cm sec^{-2} = 10^{-2} m sec^{-2}
	1 mgal = 10^{-5} m sec^{-2}
	1 μgal = 10^{-8} m sec^{-2}
Force	1 dyne = 1 g cm sec^{-2} = 10^{-5} N (Newton)
Pressure	1 dyne cm^{-2} = 1 g cm^{-1} sec^{-2} = 10^{-1} Pa (Pascal)
	1 bar = 10^{6} dyne cm^{-2} = 10^{5} Pa
	1 kbar = 10^{9} dyne cm^{-2} = 10^{8} Pa
Viscosity	1 P (poise) = 1 dyne sec cm^{-2} = 10^{-1} Pa sec
Energy	1 erg = 1 dyne cm = 10^{-7} J (joule)
Power	1 erg sec^{-1} = 10^{-7} W (watt)
Gravity gradient	1 Eötvös unit = 10^{-9} gal cm^{-1} = 10^{-4} mgal m^{-1}

Conversion factors

1 year	= 3.15567 · 10^{-7} sec
1 statute mile	= 1609 m = 0.869 nautical mile
1 nautical mile	= 1852 m = 1.151 statute mile
1 astronomical unit	= 1 AU = 1.4960 · 10^{8} km
1 tonne	= 10^{3} kg = 10^{6} g
1 kg wt.	= 9.807 · 10^{5} dyne = 9.807 N
1 kg wt. cm^{-2}	= 0.9807 bar
	= 0.9807 · 10^{6} dyne cm^{-2}
	= 0.9807 · 10^{5} Pa
1 atm. press.	= 1.013 bar
	= 1.013 · 10^{6} dyne cm^{-2}
	= 1.013 · 10^{5} Pa
1 calorie	= 4.184 · 10^{7} ergs = 4.184 J
1 radian	= 57°.30 = 2.063 · 10^{5} arc sec
1 arc sec	= 4.848 · 10^{-6} rad

APPENDIX B — NUMERICAL DATA CONCERNING THE EARTH

Equatorial radius	$a = 6{,}378.139$ km
Polar radius	$c = 6{,}356.75$ km
Mean radius (sphere of equal volume)	$\bar{r} = 6{,}370.8$ km
Volume	$v = 1.0832 \cdot 10^{21}$ m^3
Mass	$M = 5.973 \cdot 10^{24}$ kg
Gravitational constant	$G = 6.672 \cdot 10^{-8}$ g cm^3 sec^{-2}
Gravitational constant x mass	$GM = 3.98601 \cdot 10^{20}$ cm^3 sec^{-2}
Mean density	$\bar{\rho} = 5.515$ g cm^{-3}
Geometric ellipticity (flattening)	$e = 3.35282 \cdot 10^{-3} = 1/298.256$
Dynamic ellipticity	$H = 3.2732 \cdot 10^{-3} = 1/305.51$
Moments of inertia:	
About polar axis	$C = 8.0378 \cdot 10^{44}$ g cm^2
	$C = 0.33076$ Ma^2
About equatorial axis	$A = 8.0015 \cdot 10^{44}$ g cm^2
	$A = 0.32968$ Ma^2
Ellipticity coefficient	$J_2 = 1.08264 \cdot 10^{-3}$
Gravity at equator	$\gamma_{eq} = 978{,}031.69$ mgal
Gravity at pole	$\gamma_p = 983{,}241.57$ mgal
Gravitational ellipticity	$\beta = (\gamma_p - \gamma_{eq})/\gamma_{eq} = 5.327 \cdot 10^{-3}$
Geoid potential	$W_0 = 6.2637 \cdot 10^{11}$ cm^2 sec^{-2}
Solar day	$= 86{,}400$ sec
Siderial day	$= 86{,}164$ sec
Angular velocity of rotation	$\omega - 7.292115 \cdot 10^{-5}$ rad sec^{-1}
Ratio $\left(\dfrac{\text{centrifugal force}}{\text{gravity}}\right)$	$p = \omega^2 a/\gamma_{eq} = 3.46775 \cdot 10^{-3}$
Radius of earth's orbit	$R_s = 1$ AU $= 1.495979 \cdot 10^8$ km
Eccentricity of earth's orbit	$\epsilon_1 = 0.0167$
Obliquity of ecliptic	$I_s = 23^\circ.473$
Mean earth—moon distance	$R_m = 3.8441 \cdot 10^5$ km
	$= 60.34$ earth radii
Eccentricity of lunar orbit	$\epsilon_m = 0.0549$
Mean lunar orbital velocity	$\omega_L = 2.684 \cdot 10^{-6}$ rad sec^{-1}
Inclination of lunar orbit	$I_m = 5^\circ.145$
Mass of moon	$m = (1/81.3)\, M$
Mass of sun	$S = 332{,}946.8\, M$
Rate of precession of equinox	$\omega_p = 50''.37$ yr.$^{-1}$

Period of precession of equinox	= 25, 730 yr.
Age of the earth	= $4.55 \cdot 10^9$ yr.
Greatest ocean depth	= 10.550 km
Mean ocean depth	= 3.88 km
Greatest land height	= 8.84 km
Mean land height	= 0.84 km
Mass relative to that of the earth:	
Atmosphere	= $0.854 \cdot 10^{-6}$
Oceans	= 0.00023
Crust	= 0.00435
Mantle	= 0.6697
Outer core	= 0.3097
Inner core	= 0.0162

APPENDIX C — FORMULAS FOR COMPUTING EARTH-TIDE ACCELERATIONS

The correction for the vertical components of tidal attraction applied to gravity measurements at a point P on the earth's surface is given by:

$$\Delta g_r = (\Delta g_r)_m + (\Delta g_r)_s \qquad \text{(C.1)}$$

where the lunar and solar attractions are obtained from eqs. 3.30a and b. Formulas presented below for determining attractions which can be computer-programed follow the methods of Longman (1959).

Schureman (1924) obtained relations for the lunar and solar zenith angles ψ_m and ψ_s of the form:

$$\cos \psi_m = \sin \Phi \sin I_m \sin l_m + \cos \Phi \left[\cos^2\left(\frac{I_m}{2}\right) \cos(l_m - \chi_1) + \sin^2\left(\frac{I_m}{2}\right) \cos(l_m + \chi_1) \right] \qquad \text{(C.2a)}$$

$$\cos \psi_s = \sin \Phi \sin I_s \sin l_s + \cos \Phi \left[\cos^2\left(\frac{I_s}{2}\right) \cos(l_s - \chi_2) + \sin^2\left(\frac{I_s}{2}\right) \cos(l_s + \chi_2) \right] \qquad \text{(C.2b)}$$

where Φ is the geographic latitude at P; I_m the inclination of the moon's orbit to the equator; I_s the inclination of the equator to the ecliptic: 23°.452 (Schureman, 1941); l_m the longitude of the moon in its orbit measured from the ascending intersection with the equator (from point A in Fig. C.1); l_s the longitude of the sun in the ecliptic measured from the vernal equinox (from point E in Fig. C.1); χ_1 the right ascension of the meridian of point P (measured from A, Fig. C.1); χ_2 the right ascension of the meridian of P measured from the vernal equinox (E in Fig. C.1).

For the longitude of the moon in its orbit, l_m, Schureman (1941) obtained:

$$l_m = \bar{l}_m + 2\epsilon_m \sin(\bar{l}'_m - \bar{p}_m) + \tfrac{5}{4}\epsilon_m^2 \sin 2(\bar{l}'_m - \bar{p}_m) + \tfrac{15}{4}\mu\epsilon_m \sin(\bar{l}'_m - 2\bar{l}_s + \bar{p}_m) + \tfrac{11}{8}\mu^2 \sin 2(\bar{l}'_m - \bar{l}_s) \qquad \text{(C.3)}$$

where $\bar{l}_m$ is the mean longitude of the moon (in radians) in its orbit,

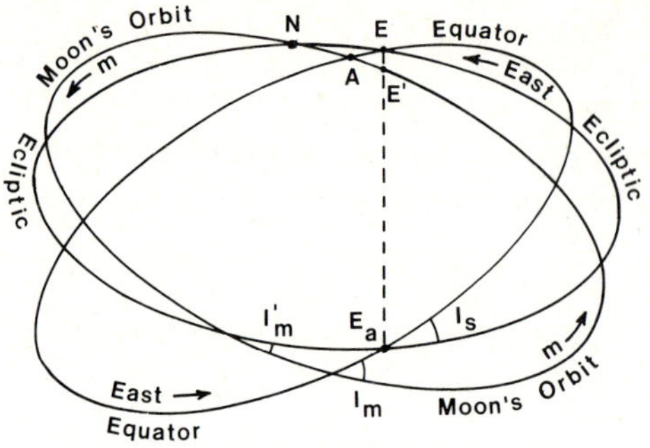

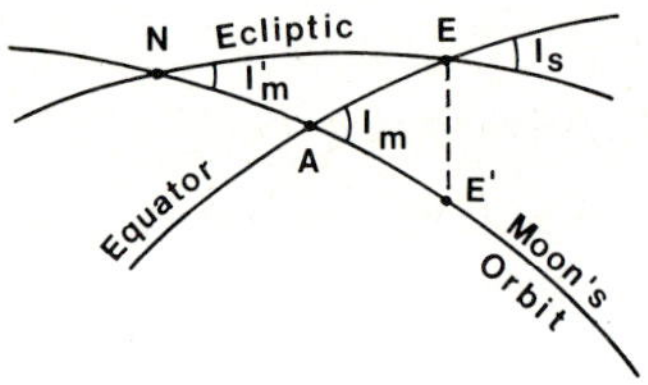

Fig. C.1. Relationships between equatorial plane, ecliptic, and moon's orbit (after Longman, 1959). A = ascending intersection of equator and moon's orbit; N = moon's ascending node; E = vernal equinox; E_a = autumnal equinox; E' = referred equinox; I_m = inclination of moon's orbit to equator; I'_m = inclination of moon's orbit to ecliptic ($5^\circ.145$); I_s = inclination of equator to ecliptic ($23^\circ.452$); $NA = \alpha$ (see eq. C.11); $AE = \nu$ (see eq. C.14); $NE' = L'_m$.

measured from A in Fig. C.1; ϵ_m the eccentricity of the moon's orbit: 0.05490 (Schureman, 1941); $\bar{l}'_m$ the mean longitude of the moon in its orbit measured from the referred equinox (E' in Fig. C.1); $\bar{p}_m$ the mean longitude of the lunar perigee; μ the ratio of the mean motion of the sun to moon: 0.074804 (Schureman, 1941); $\bar{l}_s$ the mean longitude of the sun; where values for $\bar{l}'_m$, $\bar{p}_m$, and $\bar{l}_s$ are given by:

$$\bar{l}'_m = 270^\circ 26' 14''.72 + (1{,}336 \text{ rev.} + 1{,}108{,}411''.20)T + 9''.09\,T^2 + 0''.0068\,T^3 \quad \text{(C.4)}$$

in which T is the number of Julian centuries (36,525 days) from Greenwich mean noon on December 31, 1899, and:

$$\bar{p}_m = 334^\circ 19' 40''.87 + (11 \text{ rev.} + 392{,}515''.94)T - 37''.24\,T^2 - 0''.045\,T^3 \quad \text{(C.5)}$$

$$\bar{l}_s = 279^\circ 41' 48''.04 + 129{,}602{,}768''.13\,T + 1''.089\,T^2 \quad \text{(C.6)}$$

The relation between $\bar{l}_{\mathrm{m}}$ and $\bar{l}'_{\mathrm{m}}$ is given by:

$$\bar{l}_{\mathrm{m}} = \bar{l}'_{\mathrm{m}} - L_{\mathrm{m}} \tag{C.7}$$

where L_{m} is the longitude in the moon's orbit of its ascending intersection with the celestial equator (plane of the earth's equator). From spherical trigonometry (Fig. C.1), it is seen that:

$$L_{\mathrm{m}} = L'_{\mathrm{m}} - \sin^{-1}\left(\frac{\sin I_{\mathrm{s}} \sin L'_{\mathrm{m}}}{\sin I_{\mathrm{m}}}\right) \tag{C.8}$$

where L'_{m} is the longitude of the moon's ascending node in its orbit measured from the referred equinox ($L'_{\mathrm{m}} = NE'$ in Fig. C.1). A cosine formula for the spherical triangle can be defined to assure a unique value of the inverse sine in eq. C.7. Let α be side NA in Fig. C.1. Then:

$$\cos\alpha = \cos L'_{\mathrm{m}} \cos\nu + \sin L'_{\mathrm{m}} \sin\nu \cos I_{\mathrm{s}} \tag{C.9}$$

where ν is side AE, the longitude in the celestial equator of its intersection A with the moon's orbit. Similarly, it follows that:

$$\sin\alpha = \frac{\sin I_{\mathrm{s}} \sin L'_{\mathrm{m}}}{\sin I_{\mathrm{m}}} \tag{C.10}$$

We apply the relation:

$$\tan(\alpha/2) = \frac{\sin\alpha}{(1 + \cos\alpha)}$$

to obtain:

$$\alpha = 2\tan^{-1}\left(\frac{\sin\alpha}{1 + \cos\alpha}\right) \tag{C.11}$$

which provides unique values for $0 < \alpha < 2\pi$.

The longitude of the moon's node, L'_{m}, has been shown to have a value (Schureman, 1941):

$$L'_{\mathrm{m}} = 259°10'57''.12 - (5\text{ rev.} + 489,912''.63)T + 7''.58T^2 + 0''.008\,T^3 \tag{C.12}$$

and the inclination I_{m} of the moon's orbit to the equator to be given by:

$$\cos I_{\mathrm{m}} = \cos I_{\mathrm{s}} \cos I'_{\mathrm{m}} - \sin I_{\mathrm{s}} \sin I'_{\mathrm{m}} \cos L'_{\mathrm{m}} \tag{C.13}$$

where $18° < I_{\mathrm{m}} < 29°$ is positive. Also:

$$\nu = \sin^{-1}\left(\frac{\sin I'_{\mathrm{m}} \sin L'_{\mathrm{m}}}{\sin I_{\mathrm{m}}}\right) \tag{C.14}$$

which has unique values in its range of $-15° < \nu < 15°$. Schureman (1941) obtained a value:

$$I'_m = 5°.145 \tag{C.15}$$

The angle χ_1 in eq. C.1a is given by:

$$\chi_1 = \bar{l}_s - \nu + \bar{t}_s \tag{C.16}$$

where $\bar{t}_s$ is the hour angle of mean sun measured westward from point P. At longitude λ on the earth's surface, $\bar{t}_s$ is given by:

$$\bar{t}_s = 15(t_0 - 12) - \lambda \tag{C.17}$$

where t_0 is Greenwich civil time (in hours).

The moon's zenith angle ψ_m in eq. C.2a can be computed using eqs. C.3 to C.17.

The sun's zenith angle ψ_s in eq. C.2b can be similarly obtained. The longitude of the sun in the ecliptic is measured from the vernal equinox (E in Fig. C.1) and given by:

$$l_s = \bar{l}_s + 2\epsilon_1 \sin(\bar{l}_s - \bar{p}_s) \tag{C.18}$$

where ϵ_1 is the eccentricity of earth, and $\bar{p}_s$ the mean longitude of the solar perigee. For ϵ_1 Schureman (1942) gives:

$$\epsilon_1 = 0.01675104 - 0.00004180T - 0.000000126T^2 \tag{c.19}$$

and for $\bar{p}_s$ (Schureman, 1941):

$$\bar{p}_s = 281°13'\,15''.0 + 6{,}189''.03T + 1''.63T^2 + 0''.012T^3 \tag{C.20}$$

The right ascension of the meridian of point P, χ_2, is given by:

$$\chi_2 = \bar{l}_s + \bar{t}_s \tag{C.21}$$

Eqs. C.18 to C.21 provide formulas for computing the zenith solar angle ψ_s.

Computation of tidal accelerations requires values for the variable distances between the centers of the earth and moon (R_m) and the earth and sun (R_s). Schureman (1924) obtained for the distances:

$$\frac{1}{R_m} = \frac{1}{\bar{R}_m} + \frac{1}{\bar{R}'_m}\epsilon_m \cos(\bar{l}'_m - \bar{p}_m) + \frac{1}{\bar{R}'_m}\epsilon_m^2 \cos 2(\bar{l}'_m - \bar{p}_m) + \frac{15}{8}\left(\frac{1}{\bar{R}'_m}\right)\mu\epsilon_m(\bar{l}'_m - 2\bar{l}_s + \bar{p}_m) + \frac{1}{\bar{R}'_m}\mu^2 \cos 2(\bar{l}'_m - \bar{l}_s) \tag{C.22a}$$

and:

$$\frac{1}{R_s} = \frac{1}{\bar{R}_s} + \frac{1}{\bar{R}'_s}\epsilon_1 \cos(\bar{l}_s - \bar{p}_s) \tag{C.22b}$$

in which $\bar{R}_m$ and $\bar{R}_s$ are the mean distances between centers of the earth and moon ($3.84402 \cdot 10^5$ km or 238,857 miles) and the earth and sun, and:

$$\bar{R}'_m = \bar{R}_m (1 - \epsilon_m)^2 \qquad \bar{R}'_s = \bar{R}_s (1 - \epsilon_1)^2 \tag{C.23}$$

Corrections for tidal accelerations can now be computed for a point P at latitude Φ and radial distance r (based on the 1967 reference ellipsoid, eq. 2.70), plus elevation H, from the center of the earth by obtaining values for $\bar{R}_m$ and R_s (eqs. C.22a, b) and for ψ_m and ψ_s (eqs. C.2a, b), and inserting them into eqs. 3.30a and b, using constants:

$$\left.\begin{aligned} G &= 6.673 \cdot 10^{-8}\ \mathrm{cm^3/g\ sec^2} \\ m &= 7.3537 \cdot 10^{25}\ \mathrm{g} \\ S &= 1.993 \cdot 10^{33}\ \mathrm{g} \end{aligned}\right\}\ \text{Pettit (1954)} \tag{C.24}$$

The tidal formulas above, as developed by Longman (1959), are essentially identical to those proposed by Bartels (1957) and also equations of Schureman (1924, 1941) and Pettit (1954).

Accuracies of tidal computations derived from formulas C.1 to C.24 are estimated to be within 1 μgal. Hence, it is immaterial whether the formulas by Longman or Bartels are used. It is also immaterial whether the computations are based on Hayford's (1910) spheroid, the 1924 international ellipsoid (eq. 2.68), the 1967 ellipsoid, or other reasonable spheroid approximations.

REFERENCES

Agarwal, B.N.P. and Lal, T., 1969. Calculation of the second vertical derivative of the gravity field. Pure Appl. Geophys., 76:5—16.

Agarwal, B.N.P. and Lal, T., 1971. Application of rational approximation in the calculation of the second derivative of the gravity field. Geophysics, 36:571—581.

Airy, G.B., 1855. On the computation of the effect of the attraction of mountain masses as disturbing the apparent astronomical latitude of stations in geodetic surveys. Phil. Trans. R. Soc. Lond., 145:101—104

Al-Chalabi, M., 1970. Interpretation of two-dimensional magnetic profiles by nonlinear optimization. Boll. Geofis. Teor. Appl., 12:3—20.

Al-Chalabi, M., 1971. Some studies relating to non-uniqueness in gravity and magnetic inverse problems. Geophysics, 36:835—855.

Allan, T.D., Dehlinger, P., Ganter, C., Morelli, C., Pisani, M. and Harrison, J.C., 1962. Comparison of Graf-Askania and LaCoste and Romberg surface-ship gravity meters. J. Geophys. Res. 67:5157—5162.

Allan, T.D., Charnok, H. and Morelli, C., 1964. Magnetic, gravity, and depth surveys in the Mediterranean and Red Sea. Nature, 204:1245—1248.

Ames, C.B., 1974. An Introduction to the Rotating Gravity Gradiometer for Mov-Vehicle Applications. Hughes Aircraft Co. Res. Rep't., No. 486.

Anderson, D.L., Sammis, C. and Jordon, T., 1972. Composition of the core and mantle. In: E.C. Robertson, J.F. Hays and L. Knopoff (Editors), Nature of the Solid Earth. McGraw-Hill, New York, N.Y., pp. 41—66.

Arfken, G., 1970. Mathematical Methods for Physicists, 2nd ed. Academic Press, New York, N.Y., 815 pp.

Atwater, Tanya, 1970. Implications of plate tectonics for the Cenozoic tectonics evolution of western North America. Bull. Geol. Soc. Am., 81:3513—3536.

Bartels, J., 1957. Gezeitenkräfte. In: J. Bartels (Group Editor), Handbuch der Physik, 48 (Geophysik II). Springer-Verlag, Berlin, pp. 734—774.

Bell Aerospace Co., 1971. Marine Gravity Measuring System Model 6109. Tech. Rep. No. D6109-95003 (Feb), Bell Aerospace Co., Buffalo, N.Y.

Bell, E.T., 1937. Men of Mathematics. Victor Gollancz, Ltd., London, 653 pp.

Bell, F.C., 1969. Schuler's principle and inertial navigation. Ann. N.Y. Acad. Sci., 147:493—514 (New York).

Bhattacharyya, B.K. and Navolio, M.E., 1975. Digital convolution for computing gravity and magnetic anomalies due to arbitrary bodies. Geophysics, 40:981—992.

Birch, F., 1972. The melting relations of iron and temperatures in the earth's core. Geophys. J. R. Astron. Soc., 29:373—387.

Boitnott, B.D., 1961. Instrument for airborne measuring of derivatives of the vertical component of the earth's gravity field. U.S. Patent, 3,011,347.

Boitnott, B.D., 1962. Instrument for and method of airborne gravitational geophysical exploration. U.S. Patent, 3,038,338.

Bott, M.H.P., 1960. The use of rapid digital computing methods for direct gravity interpretation of sedimentary basins. J. Geophys. Res., 3:63—67.

Bott, M.H.P., 1967. Solution of the linear inverse problem in magnetic interpretation with application to oceanic magnetic anomalies. Geophys. J. R. Astron. Soc., 13:313—323.

Bott, M.H.P., 1973. Inverse methods in the interpretation of magnetic and gravity anomalies. In: B.A. Bolt (Editor), Methods in Computational Physics, Vol. 13. Academic Press, New York, N.Y., pp. 133—162.

Boucher, G., Ruppel, B.D., Chiburis, E.F. and Dehlinger, P., 1977. Map showing free-air gravity anomalies in the southern Beaufort Sea. U.S. Geol. Survey Map MF-851, U.S. Geol. Surv., Reston, Va.

Bouguer, P., 1749. La Figure de la Terre, Paris, 364 pp.

Bower, D.R., 1966. The determination of cross-coupling errors in the measurement of gravity at sea. J. Geophys. Res., 71:487—493.

Bowie, W., 1917. Investigations of Gravity and Isostasy. U.S. Coast Geod. Surv. Spec. Publ., 40. Washington, D.C.

Bowie, W., 1924. Isostatic Investigations and Data for Gravity Stations in the United States. U.S. Coast Geod. Surv. Publ., 99. Washington, D.C.

Bowin, C.O., 1976. Caribbean gravity field and plate tectonics. Geol. Soc. Amer. Spec. Pap., 169, 79 pp.

Bowin, C.O., Aldrich, T.C. and Folinsbee, R.A., 1972. VSA gravity meter system: Tests and recent developments. J. Geophys. Res., 77:2018—2033.

Browne, B.C., 1937. The measurement of gravity at sea. Mon. Not. R. Astron. Soc., Geophys. Suppl., 4:271—279.

Bullard, E.C., 1936. Gravity measurements in East Africa. Phil. Trans. R. Soc. Lond., Ser. A, 235:445—531.

Bullen, K.E., 1963. The Introduction to the Theory of Seismology, 3rd ed. Cambridge Univ. Press, Cambridge, 381 pp.

Bullen, K.E., 1975. The Earth's Density. Halstead Press, Wiley, New York, N.Y., 420 pp.

Bullen, K.E. and Haddon, R.A., 1967. Derivation of an earth model from free earth oscillation data. Proc. Natl. Acad. Sci. (Washington, D.C.), 58:846—852.

Callandreau, M.O., 1889. Sur le développement en série du potentiel. J. Polytech.

Caputo, M., 1965. The minimum strength of the earth. J. Geophys. Res., 70:955—963.

Cathles, L.M., III, 1975. The Viscosity of the Earth's Mantle. Princeton Univ. Press, Princeton, N.J., 386 pp.

Cavendish, H., 1798. Experiments to determine the density of the earth. Phil. Trans., pp. 469—526.

Chandler, S., 1891. On the variation of latitude. Astron. J., 11:65; 83.

Chiburis, E.F. and Dehlinger, P., 1974. Operational reliability of a conventional satellite navigation system in Beaufort Sea gravity studies. Proc. Int. Symp. Application of Marine Geodesy, Marine Tech. Soc., Washington, D.C., pp. 409—410.

Chinnery, M.A., 1961. Terrain corrections for airborne gravity gradient measurements. Geophysics, 26:480—489.

Clairaut, A., 1743. Théorie de la Figure de la Terre. Paris.

Clark, J.S., 1939. An absolute determination of the acceleration due to gravity. Phil. Trans. R. Soc. Lond., Ser. A, 238:65—123.

Cleary, J. and Hales. A.L., 1966. An analysis of the travel times of P waves to North American stations in the distance range 32° to 100°. Bull. Seismol. Soc. Am., 56:467—489.

Cochran, J.R., 1973. Gravity and magnetic investigations in the Guiana Basin, Western Equatorial Atlantic. Bull. Geol. Soc. Am., 84:3249—3268.

Cochran, J.R. and Talwani, M., 1976. Gravity, magnetics, and seismicity. MESA, New York Bight Atlas, Monogr. 9. New York Sea Grant Inst., Albany, N.Y., 21 pp.

Cochran, J.R. and Talwani, M., 1977. Free-air gravity anomalies in the world's oceans and their relationships to residual elevations. Geophys. J. R. Astron. Soc., 50. In press.

Cook, A.H., 1965. The absolute determination of the acceleration due to gravity. Metrologia,1:84—114.

Cook, A.H., 1967. A new absolute determination of the acceleration due to gravity at the National Physical Laboratory, England. Phil. Trans. R. Soc. Lond., Ser. A., 261:211—252.

Cook, A.H., 1969. Gravity and the Earth. Wykeham, London, 95 pp.

Cook, A.H., 1973. Physics of the Earth and Planets. Halstead Press, Wiley, New York, N.Y., 316 pp.

Cooley, J.W. and Tukey, J.W., 1965. An algorithm for the machine calculation of complex Fourier series. Math. Comput., 19:297—301.

Coons, R.L. and Smalet, M., 1967. Surface ship gravity in oil exploration. Paper presented at Soc. Expl. Geophys., 37th Ann. Meeting, Houston, Texas.

Cordell, L. and Henderson, R.G., 1968. Iterative three-dimensional solution of gravity anomaly data using a digital computer. Geophysics, 33:596—601.

Couch, R.W., 1969. Gravity and Structure of Crust and Subcrust West of Washington and British Columbia. Ph.D. Dissertation, Oregon State Univ., Corvallis, Ore., 179 pp.

Couch, R.W. and Gemperle, M., 1977. Gravity and structure of the continental margin of southeastern Alaska. Submitted to: J. Geophys. Res.

Couch, R.W. and Huehn, B., 1977. Map and crustal section off central Chile. In preparation. (Personal communication.)

Couch, R.W. and Woodcock, S., 1977. Gravity and structure of the continental margins of southwestern Mexico and northwestern Guatemala. Submitted to: J. Geophys. Res.

Cox, A. (Editor), 1973. Plate Tectonics and Geomagnetic Reversals. Freeman, San Francisco, Calif., 702 pp.

Darwin, G.H., 1882. A numerical estimate of the rigidity of the earth. Nature, 27:22—23.

Darwin, G.H., 1898. The Tides. Houghton, Mifflin and Co., Riverside Press, Cambridge, Engl., 378 pp. (Reissued by W.H. Freeman, San Francisco, Calif., 1962.)

Dehlinger, P., 1964. Reliability at sea of gimbal-suspended gravity meters with 0.7 critically damped accelerometers. J. Geophys. Res., 69:5383—5394.

Dehlinger, P., 1969. Gravity and its relation to topography and geology in the Pacific Ocean. In: P.J. Hart (Editor), The Earth's Crust and Upper Mantle. Geophys. Monogr., Am. Geophys. Union, 13:352—363.

Dehlinger, P. and Jones B.R., 1965. Free-air gravity anomaly map of the Gulf of Mexico and its tectonic implications, 1963 ed. Geophysics, 30:102—110.

Dehlinger, P. and Yungul, S.H., 1962. Experimental determination of the reliability of the LaCoste and Romberg surface-ship gravity meter S-9. J. Geophys. Res., 67:4389—4394.

Dehlinger, P., Couch, R.W. and Gemperle, M., 1966. Surface-ship gravity measurements corrected for vertical accelerations. J. Geophys. Res., 71:6017—6023 (correction in J. Geophys. Res., 72:4180).

Dehlinger, P., Couch, R.W. and Gemperle, M., 1967. Gravity and structure of the eastern Mendocino Escarpment. J. Geophys. Res., 72:1233—1247.

Dehlinger, P., Couch, R.W. and Gemperle, M., 1968. Continental and oceanic structure from the Oregon coast westward across the Juan de Fuca Ridge. Can. J. Earth Sci., 5:1079—1090.

Dehlinger, P., Couch, R.W., McManus, D.A. and Gemperle, M., 1970. Northeast Pacific structure. In: A.E. Maxwell (Editor), The Sea, Vol. 4, Part 2. Wiley-Interscience, New York, N.Y., pp. 133—189.

Delaney, J.P., 1940. Leonardo da Vinci on isostacy. Science, 91:546—547.

De Maupertius, P.L., 1735. Sur la figure de la terre, et sur les moyens que l'astronomie et la géographie fournissent pour la détermines. Phil. Trans., Mem., 1733, Paris.

De Sitter, W., 1924. On the flattening and constitution of the earth. Bull. Astron. Inst. Neth., 55:97.

Dietz, R.S. and Menard, H.W., 1953. Hawaiian swell, deep, and arch, and subsidence of the Hawaiian Islands. J. Geol., 61:99—113.
Dobrin, M.B., 1975. Introduction to Geophysical Prospecting. 3rd ed., McGraw-Hill, New York, N.Y., 630 pp.
Doodson, A.T., 1921. The harmonic development of the tide-generating potential. Proc. R. Soc. Lond., Ser. A, 100:305—329.
Dutton, C.E., 1889. On some of the greater problems of physical geology. Bull. Wash. Phil. Soc., Ser. B, 11:51—64.
Eaton, G.P. and Rosenfeld, J.L., 1960. Gravimetric and structural investigation in central Connecticut. Int. Geol. Conf., 21st, Copenhagen, Part II, pp. 168—178.
Elkins, T.A., 1951. The second derivative method of gravity interpretation. Geophysics, 16:29—50.
Elsasser, W.M., 1969. Convection and stress application in the upper mantle. In: S.K. Runcorn (Editor), The Applications of Modern Physics to the Earth and Planetary Interiors. Wiley-Interscience, New York, N.Y., pp. 223—246.
Eötvös, R.V., 1896, Untersuchung über Gravitation und Erdmagnetismus. Annal. Phys., 59:354—400.
Euler, L., 1736. Mechanica Sive Motus Scientia Analytice Exposita. St. Petersburg, 2 vols.
Evans, P. and Crompton, W., 1946. Geological factors in gravity interpretation illustrated by evidence from India and Burma. Q. J. Geol. Soc. Lond., 102:211—249.
Evjen, H.M., 1936. The place of the vertical gradient in gravitational interpretations. Geophysics, 1:127—136.
Ewing, J.I., 1969. Seismic model of the Atlantic Ocean. In: P.J. Hart (Editor), The Earth's Crust and Upper Mantle. Geophys. Monogr., Am. Geophys. Union, 13:220—225.
Ewing, J.I., Antoine, J.W. and Ewing, M., 1960. Geophysical measurements in the western Caribbean Sea and in the Gulf of Mexico. J. Geophys. Res., 65:4087—4126.
Ewing, M., Worzel, J.L., Ericson, D.B. and Heezen, B.C., 1955. Geophysical and geological investigations of the Gulf of Mexico, I. Geophysics, 20:1—18.
Faller, J.E., 1965. Results of an absolute determination of the acceleration of gravity. J. Geophys. Res., 70:4035—4038.
Fleet, S., 1961. Clocks. Weidenfeld and Nicolson, London, 128 pp.
Fleischer, U., 1964. Schwerestörung um östlichen Mittelmeer nach Messungen mit ein Askania-Sea Gravimeter. Dtsch. Hydrogr. Z., 17:153—164.
Fowler, C.M.R., 1976. Crustal structure of the Mid-Atlantic ridgecrest at 37°N. Geophys. J. R. Astron. Soc., 47:459—491.
Frowe, E., 1947. A diving bell for underwater gravimeter operation. Geophysics, 12:1—12.
Gans, R.F., 1972.Viscosity of the earth's core. J. Geophys. Res., 77:360—366.
Gaposchkin, E.M., 1974. Earth's gravity field to the eighteenth degree and geodetic coordinates for 104 stations from satellite and terrestrial data. J. Geophys. Res., 79:5377—5411.
Gaposchkin, E.M. and Lambeck, K., 1970. 1969 Smithsonian Standard Earth (II). Smithsonian Astrophys. Obs., Spec. Rep., 315; 93 pp.
Gass, I.G., 1968. Is the Troodos Massif of Cyprus a fragment of Mesozoic ocean floor? Nature, 220:39—42.
Gass, I.G. and Masson-Smith, D., 1963. The geology and gravity anomalies of the Troodos Massif, Cyprus. Phil. Trans. R. Soc. Lond., Ser. A, 255:417—467.
Gassman, F., 1951. Proc. World Petrol. Congr., 3rd, 1951, Sect. 1. The Hague, pp. 613—621.
Gilbert, F., Dziewonski, A. and Brune, J., 1973. An informative solution to a seismological inverse problem. Proc. Natl. Acad. Sci. U.S.A., 70:1410—1413.

Gilbert, R.L.G., 1949. A dynamic gravimeter of novel design. Proc. Phys. Soc. Lond., Ser. B, 62:445--454.

Gilbert, R.L.G., 1952. Gravity observations in a borehole. Nature, 170:424—425.

Goncharov, V.P. and Mikhailov, O.V., 1963. New data concerning the topography of the Mediterranean Sea bottom. Okeanologiya, 3:1056—1061.

Goodell, R.R. and Fay, C.H., 1964. Borehole gravity meter and its application. Geophysics, 29:774—782.

Graf, A., 1958. Das Seegravimeter. Z. Instrumentenkd., 60:151—162.

Graf, A. and Schulze, R., 1961. Improvements on the sea gravimeter Gss-2. J. Geophys. Res., 66:1813—1821.

Griffin, W.R., 1949. Gravity in theory and practice. Geophysics, 14:39—56.

Gumma, W., 1974. An Interpretation of the Gravity and Magnetic Anomalies of the Rivera Fracture Zone, Eastern Pacific Ocean. M.S. Thesis, Oregon State Univ., Corvallis, Ore., 50 pp.

Gunn, R., 1943. A quantitative evaluation of the influence of the lithosphere on the anomalies of gravity. J. Franklin Inst., 236:47—65.

Gunn, R., 1947. Quantitative aspects of juxtaposed ocean deeps, mountain chains, and volcanic ranges. Geophysics, 12:238—255.

Gunn, R., 1949. Isostasy extended. J. Geol., 57:263—279.

Haalck, H., 1931. Ein statischer Schwerkraftmesser. Z. Geophys., 7:95 (ibid, 8:17, 197; 9:81, 285).

Hammer, S., 1945. Estimating ore masses in gravity prospecting. Geophysics, 10:50—62.

Hammer, S., 1963. Deep gravity interpretation by stripping. Geophysics, 28:369—378.

Hammer, S., 1976. Topographic corrections for gradients in airborne gravimetry. Geophysics, 41:1346—1352.

Hammond, J.A., 1970. J.I.L.A. Report No. 103. Univ. Colorado, Boulder, Colo.

Hammond, J.A. and Faller, J.E., 1971. A laser-interferometer system for the absolute determination of the acceleration due to gravity. In: D.N. Langenberg and B.N. Taylor (Editors), Precision Measurements and Fundamental Constants. Natl. Bur. Stand. Spec. Publ., 343: 457—463.

Hanks, T.C., 1971. The Kurile Trench — Hokkaido Rise system: Large shallow earthquakes and simple models of deformation. Geophys. J. R. Astron. Soc., 23:173—189.

Harrison, J.C., 1960. The measurement of gravity at sea. In: S.K. Runcorn (Editor), Methods and Techniques of Modern Geophysics. Wiley-Interscience, New York, N.Y., pp. 211—229.

Hartley, K.A., 1932. A new instrument for measuring very small differences in gravity. Physics, 2:123—130.

Hayford, J.F., 1909. The Figure of the Earth and Isostasy from Measurements in the United States. U.S. Coast Geod. Surv. Monogr., 178 pp.

Hayford, J.F., 1910. Supplemental Investigations in 1909 of the Figure of the Earth and Isostasy. U.S. Coast Geod. Surv. Monogr., 80 pp.

Hayford, J.F. and Bowie, W., 1912. The Effect of Topography and Isostatic Compensation upon the Intensity of Gravity. U.S. Coast Geod. Surv. Spec. Publ., 10: 132 pp.

Healy, J.H. and Warren, D.H., 1969. Explosion seismic studies in North America. In: P.J. Hart (Editor), The Earth's Crust and Upper Mantle. Geophys. Monogr., Am. Geophys. Union, 13:208—220.

Hecker, O., 1903. Bestimmung der Schwerkraft auf dem Atlantischen Ozean sowie in Rio de Janeiro, Lissabon, und Madrid. Veröff. K. Preuss. Geod. Inst., Berlin, N.F., 11.

Hecker, O., 1907. Beobachtungen an Horizontalpendeln über die Deformation des Erdkörpers unter dem Einfluss von Sonne und Mond. Veröff. K. Preuss. Geod. Inst., Berlin, N.F., 32.

Heiland, C.A., 1940. Geophysical Exploration. Prentice-Hall, New York, N.Y., 1013 pp. (Reprinted by Hafner Publ., New York, N.Y., 1963.)

Heiskanen, W.A., 1924. Untersuchungen über Schwerkraft und Isostasie. Publ. Finn. Geod. Inst., 4.

Heiskanen, W.A., 1928. Ist die Erde ein dreiachsiges Ellipsoid? Gerlands Beitr. Geophys., 19.

Heiskanen, W.A., 1938a. Investigations on the gravity formula. Publ. Isost. Inst., I.A.G. (Helsinki), 1.

Heiskanen, W.A., 1938b. New isostatic tables for the reduction of gravity values calculated on the basis of Airy's hypothesis. Publ. Isost. Inst., I.A.G. (Helsinki), 2.

Heiskanen, W.A., 1953. Isostatic reductions of gravity anomalies by the aid of high-speed computing machines. Publ. Isost. Inst., I.A.G. (Helsinki), 28.

Heiskanen, W.A. and Vening Meinesz, F.A., 1958. The Earth and its Gravity Field. McGraw-Hill, New York, N.Y., 470 pp.

Helmert, F.R., 1884. Die mathematischen und physikalischen Theorie der höheren Geodesie, 2. Teubner, Leipzig.

Helmert, F.R., 1898. Beiträge zur Theorie des Reverspendels. Veröff. K. Preuss. Geod. Inst., Berlin.

Helmert, F.R., 1901. Der normale Teil der Schwerkraft im Meresniveau. Ber. K. Preuss. Akad. Wiss., 14:328.

Helmert, F.R., 1909. Die Tiefe der Ausgleichsfläche bei der Prattschen Hypothese für das Gleichgewicht der Erdkruste und auf der Verlauf der Schwerestörung vom Innern der Kontinente und Ozeane nach den Küsten. Sitzber. K. Preuss. Akad. Wiss., 48: 1192.

Herglotz, G., 1907. Über das Benndorfsche Problem der Fortpflanzungsgeschwindigkeit der Erdbebenstrahlen. Phys. Z., 8:145—147.

Herrin, E. and Taggart, J., 1962. Regional variations in P_n velocity and their effect on the location of epicenters. Bull. Seismol. Soc. Am., 52:1037—1046.

Herron, E.M., Dewey, J.F. and Pitman, W.C., 1974. Plate tectonics model for the evolution of the Arctic. Geology (August), pp. 377—380.

Hertz, H., 1884. On equilibrium of floating elastic plates. Wiedemanns Annal., 22:449.

Hess, H.H., 1962. History of ocean basins. In: A.E.J. Engel, H.L. James, and B.F. Leonard (Editors), Petrologic Studies (A volume in honor of A.F. Buddington). Geol. Soc. Am., New York, N.Y., pp. 599—620.

Heyl, P.R. and Cook, G.S., 1936. The value of gravity at Washington. J. Res., Natl. Bur. Stand., 17:805—839.

Hide, R., 1969. Interaction between the earth's liquid core and solid mantle. Nature, 222:1055—1056.

Higgins, G.H. and Kennedy, G.C., 1971. The adiabatic gradient and the melting point gradient in the core of the earth. J. Geophys. Res., 76:1870—1878.

Holweck, F. and Lejay, P., 1930. Un instrument transportable pour la mesure rapide de la gravité. C.R. Acad. Sci., 190:1387—1388.

Hood, Peter and Ward, S.H., 1969. Airborne geophysical methods. In: H.E. Landsberg and J. van Mieghem (Editors), Advances in Geophysics, 13. Academic Press, New York, N.Y., pp. 1—112.

Howell, L.G., Heintz, K.O. and Barry, H., 1966. The development of a high precision downhole gravity meter. Geophysics, 31:764—772.

Hoyt, A., 1938. Helical ribbon spring measuring apparatus. U.S. Patent, 2,131,739.

Hubbert, M.K., 1948a. A line-integral method for computing the gravimetric effects of two-dimensional masses. Geophysics, 13:215—225.

Hubbert, M.K., 1948b. Gravitational terrain effects of two-dimensional topographic features. Geophysics, 13:226—254.

Huygens, C., 1693. Discours de la Cause de la Pesanteur. Leyden.

Isacks, B., Oliver, J. and Sykes, L., 1968. Seismology and the new global tectonics, J. Geophys. Res., 73:5855—5899.

Jacobs, J.A., 1971. Boundaries of the earth's core. Nature, Phys. Sci., 231:170—171.

Jacobs, J.A., 1975. The Earth's Core. Academic Press, New York, N.Y., 253 pp.

Jakosky, J.J., 1940. Exploration Geophysics. Times-Mirror Press, Los Angeles, Calif., 786 pp.

James, F. and Roos, M., 1969. CERN Computer 6000 Series Program Library D506.

Jeffreys, H., 1948. The figures of the earth and moon (Third Paper). Mon. Not. R. Astron. Soc., Geophys. Suppl., 5:219—248. (Reprinted in Collected Papers of Sir Harold Jeffreys on Geophysics and Other Sciences, 3. Gravity. Gordon and Breach, New York, N.Y., pp. 277—306.)

Jenkins, G.M. and Watts, D.C., 1968. Spectral Analysis and its Applications. Holden-Day, San Francisco, Calif., 525 pp.

Johnson, L.R., 1969. Array measurements of P velocities in the upper mantle. J. Geophys. Res., 72:6309—6325.

Kanasewich, E.R. and Agarwal, R.G., 1970. Analysis of combined gravity and magnetic fields in wave number domains. J. Geophys. Res., 75:5702—5712.

Kater, H., 1818. An account of experiments for determining the length of the pendulum vibrating seconds in the latitude of London. Phil. Trans. R. Soc. Lond., 108:33.

Kaula, W.M., 1963. Determination of the earth's gravitational field. Rev. Geophys., 1:507—551.

Kaula, W.M., 1964. Tidal dissipation by solid friction and the resulting orbital evolution. Rev. Geophys., 2:661—685.

Kaula, W.M., 1965. The shape of the earth. In: W.N. Hess (Editor), Introduction to Space Science. Gordon and Breach, New York, N.Y., pp. 297—317.

Kaula, W.M., 1966. Theory of Satellite Geodesy. Blaisdell, Waltham, Mass., 124 pp.

Keen, M.J., Loncarevic, B.D. and Ewing, G.N., 1970. Continental margin of eastern Canada: Georges Bank to Kane Basin. In: A.E. Maxwell (Editor), The Sea, Vol. 4, Part 2. Wiley-Interscience, New York, N.Y., pp. 251—291.

King-Hele, D.G., 1958. The effect of the earth's oblateness on the oribit of a near satellite. Proc. R. Soc. Lond., Ser. A, 247:49—72.

Knopoff, L., 1964. The convection current hypothesis. Rev. Geophys., 2:89—122.

Kolenkiewics, R., Smith, D.E. and Dunn, P.J., 1973. A re-evaluation of the tidal perturbation of the Beacon Explorer 3 spacecraft. EOS Trans. Am. Geophys. Union, 54:232. (Abstract).

Kozai, Y., 1969. Revised zonal harmonics in the geopotential. Smithsonian Astrophys. Obs., Spec. Rep., 295: 17 pp.

Krassovsky, T.N., 1938. A value of $e = 1/298.3$, determined by Krassovsky, is referenced by K. Jung in: Figur der Erde. In: J. Bartels (Group Editor), 1956. Handbuch der Physik, 47 (Geophysik I). Springer, New York, Berlin, p. 633.

Kühnen, F. and Furtwängler, P., 1906. Bestimmung der absoluten Grösse der Schwerkraft zu Potsdam. Veröff. K. Preuss. Geod. Inst., 27:397.

Kuo, J.T. and Jachens, R.C., 1975. Indirect mapping of ocean tides by solving the inverse problem for the tidal observations. Symp. Tidal Interactions, XVI Gen. Assem., Int. Union Geod. Geophys., Grenoble, France. (Separate, 36 pp.).

Kuo, J.T., Jachens, R.C., Ewing, M. and White, G., 1970. Transcontinental tidal gravity profile across the United States. Science, 168:968—971.

LaCoste, L.J.B., 1952a. Force Measuring Device. U.S. Patent, 2,589,709.

LaCoste, L.J.B., 1952b. Force Measuring Device. U.S. Patent, 2,589,710.

LaCoste, L.J.B., 1959. Surface ship gravity measurements on the Texas A and M College ship "Hidalgo". Geophysics, 24:309—322.

LaCoste, L.J.B., 1967. Measurements of gravity at sea and in the air. Rev. Geophys., 5:477—526.

LaCoste, L.J.B., 1973a. Cross-correlation method for evaluating and correlating shipboard gravity data. Geophysics, 38:701—709.

LaCoste, L.J.B., 1973b. Force Measuring Device. U.S. Patent, 3,717,036.

Lambeck, K., 1972. Gravity anomalies over ocean ridges. Geophys. J. R. Astron. Soc., 30:37—53.

Lambeck, K., 1975. Effects of tidal dissipation in the oceans on the moon's orbit and the earth's rotation. J. Geophys. Res., 80:2917—2925.

Langseth, M.G., Le Pichon, X. and Ewing, M., 1966. Crustal structure of the mid-ocean ridges, 5. Heat flow through the Atlantic Ocean floor and convection currents. J. Geophys. Res., 71:5321—5355.

Laplace, P.S., 1784. Théorie du Mouvement et de la Figure Elliptique des Planètes. Paris.

Legendre, A.M., 1785. Recherches sur l'attraction des sphéroïdes homogènes. Paris. Mémoirs.

Le Pichon, X., Francheteau, J. and Bonnin, J., 1973. Plate Tectonics. Developments in Geotectonics, 6. Elsevier, New York, Amsterdam, 300 pp.

Linsser, H., 1965. A generalized form of Nettleton's density determination. Geophys. Prospect., 13:247—258.

Longman, I.M., 1959, Formulas for computing the tidal accelerations due to the sun and moon. J. Geophys. Res., 64:2351—2355.

Love, A.E.H., 1909. The yielding of the earth and disturbing forces. Proc. R. Soc. Lond., Ser. A, 82:73—88.

Lozinskaya, A.M., 1959. The string gravimeter for measurement of gravity at sea. Bull. Acad. Sci. USSR, Geophys. Ser., 3:398—409 (English trans.).

Luenberger, D.G., 1969. Optimization by Vector Space Methods. Wiley, New York, N.Y., 326 pp.

Lundberg, H., 1957. Airborne gravity surveys. Can. Mining Metall. Bull., 60:251—259.

Lynn, W., 1975. A Geophysical Analysis of the Orozco Fracture Zone and Tectonic Evolution of the Northern Cocos Plate. M.S. Thesis, Oregon State Univ., Corvallis, Ore., 52 pp.

MacDonald, G.J.F., 1964. Tidal friction. Rev. Geophys. 2:467—541.

MacLaurin, C., 1742. A Treatise on Fluxions. Edinburgh.

Malahoff, A. and Woollard, G.P., 1970. Geophysical studies of the Hawaiian Ridge and Murray Fracture Zone. In: A.E. Maxwell (Editor), The Sea, Vol. 4, Part 2. Wiley-Interscience, New York, N.Y., pp. 73—131.

McCamy, K. and Meyer, R.P., 1966. Crustal results of fixed multiple shots in the Mississippi Embayment. In: J.S. Steinhart and T.J. Smith (Editors), The Earth Beneath The Continents (A volume in honor of Merle A. Tuve), Geophys. Monogr., Am. Geophys. Union, 10:370—381.

McKenzie, D.P., 1967. Some remarks on heat flow and gravity anomalies. J. Geophys. Res., 72:6261—6273.

McKenzie, D.P., 1968. The geophysical importance of high-temperature creep. In: History of the Earth's Crust, Proc. NASA Conf., Princeton Univ. Press, Princeton, N.J.

McKenzie, D.P., 1969. Speculations on the consequences and causes of plate motions. Geophys. J. R. Astron. Soc., 18:1—32.

McKenzie, D.P., 1972. Plate tectonics. In: E.C. Robertson, J.F. Hays and L. Knopoff (Editors), The Nature of the Solid Earth, McGraw-Hill, New York, N.Y., pp. 323—360.

McKenzie, D.P. and Bowin, C.O., 1976. The relationship between bathymetry and gravity in the Atlantic Ocean. J. Geophys. Res., 81:1903—1915.

McKenzie, D.P. and Morgan, W.J., 1969. The evolution of triple junctions. Nature, 224:125—133.

Melchior, P., 1959. Résultats de 8 mois d'enrégistrement continu de la marée à l'Observatoire Royal de Belgique (Bruxelles) par le gravimètre Askania No. 145, III ème Symp. Marées Terr., Trieste, 145:129—139.

Melchior, P., 1966. The Earth Tides. Pergamon, New York, London, 458 pp.

Merson, R.H., 1961. The motion of a satellite in an axi-symmetric gravitational field. Geophys. J. R. Astron. Soc., 4:17—52.

Miller, G.R., 1966. The flux of tidal energy out of the deep oceans. J. Geophys. Res., 71:2485—2489.

Morgan, W.J., 1968. Rises, trenches, great faults, and crustal blocks. J. Geophys. Res., 73:1959—1982.

Mott-Smith, L.M., 1938. Torsion gravimetry. U.S. Patent, 2,130,648.

Mueller, S. (Editor), 1974. The Structure of the Earth's Crust. Elsevier, Amsterdam, New York, 391 pp.

Munk, W.H. and MacDonald, G.J.F., 1960. The Rotation of the Earth. Cambridge Univ. Press, Cambridge, 323 pp.

Murrell, S.A.F., 1976. Rheology of the lithosphere — experimental indications. Tectonophysics, 36:5—24. Also in: M.H.P. Bott (Editor), Sedimentary Basins of Continental Margins and Cratons. Developments in Geotectonics, 12. Elsevier, Amsterdam, pp. 5—24.

Nadai, A., 1963. Theory of Flow and Fracture of Solids, 2. McGraw-Hill, New York, N.Y., 2nd ed., 705 pp.

Nafe, J.E. and Drake, C.L., 1963. Physical properties of marine sediments. In: M.N. Hill (Editor), The Sea, 3. Wiley-Interscience, New York, N.Y., pp. 794—815.

Nagy, D., 1966a. The gravitational attraction of a right rectangular prism. Geophysics, 31:362—371.

Nagy, D., 1966b. The prism method for terrain corrections using digital computers. Pure Appl. Geophys., 63:31—39.

Nettleton, L.L., 1940. Geophysical Prospecting for Oil. McGraw-Hill, New York, N.Y., 444 pp.

Nettleton, L.L., 1976. Gravity and Magnetics in Oil Prospecting. McGraw-Hill, New York, N.Y., 464 pp.

Nettleton, L.L., La Coste, L.J.B. and Glicken, M., 1962. Quantitative evaluation of precision of airborne gravity meter. J. Geophys. Res., 67:4395—4410.

Newcomb, S., 1892. Remarks on Mr. Chandler's law of variation of terrestrial latitudes. Astron. J., 12:49—50.

Newton, R.R., 1968. A satellite determination of tidal parameters and earth deceleration. Geophys. J. R. Astron. Soc., 14:505—539.

Newton, R.R., 1969. Secular accelerations of the earth and moon. Science, 166: 825—831.

Nuttli, O.W., 1969. Travel times and amplitudes of S waves from nuclear explosions in Nevada. Bull. Seismol. Soc. Am., 59:385—398.

O'Connell, R.J., 1977. On the scale of mantle convection. Tectonophysics, 38:119—136.

O'Connell, R.J. and Dziewonski, A.M., 1976. Excitation of the Chandler Wobble by large earthquakes. Nature. In press.

Oesterwinter, C. and Cohen, C.J., 1972. New orbital elements for moon and planets. Celest. Mech., 5:317.

Oldham, C.H.G., 1967. The sin x/x and sin y/y method for continuation of potential fields. Mining Geophysics, 2. Soc. Expl. Geophys., Tulsa, Okla., pp. 591—605.

Orlov, A. Ya., 1915. Results of observations at Yuryev, Tomsk, and Potsdam of lunisolar deformations of the earth. Tr. Astron. Observ. Novoross. un-ta, 2: 1—281 (Odessa).

Pakiser, L.C. and Robinson, R., 1966. Composition of the continental crust as estimated from seismic observations. In: J.S. Steinhart and T.S. Smith (Editors), The Earth Beneath the Continents. Geophys. Monogr., Am. Geophys. Union, 10:620—626.

Pakiser, L.C. and Steinhart, J.S., 1966. Explosion seismology in the western hemisphere. In: H. Odishaw (Editor), Solid Earth and Interface Phenomena. Research in Geophysics, 2. Mass. Inst. Tech. Press, Cambridge, Mass., pp. 123—147.

Paterson, N.R., 1961. An integration technique for airborne gravity gradient measurements. Geophysics, 26:474—479.

Pepper, T.B., 1941. The Gulf underwater gravimeter. Geophysics, 6:34—44.

Peterschmitt, E., 1956. Quelques donnés nouvelles sur les seismics profond de la mer Tyrrhénienne. Ann. Geofis. (Rome), 9:305.

Pettit, J.T., 1954. Tables for the computation of the tidal accelerations of the sun and moon. Trans. Am. Geophys. Union, 35:193—202.

Picard, J., 1675. In: Philosophical Transactions, No. 112, March 25 (in French).

Pick, M., Picha, J. and Vyskocil, V., 1973. Theory of the Earth's Gravity Field. Elsevier, Amsterdam, New York, 538 pp.

Pittman, W.C. and Talwani, M., 1972. Sea-floor spreading in the North Atlantic. Bull. Geol. Soc. Am., 83:619—646.

Pratt, J.H., 1855. On the attraction of the Himalaya Mountains and of the elevated regions beyond the plumb line in India. R. Soc. Lond., Ser. B, 145:53.

Pratt, R.M., 1963. Great Meteor Seamount, Deep Sea Res., 10:17—25.

Preston-Thomas, H., Turnbull, L.G., Green, E., Dauphinee, T.M. and Kalra, S.N., 1960. An absolute measurement of the acceleration due to gravity at Ottawa. Can. J. Phys., 38:824—852.

Rabinowitz, P.D., 1974. The boundary between oceanic and continental crust in the western North Atlantic. In: C.A. Burk and C.L. Drake (Editors), The Geology of Continental Margins. Springer-Verlag, New York, N.Y., pp. 67—84.

Rabinowitz, P.D., 1976. Geophysical study of the continental margin of southern Africa. Bull. Geol. Soc. Am., 87:1643—1653.

Rabinowitz, P.D. and Labrecque, J.L., 1977. The isostatic gravity anomaly:Key to the evolution of the ocean—continent boundary of passive continental margins. Earth Planet. Sci. Lett., 35:145—150.

Rabinowitz, P.D. and Ryan, W.B.F., 1970. Gravity anomalies and crustal shortening in the eastern Mediterranean. Tectonophysics, 10:585—608.

Rapp, R.H., 1974. Current estimates of mean earth ellipsoid parameters. Geophys. Res. Lett., 1:35—38.

Rayleigh, Lord (J.W. Strutt), 1945. The Theory of Sound, 2 vols. Dover, New York, N.Y. (Reproduced from 2nd ed., Macmillan, London, 1894.)

Richter, C.F., 1958. Elementary Seismology. Freeman, San Francisco, Calif., 768 pp.

Ringwood, A.E., 1969. Composition and evolution of the upper mantle. In: P.J. Hart (Editor), The Earth's Crust and Upper Mantle. Geophys. Monogr., Geophys. Union, 13:1—17.

Ringwood, A.E., 1972. Mineralogy of the deep mantle: current status and future developments. In: E.C. Robertson, J.F. Hays, and L. Knopoff (Editors), Nature of the Solid Earth. McGraw-Hill, New York, N.Y., pp. 67—92.

Rosenbach, O., 1953. A contribution to the computation of the second derivative from gravity data. Geophysics, 18:894—912.

Rudzki, M.P., 1905. Sur la détermination de la figure de la terre d'après les mesures de la gravité. Bull. Astron., Ser. B, 22:49.

Sakuma, A., 1971. Recent developments in the absolute measurement of gravitational acceleration. In: D.N. Langenberg and B.N. Taylor (Editors), Precision Measurement and Fundamental Constants. Natl. Bur. Stand. Spec. Publ., 343:447—456.

Schureman, P., 1924. A manual of the harmonic analysis and prediction of tides. U.S. Coast Geod. Surv. Spec. Publ., 98.

Schureman, P., 1941. A manual of the harmonic analysis and prediction of tides. U.S. Coast Geod. Surv. Spec. Publ., 98 (Revised ed.).

Schweydar, W., 1905. Untersuchung der Oscillationen der Lotlinie auf dem Astronom. Institute der Grossh. Sternwarte zu Heidelberg. Gerlands Beitr. Geophys., 7: 33—120.

Sclater, J.G. and Francheteau, J., 1970. The implications of terrestrial heat-flow observations on current tectonic and geochemical models of crust and upper mantle of the earth. Geophys. J. R. Astron. Soc., 20:509—542.

Sclater, J.G., Anderson, R.N. and Bell, M.L., 1971. Elevation of ridges and evolution of the central eastern Pacific. J. Geophys. Res., 76:7888—7915.

Seto, T., 1968. G.S.I. type surface ship gravity meter. In: Preliminary Report of the Hakuko Maru Cruise KH 68—3. Ocean Research Institute, Univ. Tokyo, Tokyo, pp. 41—46.

Shor, G.G. Jr., 1960. Crustal structure of the Hawaiian Ridge near Gardner Pinnacles. Bull. Seismol. Soc. Am., 50:563—573.

Shor, G.G. Jr., Dehlinger, P., Kirk, H.K. and French, W.S., 1968. Seismic refraction studies off Oregon and northern California. J. Geophys. Res., 73:2175—2194.

Shor, G.G. Jr., Menard, H.W. and Raitt, R.W., 1970. Structure of the Pacific Basin. In: A.E. Maxwell (Editor), The Sea, Vol. 4, Part 2. Wiley-Interscience, New York, N.Y., pp. 3—27.

Sibuet, J.C. and Le Pichon, X., 1974. Thickness of lithosphere deduced from gravity edge effects across the Mendocino fault. Nature, 252:676—679.

Smith, R.A., 1961. A uniqueness theorem concerning gravity fields. Proc. Cambridge Phil. Soc., 57:865—870.

Smith, S.W. and Jungels, P., 1970. Phase delay of the solid earth tide. Phys. Earth Planet. Inter., 2:233.

Stacey, F.D., 1977. Physics of the Earth. 2nd ed., Wiley, Santa Barbara, Calif., London, Sydney, 414 pp.

Stauder, W., 1968. Tensional character of earthquake foci beneath the Aleutian Trench with relation to sea-floor spreading. J. Geophys. Res., 73:7693—7701.

Stokes, G.G., 1849. On the variation of gravity at the surface of the earth. Trans. Cambridge Phil. Soc., 8:672.

Takahasi, R., 1929. Tilting motion of the earth crust caused by tidal loading. Bull. Earthquake Res. Inst., 6:85—108.

Talwani, M., 1966. Some recent developments in gravity measurements aboard surface ships. In: H. Orlin (Editor), Gravity Anomalies: Unsurveyed Areas. Geophys. Monogr., Am. Geophys. Union, 9:31—47.

Talwani, M., 1970, Gravity. In: A.E. Maxwell (Editor), The Sea, Vol. 4, Part 1. Wiley-Interscience, New York, N.Y., pp. 270—297.

Talwani, M., 1973. Computer usage in the computation of gravity anomalies. In: B.A. Bolt (Editor), Methods in Computational Physics, 13. Academic Press, New York, N.Y., pp. 343—389.

Talwani, M. and Eldholm, O., 1972. Continental margin off Norway: A geophysical study. Bull. Geol. Soc. Am., 83:3575—3606.

Talwani, M. and Ewing, M., 1960. Rapid computation of gravitational attraction of three-dimensional bodies of arbitrary shape. Geophysics, 25:203—225.

Talwani, M. and Hayes, D.E., 1967. Continuous gravity profiles over island-arcs and deep-sea trenches (abstract). EOS, Trans. Am. Geophys. Union, 48:217.

Talwani, M. and Le Pichon, X., 1969. Gravity field over the Atlantic Ocean. In: P.J. Hart (Editor), The Earth's Crust and Upper Mantle. Geophys. Monogr., Am. Geophys. Union, 13:341—351.

Talwani, M., Worzel, J.L. and Landisman, M., 1959. Rapid gravity computations for two dimensional bodies with application to the Mendocino submarine fracture zone. J. Geophys. Res., 64:49—59.

Talwani, M., Worzel, J.L. and Ewing, M., 1961. Gravity anomalies and crustal section across the Tonga Trench. J. Geophys. Res., 66:1265—1278.

Talwani, M., Le Pichon, X. and Ewing, M., 1965. Crustal structure of the mid-ocean ridges, 2. Computed model from gravity and seismic refraction data. J. Geophys. Res., 70:341—352.

Talwani, M., Early, W.P. and Hayes, D.E., 1966. Continuous analog computation and recording of cross-coupling and off-leveling errors. J. Geophys. Res., 71:2079—2090.

Talwani, M., Windish, C.C. and Langseth, M.G., 1971. Reykjanes Ridge crest: a detailed geophysical study. J. Geophys. Res., 76:473—517.

Tanner, J.G., 1967. An automatic method of gravity interpretation. Geophys. J. R. Astron. Soc., 13:339—347.

Taylor, G.I., 1922. Diffusion by continuous movements. Proc. Math., Ser. 2, p. 20.

Thompson, L.G.D., Bock, R.O. and Savet, P.H., 1965. Gravity gradient sensors and their applications for manned orbital spacecraft. Goddard Mem. Sympos. 3rd., Am. Astron. Soc., Washington, D.C.

Thompson, W.T., 1963. Introduction to Space Dynamics. Wiley, New York, 2nd ed., 317 pp.

Thulin, A., 1961. Détermination absolute de l'accélération due à la pesanteur au pavillon de bretenil. Trav. Mém. Bur. Int. Poids Mes., 22:A1—A91.

Thyssen-Bornemisza, S., 1970. Instrumental arrangement to measure gravity with gradients. Geophysics, 35:713—715.

Timoshenko, S., 1955. Strength of Materials, 1. Van Nostrand, Princeton, N.J., 3rd ed., 442 pp.

Todhunter, I.A., 1962. A History of the Mathematical Theories of Attraction and the Figure of the Earth. Dover, New York, Vol. I, 476 pp.; Vol II, 508 pp. (Unabridged and unaltered republication of work published by Macmillan Co., London, in 1873, 2 vols.)

Tomaschek, R. and Schaffernicht, W., 1932. Untersuchungen über die zeitlichen Änderungen der Schwerkraft, I. Messungen mit dem Bifilargravimeter. Annal. Phys., 5:787—824.

Tomoda, Y. and Aki, K., 1955. Use of function sin x/x in gravity problems. Proc. Jap. Acad., 31:443—448.

Tomoda, Y. and Kanamori, H., 1962. Tokyo surface-ship gravity meter Alpha-1. Jap. J. Geophys., 7:116—145.

Tozer, D.C., 1965. Thermal history of the earth, 1. The formation of the corc. Geophys. J. R. Astron. Soc., 9:95—112.

Tozer, D.C., 1970a. Factors determining the temperature evolution of thermally convecting earth models. Phys. Earth Planet. Inter. 2:393.

Tozer, D.C., 1970b. Temperature, conductivity, composition, and heat flow. J. Geomagn. Geoelectr., 22:35.

Tsuboi, C., 1938. A simple method of approximately determining the thickness of the earth's crust. Bull. Earthquake Res. Inst., 16:285—286.

Tsuboi, C., Tomoda, Y. and Kanamori, H., 1961. Continuous measurements of gravity on board a moving surface ship. Proc. Jap. Acad., 37:571—576.

Uotila, U.A., 1960. Investigations of the gravity field and shape of the earth. Ann. Acad. Sci. Fennicae, Ser. A, III, Geol.-Geogr., 55 (Helsinki).

Urban, T.C., Bromery, R.W., Revetta, F.A. and Diment, W.H., 1972. Simple Bouguer anomaly map of southeastern New York. New York State Museum and Science Service. Geological Survey Map and Chart Series, No. 17c. The Univ. of the State of N.Y., The State Education Dept., Albany, N.Y.

Valliant, H.D. and La Coste, L.J.B., 1976. Theory and evaluation of the La Coste and Romberg three-axis inertial platform for marine gravity. Geophysics, 41:459—467.

Van Flandern, T.C., 1975. A determination of the rate of change of G. Mon. Not. R. Astron. Soc., 170:333—342.

Vening Meinesz, F.A., 1929. Theory and practice of pendulum observations at sea, I. Publ. Neth. Geod. Comm. Waltman, Delft, 95 pp.

Vening Meinesz, F.A., 1931. Une nouvelle méthode pour la réduction isostatique régionale de l'intensité de la pesanteur. Bull. Geod., 29.

Vening Meinesz, F.A., 1939. Tables fondamentales pour la réduction isostatique régionale. Bull. Geod., 63.

Vening Meinesz, F.A., 1941a. Theory and practice of pendulum observations at sea, II. Publ. Neth. Geod. Comm. Waltman, Delft, 89 pp.

Vening Meinesz, F.A., 1941b. Gravity over the Hawaiian Archipelago and over the Madeira area. Proc. K. Ned. Akad. Wetensch., 44 pp.

Vening Meinesz, F.A., 1948. Gravity expeditions at sea 1923—1938, 4. Publ. Ned. Geod. Comm. Delftsche Uitgevers Mij., Delft, 233 pp.

Verhoogen, J., 1973. Thermal regime of the earth's core. Phys. Earth Planet. Inter., 7:47—58.

Victor, L., 1975. Structures of the Continental Margin of Central America from Northern Nicaragua to Northern Panama. M.S. Thesis, Oregon State Univ., Corvallis, Ore., 76 pp.

Volet, C., 1946. In: C.R. Acad. Sci., 222:373.

Volet, C., 1947. In: C.R. Acad. Sci., 224:1815—1816.

Wagoner, C.A., Lerch, F.J., Brownel, J.E. and Richardson, ., 1976. Improvement in the geopotential derived from satellite and surface data (GEM 7 and 8). Preprint X-921-76-20 of the Goddard Space Flight Center, Greenbelt, Md. (January), 11 pp.

Walcott, R.I., 1970a. Flexural rigidity, thickness, and viscosity of the lithosphere. J. Geophys. Res., 75:3941—3954.

Walcott, R.I., 1970b. Isostatic response to loading of the crust in Canada. Can. J. Earth Sci., 7:716—727.

Walcott, R.I., 1970c. Flexure of the lithosphere of Hawaii. Tectonophysics, 9:435—446.

Walcott, R.I., 1972. Gravity, flexure, and growth of sedimentary basins at a continental edge. Bull. Geol. Soc. Am., 83:1845—1848.

Walcott, R.I., 1973. Structure of the earth from glacio-isostatic rebound. Ann. Rev. Earth Planet. Sci., 1:15—37.

Walcott, R.I., 1976. Lithospheric flexure, analysis of gravity anomalies, and the propagation of seamount chains. In: G.H. Sutton, M.H. Manghnani, and R. Moberly (Editors), The Geophysics of the Pacific Ocean Basins and Its Margins (A volume in honor of G.P. Woollard). Geophys. Monogr., Am. Geophys. Union, 19:431—438.

Watts, A.B. (Compiler), 1975a. Gravity field of the northwest Pacific Ocean Basin and its margin: Hawaii and vicinity (one chart and summary statement). Geol. Soc. Am. Map Chart, 9.

Watts, A.B. (Compiler), 1975b. Gravity field of the northwest Pacific Ocean Basin and its margin: Aleutian Island Arc-Trench System (one chart and summary statement). Geol. Soc. Am. Map Chart, 10.

Watts, A.B. (Compiler), 1976. Gravity field of the northwest Pacific Ocean Basin and its margin: Philippine Sea (one chart and summary statement). Geol. Soc. Am. Map Chart, 12.

Watts, A.B. and Cochran, J.R., 1974. Gravity anomalies and flexure of the lithosphere along the Hawaiian—Emperor Seamount Chain. Geophys. J. R. Astron. Soc., 38:119—141.

Watts, A.B. and Talwani, M., 1974. Gravity anomalies seaward of deep-sea trenches and their tectonic implications. Geophys. J. R. Astron. Soc., 36:57—90.

Watts, A.B., Cochran, J.R. and Selzer, G., 1975. Gravity anomalies and flexure of the lithosphere: A three-dimensional study of the Great Meteor Seamount, northeast Atlantic. J. Geophys. Res., 80:1391—1398. (Correction in: J. Geophys. Res., 80:4116.)

Wegener, A.L., 1912. Petermanns Mitt., 58:185, 253, 305. In book form: Die Enstehung der Kontinente und Ozeane. Viesveg, Brunswick, 1915. Third edition translated into English by J.W. Evans, The Origin of Continents and Oceans. Methuen, London, 1924, 212 pp.

Whitsett, R.M., 1975. Gravity Measurements and Their Structural Implications for the Continental Margin of Southern Peru. Ph.D. Dissertation, Oregon State Univ., Corvallis, Ore., 82 pp.

Wiechert, E., 1907. [Über] Erdbebenwellen, I. Theoretisches über die Ausbreitung der Erdbebenwellen. Nachr. Ges. Wiss. Göttingen, Math.-Phys. Kl., pp. 415—529.

Wilcox, L.E., 1976. Airy-Woollard isostasy. In: G.H. Sutton, M.H. Manghnani, and R. Moberly (Editors), The Geophysics of the Pacific Ocean Basins and Its Margins (A volume in honor of G.P. Woollard). Geophys. Monogr., Am. Geophys. Union, 19:53—57.

Wilson, J. Tuzo, 1965. A new class of faults and their bearing on continental drift. Nature, 207:343—347.

Wing, C.G., 1969. M.I.T. vibrating string surface-ship gravimeter. J. Geophys. Res., 74:5882—5894.

Wold, R.J. and Ostenso, N.A., 1971. Gravity and bathymetry survey of the Arctic and its geodetic implications. J. Geophys. Res., 76:6253—6264.

Woodcock, S., 1975. Crustal Structure of the Tehuantepec Ridge and Adjacent Continental Margins of Southwestern Mexico and Western Guatemala. M.S. Thesis, Oregon State Univ., Corvallis, Ore., 52 pp.

Woodside, J.M., 1972. The Mid-Atlantic Ridge near 45°N: The gravity field. Can. J. Earth Sci., 9:942—961.

Woodside, J.M. and Bowin, C.O., 1970. Gravity anomalies and inferred crustal structure in the eastern Mediterranean Sea. Bull. Geol. Soc. Am., 81:1107—1122.

Woollard, G.P., 1936. An interpretation of gravity anomalies in terms of local and regional structures. Trans. Am. Geophys. Union, 1, pp. 63—74.

Woollard, G.P., 1938. The effect of geologic corrections on gravity anomalies. Trans. Am. Geophys. Union, 19:85—90.

Woollard, G.P., 1943. A transcontinental gravity and magnetic profile of North America and its relation to geologic structure. Bull. Geol. Soc. Am., 54:747—790.

Woollard, G.P., 1962. The Relation of Gravity Anomalies to Surface Elevation, Crustal Structure and Geology. Geophysical Polar Research Center, Univ. Wisconsin, Res. Rep't. 62—9, 356 pp.

Woollard, G.P., 1966. Regional isostatic relations in the United States. In: J.S. Steinhart and T.J. Smith (Editors), The Earth Beneath the Continents. Geophys. Monogr., Am. Geophys. Union, 10:557—594

Woollard, G.P., 1968. The inter-relationship of the crust, the upper mantle, and isostatic gravity anomalies in the United States. In: L. Knopoff, C.L. Drake, and P.J. Hart (Editors), The Crust and Upper Mantle of the Pacific Area. Geophys. Monogr., Am. Geophys. Union, 12:312—341.

Woollard, G.P., 1969. Regional variations in gravity. In: P.J. Hart (Editor), The Earth's Crust and Upper Mantle. Geophys. Monogr., Am. Geophys. Union, 13: 320—341.

Woollard, G.P. and Strange, W.E., 1966. The prediction of gravity. In: H. Orlin (Editor), Gravity Anomalies: Unsurveyed Areas. Geophys. Monogr., Am. Geophys. Union, 9:96—113.

Worzel, J.L., 1958. First surface gravimeter. Trans. Am. Geophys. Union, 39: 175—178.

Worzel, J.L., 1959. Continuous gravity measurements with the Graf sea gravimeter. J. Geophys. Res., 64:1299—1315.

Worzel, J.L., 1965. Pendulum Gravity Measurements at Sea: 1936—1939. Wiley, New York, N.Y., 422 pp.

Wykoff, R.D., 1936. Study of earth tides by gravitational measurements. Trans. Am. Geophys. Union, 1:46—52.

Yanovsky, B.M., 1958. Absolute determination of the acceleration due to gravity at the point VNIIM. Tr. VNIIM, 32. (Standartgiz, Moscow-Leningrad, in Russian.)

AUTHOR INDEX*

*Exclusive of references listed on pp. 295—309.

SUBJECT INDEX